AF356213

Mechanical Properties and Application of Adhesive Materials

Mechanical Properties and Application of Adhesive Materials

Editors

Marco Lamberti
Aurelien Maurel-Pantel

Basel • Beijing • Wuhan • Barcelona • Belgrade • Novi Sad • Cluj • Manchester

Editors
Marco Lamberti
Infrastructures
Italian National Agency for
New Technologies, Energy
and Sustainable Economic
Development - ENEA
Camugnano
Italy

Aurelien Maurel-Pantel
Mechanics and
Acoustics Laboratory
Aix-Marseille University
Marseille
France

Editorial Office
MDPI AG
Grosspeteranlage 5
4052 Basel, Switzerland

This is a reprint of articles from the Special Issue published online in the open access journal *Materials* (ISSN 1996-1944) (available at: https://www.mdpi.com/journal/materials/special_issues/ Adhe_Mat).

For citation purposes, cite each article independently as indicated on the article page online and as indicated below:

Lastname, A.A.; Lastname, B.B. Article Title. *Journal Name* **Year**, *Volume Number*, Page Range.

ISBN 978-3-7258-1702-3 (Hbk)
ISBN 978-3-7258-1701-6 (PDF)
doi.org/10.3390/books978-3-7258-1701-6

© 2024 by the authors. Articles in this book are Open Access and distributed under the Creative Commons Attribution (CC BY) license. The book as a whole is distributed by MDPI under the terms and conditions of the Creative Commons Attribution-NonCommercial-NoDerivs (CC BY-NC-ND) license.

Contents

About the Editors

Marco Lamberti

Dr. Marco Lamberti is a researcher and head of ENEA Brasimone Research Centre Management. His research activities include the multiscale modelling of the mechanical response of composite materials and structures; the numerical and experimental analysis of the mechanical behaviour of innovative fibre-reinforced composite structures; the mechanical behaviour of concrete structures reinforced with composite materials; the numerical-experimental analysis of GFRP adhesive connections; the hygrothermal behaviour of adhesive joints for offshore structures; and the mechanical behaviour of complex metal structures for nuclear fusion applications. He is a member of several national and international research projects developed in collaboration with several major industries, such as Kerakoll, Fiberline Ltd., and IDEOL. He is the author of numerous scientific articles published in international journals and conference proceedings.

Aurelien Maurel-Pantel

Dr. Aurelien Maurel-Pantel is an associate professor at Aix-Marseille University. He focuses his research on solid mechanics and the behaviour of materials and structures, covering various materials and structures at different scales. The work focuses on the experimental characterisation, definition, and identification of behaviour models and the finite element modelling of these structures. Furthermore, Dr. Aurelien Maurel-Pantel has focused his studies on the molecular bonding of glass for terrestrial and space optics, structural bonding using polymeric adhesives for offshore wind turbines and civil engineering, and the properties and processing of long and short fibre composites. More recently, he has been investigating 3D printing processes for short and continuous fibre composites, as well as the engineering of composites and their hybridisation with 3D printing processes. His activities maintain a link between materials, processes, and structural behaviour while maintaining a dialogue between experimentation and modelling.

Preface

Adhesive bonding is considered to be one of the key technologies for the development of many applications, in particular for the lightweighting of structures in the automotive, aerospace, medical, mechanical, civil, and transport sectors, in a future where resources are limited and environmental concerns are growing. The potential for the use of multi-material joints is considered to be greater for this type of joining technology than for most other techniques. In the transport sector, the combination of lightweight materials such as composites, aluminium, steel, and titanium alloys can reduce the weight of structural elements without compromising rigidity or safety. In civil engineering, bonding is used to ensure the durability of structures and to reinforce concrete structures in order to reduce the environmental impact of new buildings. Bonding also reduces tolerance requirements during the manufacturing process, allowing the adhesive to be used with varying joint thicknesses, making manufacturing processes more cost-effective. Bonding technology is also used in microelectronics and optical devices to bond ceramics such as fused silica or silicon. These applications are based on direct bonding technology and extend to more exotic materials, such as crystals and polymers. Nowadays, there is no doubt that adhesive materials have penetrated many aspects of science and technology, from basic research to technological integration and industrial applications. Simulation tools have also been developed to predict the static and dynamic strength and durability of these structures using various methods, such as cohesive zone and interface models. This field is rapidly progressing towards new discoveries in the following areas: interfacial interactions, surface chemistry, test methods, accumulation of test data on physical and mechanical properties, environmental effects, durability, thermo-hydric ageing, new adhesive materials, sealants, joint design, manufacturing technology, theoretical models, analytical models, finite element models, and analysis.

Marco Lamberti and Aurelien Maurel-Pantel
Editors

Article

A Numerical Model for Understanding the Development of Adhesion during Drying of Cellulose Model Surfaces

Magdalena Kaplan and Sören Östlund *

Solid Mechanics, Department of Engineering Mechanics, KTH Royal Institute of Technology,
100 44 Stockholm, Sweden
* Correspondence: soren@kth.se

Abstract: Adhesion is crucial for the development of mechanical properties in fibre-network materials, such as paper or other cellulose fibre biocomposites. The stress transfer within the network is possible through the fibre–fibre joints, which develop their strength during drying. Model surfaces are useful for studying the adhesive strength of joints by excluding other parameters influencing global performance, such as geometry, fibre fibrillation, or surface roughness. Here, a numerical model describes the development of adhesion between a cellulose bead and a rigid surface using an axisymmetric formulation, including moisture diffusion, hygroexpansion, and cohesive surfaces. It is useful for studying the development of stresses during drying. A calibration of model parameters against previously published contact and geometry measurements shows that the model can replicate the observed behaviour. A parameter study shows the influence of cohesive and material parameters on the contact area. The developed model opens possibilities for further studies on model surfaces, with quantification of the adhesion during pull-off measurements.

Keywords: adhesion modelling; cohesive interactions; cellulose model surfaces; fibre–fibre joints

Citation: Kaplan, M.; Östlund, S. A Numerical Model for Understanding the Development of Adhesion during Drying of Cellulose Model Surfaces. *Materials* **2023**, *16*, 1327. https:// doi.org/10.3390/ma16041327

Academic Editors: Marco Lamberti and Aurelien Maurel-Pantel

Received: 9 January 2023
Revised: 27 January 2023
Accepted: 2 February 2023
Published: 4 February 2023

Copyright: © 2023 by the authors. Licensee MDPI, Basel, Switzerland. This article is an open access article distributed under the terms and conditions of the Creative Commons Attribution (CC BY) license (https:// creativecommons.org/licenses/by/ 4.0/).

1. Introduction

Fibre network materials are a composite type with two constituting parts: the individual fibres and the interactions between them. When subjecting a network material to loading, forces transfer between the fibres through the points of contact, or fibre joints. The strength of the fibre joints thus becomes decisive for the strength of the composite.

Various kinds of fibres (most commonly cellulose-based) can make up these materials. Amongst these, many consist of paper pulp fibres. In paper making, pulp fibres are wetted to form a slurry, then pressed and dried to form the desired product or material. The removal of moisture changes the properties of individual fibres, but most importantly the network joints develop when the slurry transforms into a cohesive material. This means that the joints in the network develop their strength during drying [1].

To achieve the desired properties of fibre biocomposites, understanding the development of the joints is necessary. One course of action is manipulating pulp fibres and studying the properties of complete networks [2]. Fibre network modelling is another useful tool, where specific properties of fibres or joints can be varied and evaluated at a global scale [3–5]. Testing and evaluating laboratory-made fibre joints is also an option used in previous studies [6,7]. However, it may be difficult to isolate the influence of joint strength in all those cases, as many other parameters can influence performance; for example, the number of fibre contacts, fibrillation or surface roughness, and strength of individual fibres can all impact macroscopic behaviour [8,9]. Cellulose model surfaces have previously been useful in studying the effect of specific modifications on adhesive properties [10,11]. The present study uses model surfaces in the shape of solid homogenous beads, described and characterised in previous publications [12–16].

Several analytic models have been developed to describe adhesive behaviour between two surfaces. The JKR- [17] and DMT-models [18] are among the most known and are based on Hertz's [19] theory of deformation between elastic bodies. The JKR-model accounts for large and soft bodies, where the external loading is small, and attractive forces caused by the surface energy enlarge the contact area. On the other hand, the DMT-model applies to stiff and small bodies, where the contact area expands due to van der Waals forces, but the contribution to the force needed to separate the surfaces vanishes. Maugis [20] later developed a model bridging the two extreme behaviours by assuming a constant stress at the edge of the contact zone and determining the stress concentration factor.

More recent studies cover the process of contact formation and how boundary conditions influence adhesion. Thouless and Jensen [21] conclude that residual stresses introduced during contact formation affect the loading mode and have significant effects on the adhesive properties. Olsson and Larsson [22] developed a model for adhesion between stiff, spherical particles subjected to compressive loading, based partly on Brinell indentation, and verified it using numerical simulations. Peng et al. [23] showed that the measured adhesion strength also depends on the geometry and size of the contacting objects. Other studies have also recently reported on other factors affecting adhesion, such as shear stresses, electric or magnetic potential, and rate dependency [24–26].

The drying process of single droplets has been widely investigated in other fields, such as pharmaceuticals and food technology [27]. Perdana et al. analysed the changes in geometry and enzyme activity for a single droplet, experimentally and numerically using an effective diffusion approach [28,29]. In contrast, the drying of cellulose beads is not studied as extensively and previous studies, especially, lack a closer investigation of the development of adhesive properties during drying. A suitable tool for studying the drying process is numerical modelling. Publications in the literature use numerical models as a way of describing and investigating the adhesive properties of fibre networks [3]. In particular, cohesive zone models, as pioneered by Dugdale [30] and Barenblatt [31] are useful for describing adhesive behaviour, where the contact stress development defines the strength of the joint. Lin et al. [32] experimentally and numerically characterised the moisture-dependent constitutive behaviour of paper fibres and implemented degradation of the cohesive joints similar to hydrogen embrittlement of steel. Previous studies have also investigated the dependence of both moisture and temperature on adhesive properties for other materials [33–35]. This work uses a cohesive zone model to simulate the behaviour of a cellulose bead during the drying process in a finite element (FE) model, quantifying the moisture-dependent adhesive properties.

This paper presents a numerical model replicating the geometry development of the cellulose bead model surface during drying. It is organised as follows: first, the method section presents the modelling objective and the numerical methods, including the geometry of the FE model, the assumptions made of the material description, and boundary conditions. An explanation of the cohesive interaction definition used for the adhesive description follows. The method section further describes the parameter fitting procedure and the parameter study. Finally, the results section presents the determined material and cohesive properties dependent on moisture. Analysis and discussion of the significance of the work follow, and the paper concludes by stating the future applications and further developments of the model. The novelty of the presented work lies in the determination of stresses developed during drying and how these affect the development and degradation of the joint strength. It is also possible to study the effect of individual material parameters and it opens the possibility to study complex phenomena in fibre-joint formation by understanding which parameters cause them, without including the additional difficulties resulting from the complex geometries of individual fibre–fibre joints. The understanding of the mechanics during drying complements the role of surface chemistry and is a step on the way towards fully understanding fibre–fibre joint formation.

2. Materials and Methods

The numerical investigation, including model development and pre- and postprocessing, used the FE-software ABAQUS 2019 [36].

2.1. Modelling Objective

One method of studying adhesion between two entities is the use of model surfaces. They allow for isolating and quantifying adhesive properties alone, without the influence of other factors on global performance. For example, fibre network materials owe their properties partly to the adhesion between individual fibre–fibre joints, but also to individual fibre properties, surface fibrillation, friction, entanglements, geometry, etc. Using an idealised geometry with a smooth surface makes it possible to examine only the adhesion properties. Adhesion testing measures the force required to separate contacting surfaces, in this case the bead and an underlying flat surface. The measured force depends on the process leading up to the contact formation; when the bead dries, stresses will develop, and the observed adhesive properties will change.

The modelling object used in this study was a nanometre-smooth, water-swollen cellulose bead made from regenerated cellulose. The resulting material is amorphous and isotropic [12]. The bead was set to dry on a hard surface covered with Kapton tape, and the geometries of the bead and contact zone were measured over the course of drying by Li et al. [15]. Section 2.3 explains the experimental data used further.

2.2. Numerical Model

The developed model represents the drying process, including moisture diffusion and shrinkage. The model uses a two-dimensional axisymmetric geometry, assuming rotational symmetry because of the spherical shape of the bead. Figure 1a shows the geometry of the bead and underlying rigid surface; the radius of the bead R corresponds to the wet bead radius, which is set to 0.63 mm according to experimental measurements from Li et al. [15]. The bead model was discretised using 4-noded quadrilateral elements, which is suitable for a contact analysis. The element size at the contact zone was small, with a dimension of about $R/250$. A transition zone allowed the element size far from the contact zone to be about ten times larger. Figure 1b shows the mesh and close-up of the transition region. The class of elements used depends on the analysis type, as explained further in Section 2.2.1. Comparison of a simple test case to the Hertz contact model [19] verified the validity of the mesh discretisation. A simplified model with linear elastic material on both bead and surface, and an applied compressive force yielded results consistent with the analytic solution.

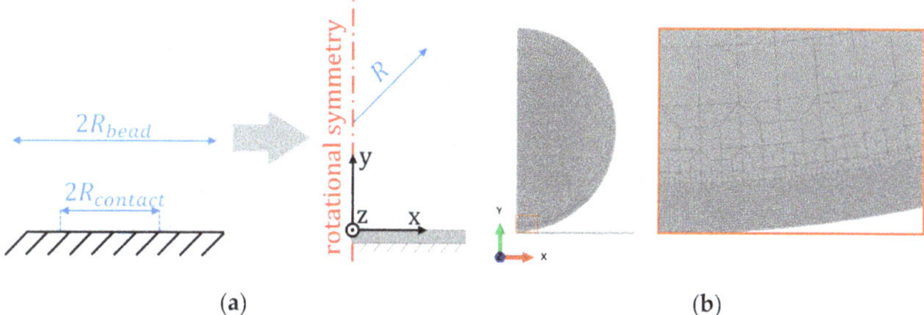

Figure 1. Model symmetries (**a**) and final FE-model with mesh used, including a close-up of transition elements (**b**).

A rotational symmetry line with a zero-displacement constraint in the x-direction prevented both the bead and surface from moving horizontally. Additionally, a fixed constraint at the bottom edge of the underlying surface inhibited it from movement in all degrees of freedom.

2.2.1. Boundary Conditions and Analysis Steps

A sequential procedure allowed for solving the moisture-expansion problem in two separate analyses. First, by solving the moisture field in a coupled stress-diffusion problem, where a moisture boundary condition placed on the external edge of the bead allowed the moisture to diffuse into the bulk of the bead. The diffusion problem used 4-node axisymmetric hybrid thermally coupled quadrilateral, bilinear displacement and temperature, constant pressure elements (CAX4HT) [36].

The second analysis, a stress procedure with cohesive interactions, consisted of three steps: compaction, loading, and drying, as summarised in Figure 2. The first two are preparation steps to get the correct geometry before the drying begins. In experiments, when placing the bead on the surface, it immediately deforms due to capillary forces pulling it towards the surface. A vertical pressure on the lower edge of the bead replaced the capillary forces, deforming it against the surface (compaction). After releasing the pressure (unloading), the cohesive interaction prevents the bead from separating from the surface and the initial deformation emerges.

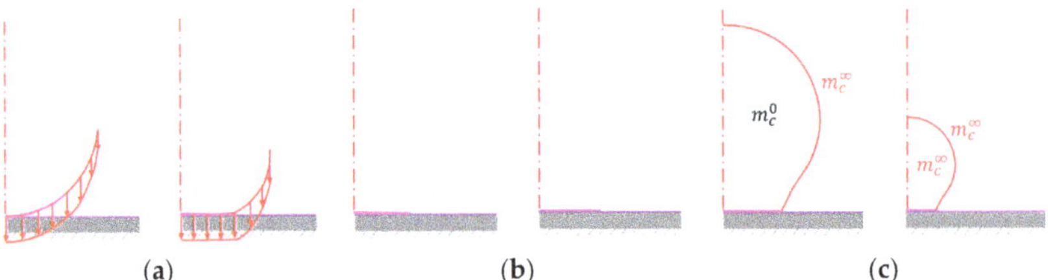

Figure 2. The three steps of the stress-shrinkage analysis with illustrated boundary conditions: (**a**) compaction, (**b**) unloading, and (**c**) shrinkage.

The drying step used the resulting moisture field from the stress-diffusion analysis as a boundary condition. As the moisture changes, it causes hygroscopic strains to develop, and eventually, the bead starts to shrink. The ambient moisture condition was replaced by a boundary condition at the initially free edge of the bead; the rate of applied moisture was fit to give the same radius evolution as the experimental data, with the final moisture value corresponding to the ambient moisture. The free edge will extend when the contact area begins to fail making the boundary condition application a simplification. The elements used in the stress analysis were 4-node bilinear elements, hybrid with constant pressure (CAX4H) [36].

2.2.2. Material Models

Li et al. [14] measured the elastic modulus of the cellulose beads using atomic force microscopy (AFM). The resulting stiffness was 0.45 MPa in the wet state and 12 MPa in the dry state. These measurements are the basis of the constitutive model used in this study.

Elastic–plastic material models describe the behaviour of many cellulose materials well. Marin et al. showed that a linear variation in moisture ratio agreed with the constitutive behaviour of paperboard [37]. However, this concerns low moisture levels: with relative humidity in the range of 20–90%, resulting in moisture ratios of less than 0.2. In the wet state, paper/cellulose holds much greater amounts of water, and the mechanical properties are drastically different. Like the wet surface of a pulp fibre [38], the cellulose bead is a hydrogel in the wet state [39]. The chosen material model must thus account for a soft, gel-like behaviour in the wet state and elastic in the dry state. A hyperelastic model captures the soft behaviour in the wet state and allows fitting of the input parameters as moisture dependent functions to account for the behaviour in the dry state. An isotropic

Mooney-Rivlin [40,41] model with three parameters achieves this using the strain energy potential function

$$W = C_{10}\left(\bar{I}_1 - 3\right) + C_{01}\left(\bar{I}_2 - 3\right) + \frac{1}{D_1}\left(J^{el} - 1\right)^2,\tag{1}$$

where $\bar{I}_1$ and $\bar{I}_2$ are the first and second deviatoric strain invariants, and J^{el} is the elastic volume ratio. Linear elastic materials parameters approximated the coefficients C_{10} and C_{01}, and D_1 using the relations

$$C_{10} = \frac{G}{2\left(1 + \frac{C_{01}}{C_{10}}\right)},\tag{2}$$

and

$$D_1 - 2/K\tag{3}$$

Here, G is the shear modulus defined by the elastic modulus of the material E and Poisson's ratio v as:

$$G = \frac{E}{2(1+v)},\tag{4}$$

and K is the bulk modulus, also defined by E and v as

$$K = \frac{E}{3(1-2v)}.\tag{5}$$

The hyperelastic model aims at mimicking the linear-elastic material behaviour as well as possible. Rewriting Equation (2) with the ratio

$$x = C_{01}/C_{10},\tag{6}$$

allows for a parameter investigation. The linear-elastic and hyperelastic models were compared using a simple model with one two-dimensional quadrilateral element. Applying uniaxial and biaxial loading to the element allowed for a comparison of the resulting stress-strain plots, as shown in Figure 3. The value of $x = 0.5$ resulted in the best agreement with the linear-elastic curve. This ratio gave the hyperelastic coefficients throughout the simulations. Poisson's ratio was assumed to be 0.3.

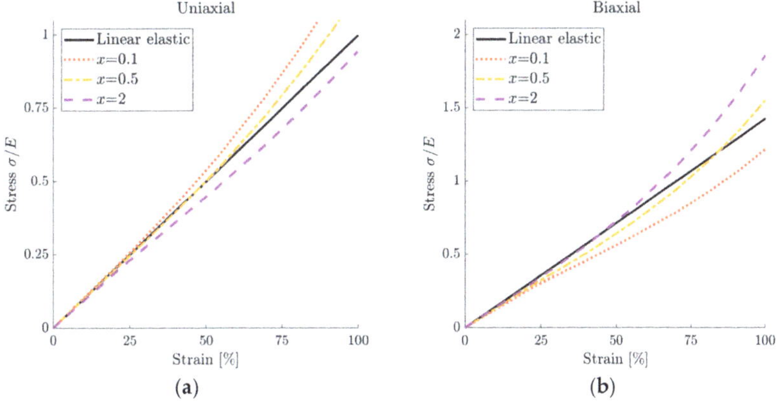

(a)
(b)

Figure 3. Comparison of a hyperelastic material model to linear-elastic for a simple one-element case in (a) uniaxial loading and (b) biaxial loading.

The values for the elastic modulus E measured by Li et al. [14] determined the moisture dependence of the hyperelastic parameters. The moisture ratio m_r depends on the mass of water m_{water} and dry content m_{dry} as

$$m_r = \frac{m_{water}}{m_{dry}}. \tag{7}$$

The moisture content m_c has a non-linear relationship with the moisture ratio.

$$m_c = \frac{m_{water}}{m_{water} + m_{dry}} = \frac{m_r}{1 + m_r}. \tag{8}$$

Assuming that the relationship between E and m_r is linear, a non-linear relationship between E and m_c emerges. The parameter study evaluates this assumption, comparing it to another dependency, see Section 2.4. Moisture transport within the bead followed the moisture diffusion equation, describing the development of m_c over time as

$$\frac{\partial m_c}{\partial t} = \alpha \nabla^2 m_c, \tag{9}$$

where α is the diffusion coefficient. The hygroscopic strain ε_h depends linearly on m_c as

$$\varepsilon_h = \beta \Delta m_c, \tag{10}$$

where β is the expansion coefficient. The assumed moisture content of the bead was $m_c^0 = 0.9$ at the onset of drying and $m_c^\infty = 0.1$ after drying. Previous measurements of the water content in similar beads in the wet state gave the initial value [42]. The final value of the moisture content was approximated by water contents in dry paper sheets, usually around 4–6% depending on the type of paper and ambient moisture content [43].

A linear-elastic material model describes the constitutive behaviour of the underlying surface, with an elastic modulus >10 times the stiffness of the bead and the same Poisson's ratio. Together with the fixed constraint applied at the bottom edge, the surface does not deform noticeably in relation to the deformation of the bead.

2.2.3. Contact Description

The contact between the bead and the surface is defined by a surface pair with a node-to-surface formulation. As the softer body, the bead is the secondary surface, while the underlying body is the primary surface.

The adhesion in the contact zone is modelled by a cohesive interaction. It is similar to the implementation of cohesive elements [44], but the surface has no volume. The cohesive behaviour describes the loads that arise when the surfaces attempt to separate. Separation can occur in two directions: normal and shear. The normal direction only considers tensile separation, as compressive stresses only push the surfaces closer together. A traction-separation law describes the relationship between the separation of the surfaces and the resulting tractions. This case uses a bilinear law consisting of three unique parameters, as seen in Figure 4a.

Here, the fracture energy G_c and the damage initiation traction defined the cohesive degradation. Cohesive displacements in one direction do not cause cohesive stresses in the other, making the cohesive stiffness matrix uncoupled. Initially, the bilinear traction separation law defines the cohesive traction t_i ($i = n,s$, either normal or shear mode) as dependent on the cohesive stiffness K_i and the separation δ_i

$$t_i = K_i \delta_i, \tag{11}$$

until it reaches the damage initiation traction t_i^0

$$\max\left(\frac{t_n}{t_n^0}, \frac{t_s}{t_s^0}\right) = 1, \tag{12}$$

where t_n and t_s, are the tractions in the normal and shear directions. The brackets $\langle \cdot \rangle$ are the Macauley operator, only accounting for positive values (damage does not evolve in compression). After damage initiation, the traction drops until it reaches the failure separation δ^f, or releases the fracture energy G_c. The damage parameter D degrades the tractions after passing the peak as

$$t_n = \begin{cases} (1-D)\bar{t}_n & , \ (\bar{t}_n \geq 0), \\ \bar{t}_n & , \ (\bar{t}_n < 0), \end{cases} \tag{13}$$
$$t_s = (1-D)\bar{t}_s,$$

where the nominal tractions $\bar{t}_i$ are tractions calculated from the nominal cohesive stiffness, see Equation (11). An effective measure accounts for the contributions from the normal and shear directions

$$(\cdot)_m = \sqrt{(\cdot)_n{}^2 + (\cdot)_s^2}, \tag{14}$$

where $(\cdot)$ is the relevant property. The damage parameter depends on the effective displacement,

$$D = \frac{\delta_m^f \left(\delta_m^{max} - \delta_m^0\right)}{\delta_m^{max} \left(\delta_m^f - \delta_m^0\right)}. \tag{15}$$

where δ_m^{max} is the maximum value of the effective displacement δ_m to have occurred (D can not decrease) and δ_m^0 is the damage initiation value. The fracture energy and effective traction at damage initiation t_m^0 (defined according to Equation (14)) define the effective displacement at failure

$$\delta_m^f = 2G_c / t_m^0. \tag{16}$$

The cohesive model includes moisture dependence by prescribing different parameters for different moisture levels, in between which the parameters vary linearly, as illustrated in Figure 4b. The damage evolution process (post-peak behaviour) can have a different softening behaviour than linear, see Figure 4c, but previous studies show that this has little influence on the results for loading in tension unless the adhesive material is highly ductile [45,46]. The parameter study considers this by varying the softening law, see Section 2.4.

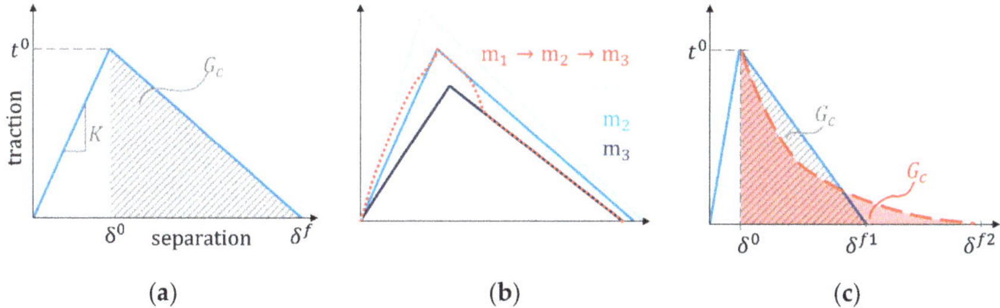

Figure 4. Schematic of (**a**) traction–separation law, (**b**) contact constitutive behaviour under moisture change during loading, and (**c**) example of cohesive law with similar fracture energy and damage initiation tractions but different softening behaviours.

2.2.4. Numerical Aspects

Examining the energy outputs in the model is a way to ensure the credibility of the modelling procedures. The energy outputs were all positive, meaning that the model dissipates and does not generate energy. Running the simulations with a halved element size (in the finest region at the contact zone) assured a mesh-independent solution. It gave the same result in terms of contact stresses and deformation. However, changing the element size introduced convergence issues for the solution. Since the elements in the bead shrink significantly while the underlying surface does not shrink at all, it is believed that a size mismatch in the contact algorithm causes the problems. Moreover, this made the solution sensitive to parameter variations. Using other parameters or mesh would require a different mesh strategy, such as an adaptive procedure. Unlike the current implementation, an adaptive procedure allows node positions to adjust to the deformation thus preventing excessive distortion and size mismatch.

2.3. Experimental Comparison

A trial-and-error procedure comparing the experimental geometry as measured by Li et al. [15] to the simulation results resulted in the material parameters summarised in Table 1. Figure 5 shows the experimental data regarding the radius of the bead and the radius of the contact area between the bead and the underlying surface.

Table 1. Material parameters in the model determined from experimental comparison.

Notation	Unit	Description
α	mm^2/s	Diffusion coefficient (Equation (9)
β	-	Expansion coefficient (Equation (10))
K_n	MPa/mm	Cohesive stiffness in the normal direction (Equation (11))
K_s	MPa/mm	Cohesive stiffness in the shear direction (Equation (11))
t_n^0	MPa	Maximum cohesive traction in the normal direction (Equation (12))
t_s^0	MPa	Maximum cohesive traction in the shear direction (Equation (12))
G_c	N/mm	Cohesive fracture energy (Equation (16))

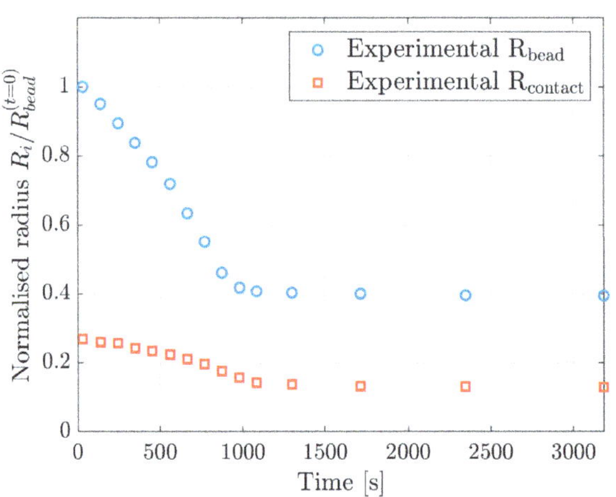

Figure 5. Drying geometry of water-swollen cellulose beads replotted from Li et al. [15]. The radii are normalised against the initial radius of the bead.

2.4. Parameter Study

A parameter study followed the calibration of the fitting parameters. Introducing variations to the model parameters showed their influence on the overall response. The influence is summarised by observing the effect on the resulting bead geometry, and by comparing the stresses at select locations in the model for the different variations.

The first step in the parameter investigation examined the effect of changing the constitutive behaviour of the bulk material by altering the evolution of the elastic modulus with moisture, so that the change was slow at first and then accelerated by the end. An exponential variation in m_c replaced the linear variation in m_r, see Figure 6. The hyperelastic parameters were determined with the same procedure as before, using the resulting elastic modulus, as previously described in Section 2.2.2.

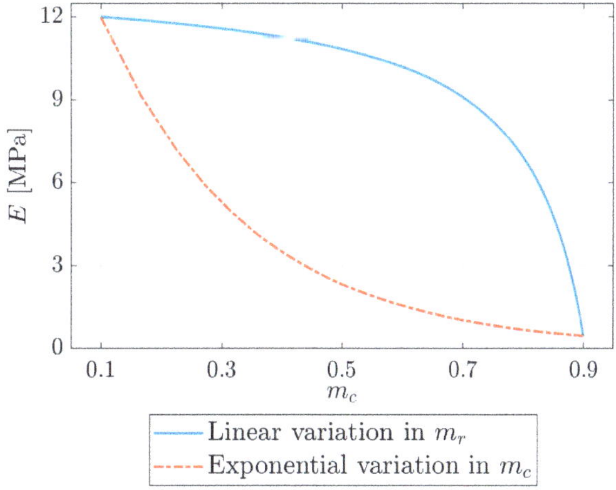

Figure 6. Development of the elastic modulus E with moisture content m_c for the two tested cases.

The second part of the parameter study examined the effect of changing the cohesive law. Changing the softening behaviour, while keeping the fitting parameters constant, allowed for studying its impact on the solution (see Figure 4c for an example of different softening behaviours). Replacing the linear softening as defined in Equation (15) with an exponential gives the damage variable as

$$D = \int_{\delta_m^0}^{\delta_m^f} \frac{t_m}{G_c - G_0} d\delta, \tag{17}$$

where G_0 is the elastic energy at the onset of damage.

3. Results

3.1. Parameter Fitting to Experimental Measurements

Figure 7a shows the simulated bead and contact radii resulting from the parameter fit, together with the experimental data. Figure 7b presents the resulting damage variable D, as defined in Equation (15), plotted against the deformed shape of the bead. Table 2 shows the determined fitting parameters. Figure 8 presents the resulting bead contact radius when varying some of the cohesive fitting parameters. The variation of remaining cohesive parameters (cohesive stiffness and strength in the normal direction, K_n and t_n^0, respectively) gave no difference in results when changing the parameters by up to 50%.

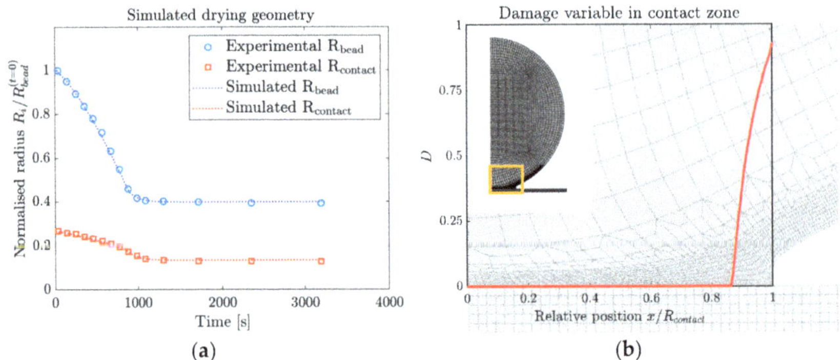

Figure 7. Results from simulations: (**a**) geometry evolution obtained by fitting of drying and diffusion coefficients, compared to experimental data replotted from Li et.al. [15] and (**b**) damage variable D (red) plotted for the last time step in the deformed model.

Table 2. Resulting material parameters in the model given for the initial and final values of the moisture content, $\mathbf{m_c^0}$ and $\mathbf{m_c^\infty}$, respectively.

Notation	Unit	Value	
		$\mathbf{m_c^0 = 0.9}$	$\mathbf{m_c^\infty = 0.1}$
α	mm^2/s	5.0×10^{-4}	5.0×10^{-5}
β	-	0.50	0.75
K_n	MPa/mm	8.0×10^6	8.0×10^6
K_s	MPa/mm	10	300
t_n^0	MPa	1.0×10^3	1.0×10^3
t_s^0	MPa	6.0	6.0
G_c	N/mm	0.27	0.27

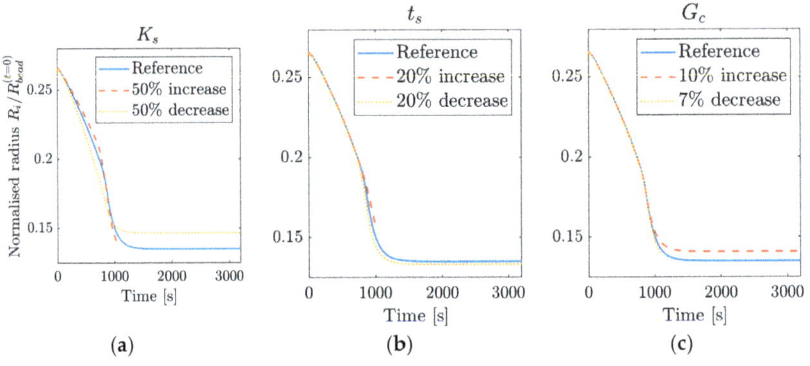

Figure 8. Parameter variations for (**a**) the cohesive stiffness in the shear direction K_s, (**b**) the cohesive strength in the shear direction t_s^0, and (**c**) the fracture energy G_c. The reference curves correspond to the values given in Table 2.

3.2. Effect of Parameter Variation

Figure 9 compares previously simulated reference curves (as shown in Figure 7a) to the resulting drying geometry with the customised development of the elastic modulus and exponential damage softening. In Figure 9a, both the bead and contact radii are shown for the different simulations. Figure 9b shows a close-up of the contact radius results and in addition, comparison to the experimental data. Figure 10 presents the normal stresses in the tangential (x-direction) and perpendicular (y-direction) directions for three different elements in the bead, placed slightly above the contact zone. Figure 10a–c compares the stress development over time in each element for the three different parameter sets (see Section 2.4). The positions of the elements are shown before the onset of shrinkage in Figure 10d and after finalised shrinkage in Figure 10e; the first element (from left to right) is the contact zone, the second is close to the edge of the contact zone, and the third one is outside the contact zone.

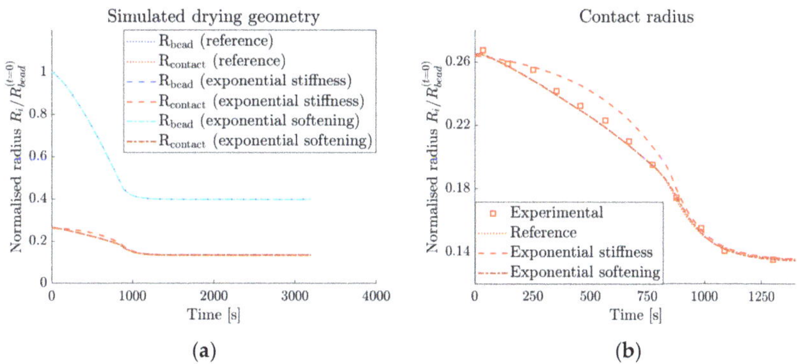

Figure 9. Simulation results for (**a**) bead and contact radii compared for reference simulations (as described in Section 3.1) and custom stiffness and exponential damage softening (as described in Section 2.4), and (**b**) a close-up of simulated contact radii compared to experimental data.

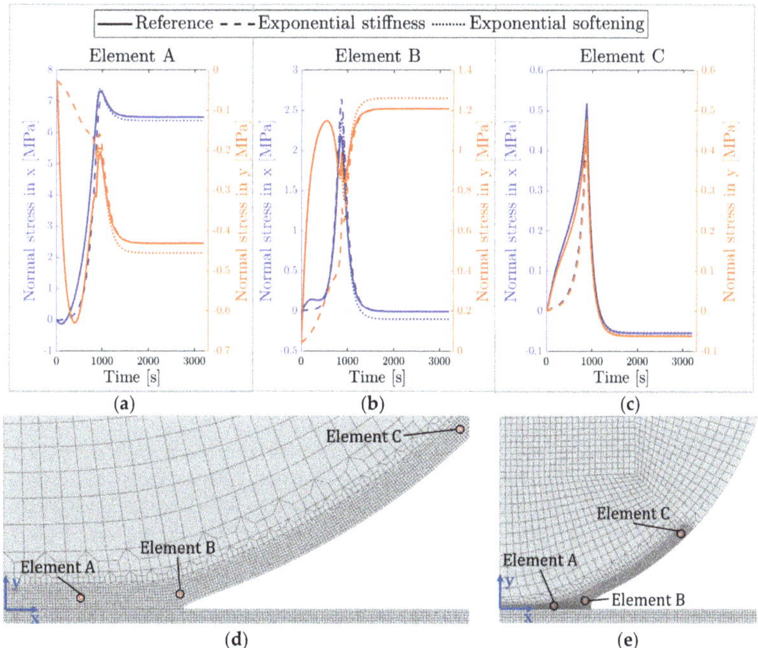

Figure 10. Element investigation: (**a**–**c**) stresses in two directions in three elements of the model for the parameter variations, and position of the elements in the model (**d**) before the onset of shrinkage and (**e**) after finalised shrinkage.

4. Discussion

The comparison between experimental measurements and simulated results (see Figure 7a) shows that the model is successful in replicating the observed drying geometry. The values presented for the fitting parameters in Table 2 show that varying all parameters linearly with moisture content is sufficient, suggesting that the model is well-suited for the application. Examining the damage value after the completion of drying, as shown in Figure 7b, shows that the edge of the contact zone is close to complete degradation, with a maximum value of $D = 0.93$.

The resulting material parameters, as presented in Table 2, give insight into what is happening with the material as it dries. The diffusion coefficient decreases with decreased moisture, while the expansion coefficient instead increases. As opposed to the onset of drying, the moisture gradient is higher in the late stages. Moreover, the shrinkage is also greater and thus, the developed stresses will be larger. Looking at the cohesive material parameters, the high tensile stiffness K_n allows little to no separation between the contacting surfaces, while the lower value of K_s allows some slippage in the shear direction. However, the shear stiffness increases over the course of drying, while the maximum shear traction t_s^0 and the fracture energy G_c remain constant. Consequently, the joint becomes more brittle as it dries, which translates well to fibre networks; the material can be highly ductile in the wet state, but brittle when dry.

At closer examination, only contact stresses in the shear direction contribute to the degradation of the cohesive interaction. Consequently, the result is insensitive to higher values of cohesive damage initiation traction in the normal direction t_n^0. Another set of data would thus be necessary to determine the normal strength of the adhesive zone. The model could then be applied to pull-off measurements, where the force required to pull the bead off the surface is measured. The method developed here presents a good starting point for such an investigation, with the shear parameters already determined. Since the shear mode dominates this part of the analysis, further analysis should also investigate coupled effects, where the damage initiation and evolution depend on the normal and shear contributions nonuniformly.

As shown in Figure 8, the initial contact radius is insensitive to variations in cohesive parameters. However, several cohesive parameters affect the final contact radius. It raises the question of the uniqueness of the determined parameter set, which also requires further examination.

As expected, changing the stiffness evolution of the bead does not affect its shape during drying, as shown in Figure 9a. The hygroscopic strains are only dependent on the moisture expansion coefficient and the moisture content, see Equation (10). Changing the damage evolution behaviour also has no effect. However, changing the stiffness evolution does visibly affect the development of the contact radius, see Figure 9b. The lower stiffness in the initial stages (see Figure 6) will cause the cohesive tractions to be lower as well and damage will develop slower, therefore delaying decohesion (radius decrease). Since the final stiffness values are the same in both cases, the final values for the contact radius are the same, as the exponential variation increases the stiffness rapidly by the end. To understand how different stiffness values during drying impact the damage evolution in the normal direction, there needs to be further investigation. A possible course of action is, for example, studying pull-off measurements in intermediate states, prior to the complete drying of the bead. Subsequently, the experimental results can be compared to simulations with different stiffness evolution.

Changing the damage evolution shape has a negligible effect on the contact radius, as shown in Figure 9b. However, there is an effect on the residual stresses after completed drying. The curves in Figure 10a–c show that the peak stresses occur in all locations at almost the same time. The observed peak corresponds to the time when the moisture difference between the bulk of the bead and the ambient conditions stabilises. With one exception, the absolute value of the stresses decreases after this point, see Figure 10a–c. Presumably, the vanishing moisture gradient causes this drop. However, the stress in the y-direction close to the contact zone (element B) experiences an increase, see Figure 10b. A possible explanation is that the surrounding elements compress the edge of the contact zone in the presence of the moisture gradient, and when it disappears, the tensile stresses increase.

The changing of stiffness evolution introduces differences in the stresses compared to the reference model, which is apparent in all three examined elements. Close to the edge of the contact zone and outside of it, corresponding to element B and C (see Figure 10), respectively, the numerical values are similar, but the delayed stiffening also delays the stress development. Inside the contact zone the stress curves in the y-direction look initially very different. The slower development of stresses delays the development of damage. However, this difference in stress is not as large in the x-direction, where all three curves agree very well.

Analysing the effect of changing the softening behaviour from linear (as in the reference simulation) to exponential, proves that the effect on the stress evolution is small. However, there is a visible difference in the final stress value in the elements within the contact zone. The damage parameter D, as defined in Equations (15) and (17), will have different values in the two cases. The tractions are thus not equally reduced, causing the resulting residual stresses to differ. The difference is mostly small (1–6%), except in the case of the parallel stress close to the edge of the contact zone. The residual stress in the reference case is almost zero, but with the exponential damage evolution it results in a slight compressive stress.

Being able to examine what happens in the adhesive zone in detail makes the developed model useful for future evaluations of adhesion in model surfaces. As mentioned previously, pull-off tests will be of particular interest to further the understanding of other loading modes and adhesive failure. In a recent study Li et al. measured the adhesion between cellulose beads and a cellulose thin film [47]. In the pull-off test, the force required to separate surfaces is measured, but it says little about the stress state in the contact zone. The proposed model makes it possible to link global quantities, such as the pull-off force or drying geometry, to the stress state in the contact zone and degree of decohesion.

Another interesting application is studying the adhesion in the intermediate state, before complete drying, to understand how the moisture gradient affects the adhesion. As previously shown, mechanical properties and the internal structure change during drying [14,15,42], which is why it would be interesting to investigate the differences and similarities to changes in adhesive properties: the model can estimate cohesive properties and residual stresses. In addition to investigating the surface chemistry and how it contributes to the adhesion, the impact of drying can be explored, and the findings transferred to explain what happens when a fibre–fibre joint is formed.

For stronger interactions, e.g., when cellulose interacts with cellulose, the model could capture other phenomena important in contact creation. For example, in the case of a cellulose bead placed on a deformable film, wrinkling or other significant deformation of the film may occur. This model allows for studying this virtually, before doing complicated and time-consuming experiments. It is valuable both for designing and interpreting experiments.

However, the model might need further development to capture all major phenomena. For example, when the bead shrinks, the introduced deformations are very large. Because of the adhesive forces, the contact radius does not shrink proportionally to the bead radius. In cases with even larger adhesive forces, like cellulose–cellulose interactions, this mismatch will grow. Increased stresses close to the edge of the contact zone, greater strain gradients in the bead, and mesh distortions will follow, making the problem numerically more complex to solve. In this case, a different mesh approach would be suitable. The current approach uses the Lagrangian method, where the mesh nodes follow the deformation of the body; instead, an adaptive mesh could be introduced, where the distortion is limited.

Another aspect of the model that might need further development is the material assumptions. The hyperelastic model allowed for modelling soft behaviour with large deformations, although little is known about the constitutive behaviour of the beads. In particular, the material dependence on moisture and by extension, the moisture field in the bead during drying are interesting for future examination. By comparing the model to new experimental data, the material assumptions can be further evaluated and refined if needed.

Similarly, the description of the adhesive forces needs evaluation. The cellulose model surfaces intend to reflect simplified surfaces of fibres and capture interactions between them. However, which forces act in a fibre joint is still not completely determined. Molecular bonds, friction, and capillary forces are examples of physical phenomena believed to play a role in the fibre joint strength. Here, a cohesive description based on two principal directions, normal and tangential to the surface, with irreversible damage development replaces these. The model with the fitted parameters was fit to successfully replicate the drying geometry, but further investigation is needed to determine if the proposed description sufficiently replaces the adhesive forces.

To summarise, this paper presents a useful model for the investigation of adhesive forces between a bead and a surface. It is shown to be successful in replicating the observed drying geometry of a cellulose model bead and a rigid surface. Evaluation and comparison to pull-off experiments will further the understanding of the drying process in the formation of fibre–fibre joints. In particular, the model makes it possible to study an, at present, unmeasurable property: the development of stresses in the contact zone. The well-defined geometry of the cellulose bead model surface makes it possible to isolate adhesion from other phenomena and the findings can be implemented on more complex geometries, for example, fibre joints. The model also offers the possibility of implementation in fibre network models, with a more authentic description of the joint mechanics.

5. Conclusions

This paper presents a model for understanding the development of adhesion of cellulose beads during drying. The developed model is useful in furthering the knowledge of fibre joint formation in fibre network materials, and the most important conclusions are presented below.

- The developed model successfully replicates the drying geometry observed from experiments with parameters varying linearly with moisture.
- Several fitting parameters have similar effects on the solution; therefore, determination of a unique solution needs more examination.
- The validity of the model assumptions needs further investigation, especially in terms of moisture dependence and material parameters.
- It is a valuable tool for understanding the development of stresses during drying, and implementation in pull-off tests of model surfaces can further the insights.
- By determining the cohesive material parameters, it is shown that the shear mode completely dominates the decohesion during drying.

Author Contributions: Conceptualization, M.K. and S.Ö.; methodology, M.K. and S.Ö.; software, M.K.; formal analysis, M.K.; investigation, M.K.; writing—original draft preparation, M.K.; writing—review and editing, S.Ö.; visualization, M.K.; supervision, S.Ö.; funding acquisition, S.Ö. All authors have read and agreed to the published version of the manuscript.

Funding: This research was funded by Knut and Alice Wallenberg Foundation.

Institutional Review Board Statement: Not applicable.

Informed Consent Statement: Not applicable.

Data Availability Statement: The experimental data used is found in the literature, simulated data can be acquired upon request from the corresponding author.

Acknowledgments: The authors would like to acknowledge current and former colleagues at Solid Mechanics, Department of Engineering Mechanics, and Fibre Technology, Department of Fibre and Polymer Technology, KTH Royal Institute of Technology. Among those, in particular, Artem Kulachenko and Erik Olsson for help with software implementation and troubleshooting.

Conflicts of Interest: The authors declare no conflict of interest.

References

1. Delgado-Fornué, E.; Contreras, H.J.; Toriz, G.; Allan, G.G. Adhesion between Cellulosic Fibers in Paper. *J. Adhes. Sci. Technol.* **2011**, *25*, 597–614. [CrossRef]
2. Torgnysdotter, A.; Wågberg, L. Tailoring of Fibre/Fibre Joints in Order to Avoid the Negative Impacts of Drying on Paper Properties. *Nord. Pulp Pap. Res. J.* **2006**, *21*, 411–418. [CrossRef]
3. Borodulina, S.; Kulachenko, A.; Nygårds, M.; Galland, S. Stress-Strain Curve of Paper Revisited. *Nord. Pulp Pap. Res. J.* **2012**, *27*, 318–328. [CrossRef]
4. Goutianos, S.; Mao, R.; Peijs, T. Effect of Inter-Fibre Bonding on the Fracture of Fibrous Networks with Strong Interactions. *Int. J. Solids Struct.* **2018**, *136–137*, 271–278. [CrossRef]
5. Li, Y.; Yu, Z.; Reese, S.; Simon, J.-W. Evaluation of the Out-of-Plane Response of Fiber Networks with a Representative Volume Element Model. *Tappi J.* **2018**, *17*, 329–339. [CrossRef]
6. Jajcinovic, M.; Fischer, W.J.; Hirn, U.; Bauer, W. Strength of Individual Hardwood Fibres and Fibre to Fibre Joints. *Cellulose* **2016**, *23*, 2049–2060. [CrossRef]
7. Magnusson, M.S.; Zhang, X.; Östlund, S. Experimental Evaluation of the Interfibre Joint Strength of Papermaking Fibres in Terms of Manufacturing Parameters and in Two Different Loading Directions. *Exp. Mech.* **2013**, *53*, 1621–1634. [CrossRef]
8. Hirn, U.; Schennach, R. Fiber-Fiber Bond Formation and Failure: Mechanisms and Analytical Techniques. In Proceedings of the Advances in Pulp and paper Research, Oxford, UK, September 2017; pp. 839–863. Available online: https://bioresources.cnr.ncsu.edu/features/proceedings-of-the-pulp-and-paper-fundamental-research-symposia/ (accessed on 1 February 2023).
9. Huang, F.; Li, K.; Kulachenko, A. Measurement of Interfiber Friction Force for Pulp Fibers by Atomic Force Microscopy. *J. Mater. Sci.* **2009**, *44*, 3770–3776. [CrossRef]
10. Eriksson, M.; Notley, S.M.; Wågberg, L. Cellulose Thin Films: Degree of Cellulose Ordering and Its Influence on Adhesion. *Biomacromolecules* **2007**, *8*, 912–919. [CrossRef]
11. Sczech, R.; Riegler, H. Molecularly Smooth Cellulose Surfaces for Adhesion Studies. *J. Colloid Interface Sci.* **2006**, *301*, 376–385. [CrossRef] [PubMed]
12. Carrick, C.; Pendergraph, S.A.; Wågberg, L. Nanometer Smooth, Macroscopic Spherical Cellulose Probes for Contact Adhesion Measurements. *ACS Appl. Mater. Interfaces* **2014**, *6*, 20928–20935. [CrossRef] [PubMed]
13. Hellwig, J.; Karlsson, R.-M.P.; Wågberg, L.; Pettersson, T. Measuring Elasticity of Wet Cellulose Beads with an AFM Colloidal Probe Using a Linearized DMT Model. *Anal. Methods* **2017**, *9*, 4019–4022. [CrossRef]
14. Li, H.; Mystek, K.; Wågberg, L.; Pettersson, T. Development of Mechanical Properties of Regenerated Cellulose Beads during Drying as Investigated by Atomic Force Microscopy. *Soft Matter* **2020**, *16*, 6457–6462. [CrossRef] [PubMed]
15. Li, H.; Kruteva, M.; Mystek, K.; Dulle, M.; Ji, W.; Pettersson, T.; Wågberg, L. Macro- and Microstructural Evolution during Drying of Regenerated Cellulose Beads. *ACS Nano* **2020**, *14*, 6774–6784. [CrossRef]
16. Li, H.; Kruteva, M.; Dulle, M.; Wang, Z.; Mystek, K.; Ji, W.; Pettersson, T.; Wågberg, L. Understanding the Drying Behavior of Regenerated Cellulose Gel Beads: The Effects of Concentration and Nonsolvents. *ACS Nano* **2022**, *16*, 2608–2620. [CrossRef] [PubMed]
17. Johnson, K.L.; Kendall, K.; Roberts, A.D. Surface Energy and the Contact of Elastic Solids. *Proc. R. Soc. London. A Math. Phys. Sci.* **1971**, *324*, 301–313. [CrossRef]
18. Derjaguin, B.V.; Muller, V.M.; Toporov, Y.P. Effect of Contact Deformations on the Adhesion of Particles. *J. Colloid Interface Sci.* **1975**, *53*, 314–326. [CrossRef]
19. Hertz, H. Ueber Die Berührung Fester Elastischer Körper. *Crll* **1882**, *1882*, 156–171. [CrossRef]
20. Maugis, D. Adhesion of Spheres: The JKR-DMT Transition Using a Dugdale Model. *J. Colloid Interface Sci.* **1992**, *150*, 243–269. [CrossRef]
21. Thouless, M.D.; Jensen, H.M. The Effect of Residual Stresses on Adhesion Measurements. *J. Adhes. Sci. Technol.* **1994**, *8*, 579–586. [CrossRef]
22. Olsson, E.; Larsson, P.-L. On Force–Displacement Relations at Contact between Elastic–Plastic Adhesive Bodies. *J. Mech. Phys. Solids* **2013**, *61*, 1185–1201. [CrossRef]
23. Peng, B.; Feng, X.-Q.; Li, Q. Decohesion of a Rigid Flat Punch from an Elastic Layer of Finite Thickness. *J. Mech. Phys. Solids* **2020**, *139*, 103937. [CrossRef]
24. Menga, N.; Carbone, G.; Dini, D. Do Uniform Tangential Interfacial Stresses Enhance Adhesion? *J. Mech. Phys. Solids* **2018**, *112*, 145–156. [CrossRef]
25. Wu, F.; Li, C. Theory of Adhesive Contact on Multi-Ferroic Composite Materials: Conical Indenter. *Int. J. Solids Struct.* **2021**, *233*, 111217. [CrossRef]
26. Yang, T.; Liechti, K.M.; Huang, R. A Multiscale Cohesive Zone Model for Rate-Dependent Fracture of Interfaces. *J. Mech. Phys. Solids* **2020**, *145*, 104142. [CrossRef]
27. de Souza Lima, R.; Ré, M.-I.; Arlabosse, P. Drying Droplet as a Template for Solid Formation: A Review. *Powder Technol.* **2020**, *359*, 161–171. [CrossRef]

28. Perdana, J.; Fox, M.B.; Schutyser, M.A.I.; Boom, R.M. Mimicking Spray Drying by Drying of Single Droplets Deposited on a Flat Surface. *Food Bioprocess Technol.* **2013**, *6*, 964–977. [CrossRef]
29. Perdana, J.; Aguirre Zubia, A.; Kutahya, O.; Schutyser, M.; Fox, M. Spray Drying of Lactobacillus Plantarum WCFS1 Guided by Predictive Modeling. *Dry. Technol.* **2015**, *33*, 1789–1797. [CrossRef]
30. Dugdale, D.S. Yielding of Steel Sheets Containing Slits. *J. Mech. Phys. Solids* **1960**, *8*, 100–104. [CrossRef]
31. Barenblatt, G.I. The Mathematical Theory of Equilibrium Cracks in Brittle Fracture. In *Advances in Applied Mechanics*; 1962; Volume 7, pp. 55–129. Available online: https://www.sciencedirect.com/science/article/pii/S0065215608701212?via%3Dihub#aep-abstract-id9 (accessed on 1 February 2023).
32. Lin, B.; Auernhammer, J.; Schäfer, J.-L.; Meckel, T.; Stark, R.; Biesalski, M.; Xu, B.-X. Humidity Influence on Mechanics of Paper Materials: Joint Numerical and Experimental Study on Fiber and Fiber Network Scale. *Cellulose* **2022**, *29*, 1129–1148. [CrossRef]
33. Peng, X.-L.; Huang, G.-Y. Effect of Temperature Difference on the Adhesive Contact between Two Spheres. *Int. J. Eng. Sci.* **2017**, *116*, 25–34. [CrossRef]
34. Günther, N.; Griese, M.; Stammen, E.; Dilger, K. Modeling of Adhesive Layers with Temperature-Dependent Cohesive Zone Elements for Predicting Adhesive Failure during the Drying Process of Cathodic Dip Painting. *Proc. Inst. Mech. Eng. Part L J. Mater. Des. Appl.* **2018**, *233*, 146442071880600. [CrossRef]
35. Qin, R.; Lau, D. Evaluation of the Moisture Effect on the Material Interface Using Multiscale Modeling. *Multiscale Sci. Eng.* **2019**, *1*, 108–118. [CrossRef]
36. *ABAQUS*; Dassault Systémes Simulia Corp.: Providence, RI, USA, 2019.
37. Marin, G.; Nygårds, M.; Östlund, S. Elastic-Plastic Model for the Mechanical Properties of Paperboard as a Function of Moisture. *Nord. Pulp Pap. Res. J.* **2020**, *35*, 353–361. [CrossRef]
38. Pelton, R. A Model of the External Surface of Wood Pulp Fibers. *Nord. Pulp Pap. Res. J.* **1993**, *8*, 113–119. [CrossRef]
39. Karlsson, R.-M.P.; Larsson, P.T.; Yu, S.; Pendergraph, S.A.; Pettersson, T.; Hellwig, J.; Wågberg, L. Carbohydrate Gel Beads as Model Probes for Quantifying Non-Ionic and Ionic Contributions behind the Swelling of Delignified Plant Fibers. *J. Colloid Interface Sci.* **2018**, *519*, 119–129. [CrossRef]
40. Mooney, M. A Theory of Large Elastic Deformation. *J. Appl. Phys.* **1940**, *11*, 582–592. [CrossRef]
41. Rivlin, R.S. Large Elastic Deformations of Isotropic Materials IV. Further Developments of the General Theory. *Philos. Trans. R. Soc. London Ser. A Math. Phys. Sci.* **1948**, *241*, 379–397. [CrossRef]
42. Mystek, K.; Reid, M.S.; Larsson, P.A.; Wågberg, L. In Situ Modification of Regenerated Cellulose Beads: Creating All-Cellulose Composites. *Ind. Eng. Chem. Res.* **2020**, *59*, 2968–2976. [CrossRef]
43. Tillmann, O. Paper and Board Grades and Their Properties. In *Handbook of Paper and Board*; Holik, H., Ed.; Wiley: New York, NY, USA, 2006; pp. 446–466. ISBN 9783527309979.
44. Spring, D.W.; Paulino, G.H. A Growing Library of Three-Dimensional Cohesive Elements for Use in ABAQUS. *Eng. Fract. Mech.* **2014**, *126*, 190–216. [CrossRef]
45. Camanho, P.P.; Davila, C.G.; de Moura, M.F. Numerical Simulation of Mixed-Mode Progressive Delamination in Composite Materials. *J. Compos. Mater.* **2003**, *37*, 1415–1438. [CrossRef]
46. Fernandes, R.L.; Campilho, R.D.S.G. Testing Different Cohesive Law Shapes to Predict Damage Growth in Bonded Joints Loaded in Pure Tension. *J. Adhes.* **2017**, *93*, 57–76. [CrossRef]
47. Li, H.; Roth, S.V.; Freychet, G.; Zhernenkov, M.; Asta, N.; Wågberg, L.; Pettersson, T. Structure Development of the Interphase between Drying Cellulose Materials Revealed by In Situ Grazing-Incidence Small-Angle X-ray Scattering. *Biomacromolecules* **2021**, *22*, 4274–4283. [CrossRef] [PubMed]

Disclaimer/Publisher's Note: The statements, opinions and data contained in all publications are solely those of the individual author(s) and contributor(s) and not of MDPI and/or the editor(s). MDPI and/or the editor(s) disclaim responsibility for any injury to people or property resulting from any ideas, methods, instructions or products referred to in the content.

Article

Creep Crack Growth Behavior during Hot Water Immersion of an Epoxy Adhesive Using a Spring-Loaded Double Cantilever Beam Test Method

Kota Nakamura [1], Yu Sekiguchi [2,*], Kazumasa Shimamoto [3], Keiji Houjou [3], Haruhisa Akiyama [3] and Chiaki Sato [2]

[1] Department of Mechanical Engineering, Tokyo Institute of Technology, 4259 Nagatsuta-cho, Midori-ku, Yokohama 226-8503, Japan

[2] Institute of Innovative Research, Tokyo Institute of Technology, 4259 Nagatsuta-cho, Midori-ku, Yokohama 226-8503, Japan

[3] Nanomaterials Research Institute, National Institute of Advanced Industrial Science and Technology (AIST), 1-1-1 Higashi, Tsukuba 305-8565, Japan

* Correspondence: sekiguchi.y.aa@m.titech.ac.jp

Abstract: Double cantilever beam (DCB) tests were conducted by immersing the specimens in temperature-controlled water while applying a creep load using a spring. By introducing a data reduction scheme to the spring-loaded DCB test method, it was confirmed that only a single parameter measurement was sufficient to calculate the energy release rate (ERR). Aluminum alloy substrates bonded with an epoxy adhesive were used, and DCB tests were performed by changing the initial load values, spring constants, and immersion temperatures for two types of surface treatment. The initial applied load and spring constant had no effect on the ERR threshold. In contrast, the threshold decreased with the increasing immersion temperature, but even in the worst case, it was 15% of the critical ERR in the static tests. Using the creep crack growth relationship, it was revealed that there were three phases of creep immersion crack growth in the adhesive joints, and each phase was affected by the temperature. The spring-loaded DCB test method has great potential for investigating the combined effects of creep, moisture, and temperature, and this study has demonstrated the validity of the test method. The long-term durability of adhesive joints becomes increasingly important, and this test method is expected to become widespread.

Keywords: adhesive bonding; hydrothermal creep; degradation; creep fracture toughness; aging

Citation: Nakamura, K.; Sekiguchi, Y.; Shimamoto, K.; Houjou, K.; Akiyama, H.; Sato, C. Creep Crack Growth Behavior during Hot Water Immersion of an Epoxy Adhesive Using a Spring-Loaded Double Cantilever Beam Test Method. *Materials* **2023**, *16*, 607. https://doi.org/10.3390/ma16020607

Academic Editors: Marco Lamberti and Aurelien Maurel-Pantel

Received: 27 December 2022
Revised: 4 January 2023
Accepted: 6 January 2023
Published: 8 January 2023

Copyright: © 2023 by the authors. Licensee MDPI, Basel, Switzerland. This article is an open access article distributed under the terms and conditions of the Creative Commons Attribution (CC BY) license (https://creativecommons.org/licenses/by/4.0/).

1. Introduction

Joining technology is essential for the assembly of parts. In particular, adhesive bonding is superior to other joining methods in terms of stress dispersion, bonding of dissimilar materials, and weight reduction. Therefore, they are widely used for electrical parts, vehicles, and buildings. As the use of adhesives increases, reliability issues become more important. Many factors are related to the degradation of adhesive joints, but moisture is one of the most well-known and important factors [1,2]. Because adhesives are mainly composed of polymers, they absorb water. Water penetrating the adhesive not only changes the material properties of the adhesive [3,4] but also attacks the interface and weakens the bond [5–7]. Creep and fatigue loads are also important for the durability of adhesive bonding [8–10]. The creep strength and fatigue strength are much lower than the static strength but even lower under humid conditions. Therefore, it is important to study the combination of moisture and creep or fatigue loads [11]. In general, specimens are first immersed or subjected to high humidity, and then fatigue tests are conducted under ambient conditions [12–15]. Conversely, some studies have found that the deterioration of joints by immersion in water is accelerated by external loads [3,16–19]. Therefore, immersion in water during creep/fatigue testing could alter the results.

The mechanical properties of adhesive joints are generally evaluated based on their strength and toughness. The best-known method for evaluating strength is the single-lap joint (SLJ) test. Springs are sometimes used when applying creep loads to SLJ specimens under humid conditions [20,21]. Spring-loaded SLJ specimens have also been immersed in hot water or exposed on ship decks [16,22]. The double cantilever beam (DCB) test is commonly used to evaluate fracture toughness [23,24]. In the static fracture test, the DCB specimen is set to a mechanical tensile testing machine and loaded in mode I at a constant opening speed. Three parameters, the crack length, load, and opening displacement, are required to calculate the fracture toughness of the adhesives in the DCB tests. However, the optical measurement of the crack length causes large errors in the calculation of fracture toughness. Therefore, new approaches for crack measurement methods using digital image correlation and mechanoluminescence have been proposed [25,26]. Moreover, a method to reduce one of the three parameters has been proposed: the compliance-based beam method (CBBM) [27]. To apply a creep load to the DCB specimens, a tensile testing machine can be used in the same way as for static tests [8]. However, in this case, it is difficult to maintain a specimen exposed to humid conditions. Another method of testing creep crack resistance is to insert a wedge into the DCB specimen, which is the Boeing wedge test [28–31]. In this case, creep-loaded specimens can be exposed to different environmental conditions to investigate their degradation [32–35]. Conversely, optical crack measurement is required to calculate the fracture toughness, leading to concerns regarding increased calculation errors. Studies have been conducted to accurately measure the crack length in wedge tests [36,37]; however, these techniques are not a panacea. To overcome these weaknesses, a spring-loaded DCB test has been proposed for hot water immersion tests but remains largely untested to date [38]. Similar to the spring-loaded SLJ test, the spring exerts a creep load on the specimen; however, the displacement increases and the load decreases as the crack propagates. By measuring the load and displacement over time and applying the CBBM, a change in the fracture toughness can be obtained. Therefore, it has the potential to evaluate crack resistance under a combination of creep load and immersion without measuring the crack length. Although no studies have focused on the effectiveness and accuracy of the spring-loaded DCB test method so far, this test method deserves more attention.

In this study, spring-loaded DCB tests were conducted under hot and wet conditions. First, the effects of initial load and spring constant on the creep crack growth (CCG) were investigated to verify the accuracy of the calculation process of the energy release rate and the experimental equipment in the test method. In addition, the effects of the water temperature and surface preparation on CCG behavior were investigated. Three temperature conditions (32, 63, and 90 °C) and two surface treatments (sandblasting and pickling) were examined, and the results are discussed in terms of the CCG rate.

2. Methods

2.1. Compliance-Based Fracture Energy Calculation

The fracture behavior of adhesive joints has been widely discussed based on linear elastic fracture mechanics, and the mode I critical energy release rate, G_{IC}, is given by:

$$G_{IC} = \frac{P^2}{2b}\frac{dC}{da} \tag{1}$$

where P is the applied load, b is the specimen width, $C = \delta/P$ is the compliance, δ is the opening displacement, and a is the crack length. Therefore, in DCB tests, the crack length, displacement, and load must be measured to calculate the energy release rate. However, because it is known that simple beam theory (SBT) provides the relationship between these three parameters, a parameter reduction scheme using SBT (i.e., CBBM) was validated for measuring the fracture toughness of various adhesives [39–41]. Considering the shear effect in the SBT, the compliance can be expressed as follows:

$$C = \frac{\delta}{P} = \frac{8a_e^3}{Ebh^3} + \frac{12a_e}{5bhG} \tag{2}$$

where a_e is the equivalent crack length, which is the sum of the crack length a and the crack length correction Δa; E, G, and h are the longitudinal and transverse moduli of the elasticity and thickness of the substrate, respectively. Substituting Equation (2) into Equation (1):

$$G_{IC} = \frac{6P^2}{b^2h} \left(\frac{2a_e^2}{h^2E} + \frac{1}{5G} \right) \tag{3}$$

is obtained. By solving Equation (2) for a_e and substituting it into Equation (3), the energy release rate is expressed as a function of the load and displacement [40].

2.2. Load–Displacement Curve in the Spring-Loaded Testing Method

A schematic of the loading procedures in the spring-loaded DCB test method is shown in Figure 1. The wing nut was tightened to compress the spring and apply an initial load to the specimen. Here, the amount of spring contraction, Δ_{spring}, is given by:

$$\Delta_{\text{spring}} = \frac{P}{k_{\text{spring}}} \tag{4}$$

where k_{spring} is the spring constant. The spring-loaded DCB test includes two stages: load applying stage (stage 1) and crack propagation stage (stage 2). In the first stage, the crack length is considered constant as $a = a_1$, where a_1 is the crack length at the beginning of the first stage. Therefore, the compliance in the first stage C_1 keeps constant. From Equations (2) and (4), the opening displacement in the first stage can be expressed as

$$\delta = PC_1 = k_{\text{spring}} \Delta_{\text{spring}} C_1 \tag{5}$$

Thus, there is a linear relationship between δ and Δ_{spring}. In the second stage, the cracks gradually propagate. Therefore, the compliance and displacement increases and the load decreases. Considering the change in the magnitude of the spring contraction, the relationship between the load and displacement is as follows:

$$(P - P_2) = -k_{\text{spring}}(\delta - \delta_2) \tag{6}$$

where P_2 and δ_2 are the load and displacement at the stage change point, i.e., the initial values of the second stage.

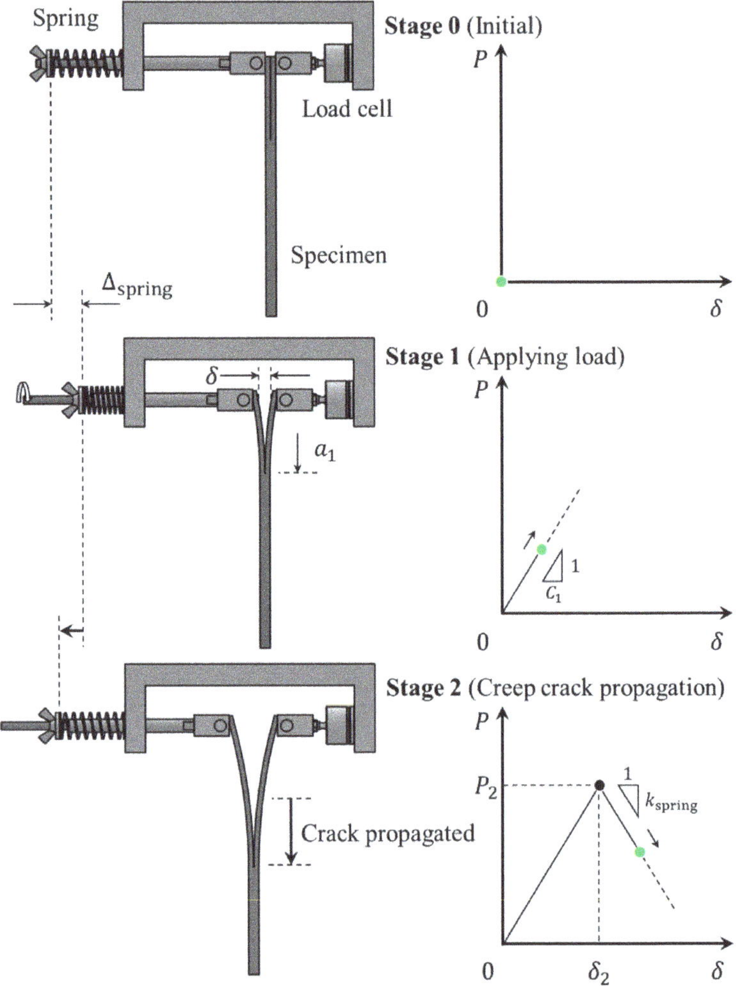

Figure 1. Schematic diagram of a spring-loaded DCB test: stage 0—before a test; stage 1—load applying process by tightening a nut; stage 2—creep load applying process.

3. Experimental

3.1. Materials and Specimens

An aluminum alloy (A6061-T6) with a length of $l = 188$ mm, width of $b = 25$ mm, and thickness of $h = 4$ mm was used as the substrate of the DCB specimens. The thermoset epoxy adhesive used (Cemedine Co., Ltd., Tokyo, Japan) consisted of bisphenol A epoxy resin, which contained carboxyl-terminated butadiene acrylonitrile rubber (CTBN), fumed silica, $CaCO_3$, and CaO as a base resin; dicyandiamide as a curing agent; and 3-(3,4-dichlorophenyl)-1,1'-dimethylurea as a curing accelerator. CTBN was contained to improve its fracture toughness and elongation. Two different surface treatments were applied to the aluminum alloy substrates: sandblasting and pickling. The process of the first method was sandblasting with an air pressure of 0.7 MPa using Al_2O_3 abrasive grains, followed by degreasing with acetone. In the second method, the substrates were degreased with acetone, immersed in an alkaline solution at 60 °C for 30 s, washed with purified water, dried, and immersed in an acidic solution at 60 °C for 30 s. The substrates were bonded with the adhesive and cured for 1 h at 180 °C in an electric furnace. The thickness of the adhesive layer was controlled by inserting a 0.3 mm thick polytetrafluoroethylene (PTFE) sheet at both ends of the adhesive layer. The PTFE sheet also produced an initial crack: $a_0 = 50$ mm. The geometry of the DCB specimens used in this study is shown in Figure 2. In addition, before the creep test, a crack of approximately 5 mm in length was generated using a universal tensile tester (AGS-X 10kN, Shimadzu Corp., Kyoto, Japan) to produce a sharp initial crack, i.e., $a_1 \approx 55$ mm.

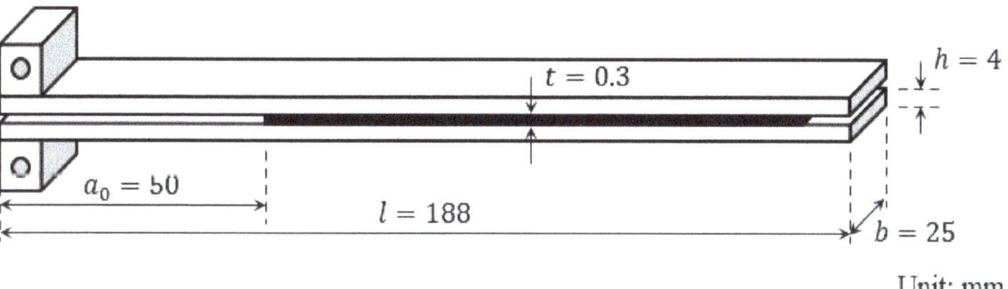

Figure 2. Geometry of a DCB test specimen.

3.2. Experimental Setup

A schematic diagram of the spring-loaded experimental setup is shown in Figure 3. The applied load and opening displacement were measured using a load cell (LUX-B-1kN, Kyowa Electronic Instruments Co., Ltd., Tokyo, Japan) and a displacement sensor (DTK-A-50, Kyowa Electronic Instruments Co., Ltd., Tokyo, Japan). These output data were converted to voltage via strain amplifiers (DA-18A, Tokyo Measuring Instruments Laboratory Co., Ltd., Tokyo, Japan), digital data via an analog input unit (AI-1608GY-USB, CONTEC Co., Ltd., Osaka, Japan), and then recorded on a PC using in-house developed data acquisition software. The temperature of the purified water was controlled using a water bath.

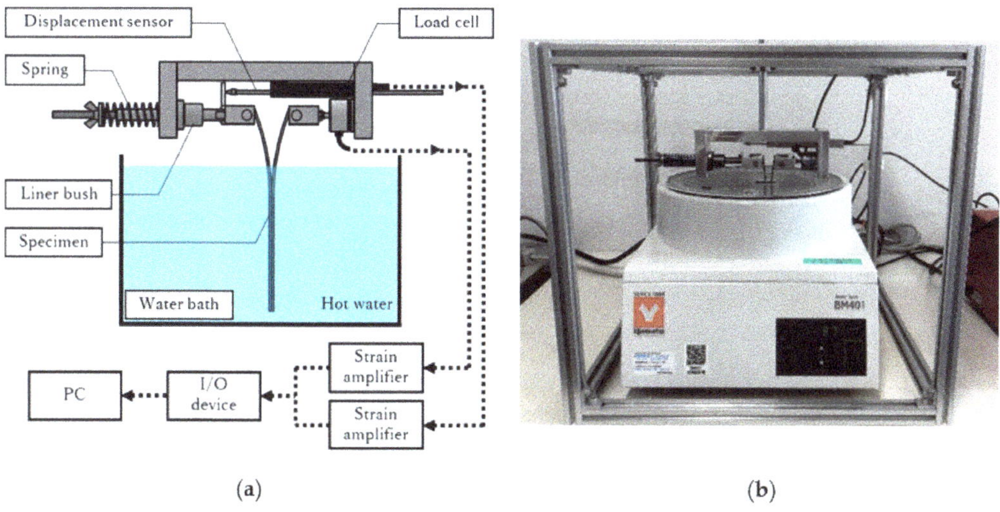

Figure 3. (**a**) Schematic diagram and (**b**) photograph of the spring-loaded DCB test setup.

3.3. Spring-Loaded DCB Test

The tests were conducted under two test conditions with different objectives. First, the accuracy of the experimental system and calculation method was investigated by changing the spring constant and initial loading conditions. The test conditions are listed in Table 1. Next, the effects of the surface treatment and water immersion temperature, T (32, 63, and 90 °C), on the crack growth behavior were investigated keeping the initial energy release rate, G_{ini}, at approximately 500 J/m^2, which is approximately 75% of the critical energy release rate ($G_{IC} = 679$ J/m^2). The test conditions are listed in Table 2.

Table 1. Experimental conditions for studying the effects of the spring constant and initial energy release rate.

No.	k_{spring} (N/mm)	G_{ini} (J/m^2)	Surface Treatment	T (°C)
1	25	500		
2	49	500		
3	98	498	Sandblasting	90
4	49	621		
5	49	372		
6	49	253		

Table 2. Experimental conditions for studying the effects of the surface treatment and immersion temperature.

Specimen	k_{spring} (N/mm)	G_{ini} (J/m^2)	Surface Treatment	T (°C)
SB1		475		32
SB2	25	504	Sandblasting	63
SB3		500		90
AC1		465		32
AC2	25	492	Pickling	63
AC3		539		90

4. Results and Discussion

4.1. Verification of the Spring-Loaded DCB Test

A linear bush was inserted to reduce the friction and keep the shaft coaxial with the load cell (see Figure 3). If the friction is negligible and the shaft moves smoothly, Equations (5) and (6) are satisfied, and the load can be calculated from the displacement, and vice versa. In this way, a further parameter reduction is possible, and the energy release rate can be calculated from the load or displacement only. The results with varying spring constants are shown in Figure 4 for the load–displacement relationship and in Figure 5 for the crack length and energy release rate as a function of time. The lines indicate the results obtained with measured displacements only, whereas the marks indicate the results obtained with the measured loads and displacements. Because the difference between the lines and marks was small, it shows that the friction was sufficiently low, and the equipment system worked well. Thus, Equations (5) and (6) are valid, and it is possible to calculate the energy release rate by measuring only one parameter. However, to avoid unexpected situations, both the loads and displacements were recorded in all experiments, and the energy release rate was calculated from these two values in subsequent experiments.

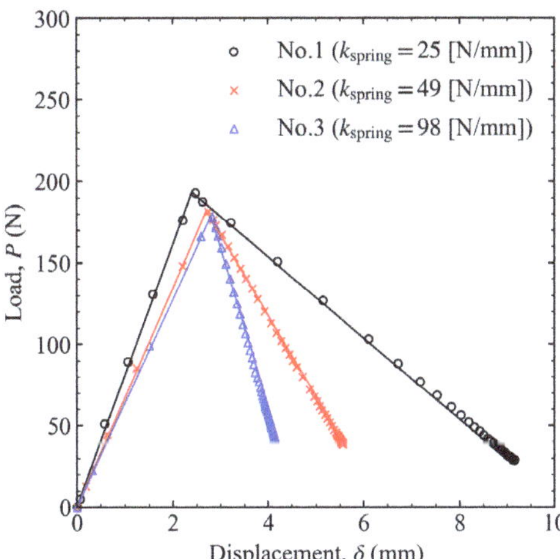

Figure 4. Relationship between load and displacement when the spring constant was changed to 25, 49, and 98 N/m.

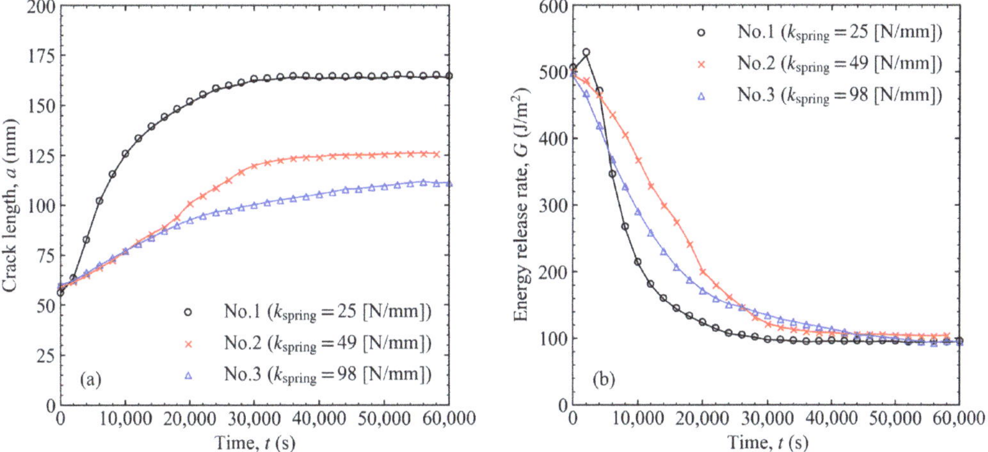

Figure 5. (**a**) Crack length and (**b**) energy release rate variation over time when the spring constant was changed to 25, 49, and 98 N/m.

When the spring constant was changed, the slopes of the load–displacement relationship in the first and second stages varied, as shown in Figure 4. The smaller the spring constant, the more cracks propagated, as shown in Figure 5a. Conversely, the energy release rate converged to a certain value regardless of the spring constant, as shown in Figure 5b. This value is called the threshold for the energy release rate (G_{th}). This was approximately 15% of G_{IC}. Although G_{ini} was set to almost the same value, different spring constants resulted in different changes in the energy release rate. At the smallest spring constant, the energy release rate initially increased and then rapidly decreased to approach G_{th}. However, at a larger spring constant, the energy release rate decreased monotonically to G_{th}. The difference in the initial G variation was related to the relationship between the line in Equation (6) and the curve of $G = G_{ini}$. In Figure 6, I_2 is the change point from stages 1 to 2. When the spring constant was small, the line was positioned above the curve of $G = G_{ini}$, between I_2 and I_4, as shown in Figure 6a. In such cases, G increased between I_2 and I_3

and then decreased. I_3 is the point where the line and the G constant curve intersected at a single point. Conversely, when the spring constant was sufficiently large, the line was below the curve after intersecting the curve at I_2, as shown in Figure 6b. In this case, G decreased monotonically.

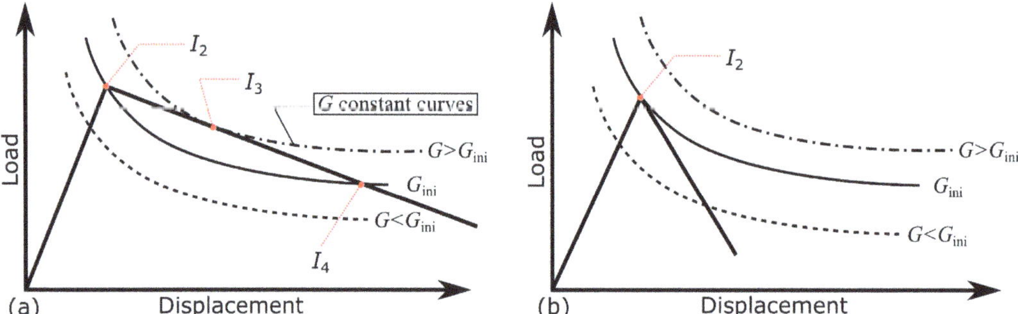

Figure 6. Schematic diagram of the relationship between the experimentally determined load–displacement relationship and energy release rate constant curve in the case for (**a**) a small spring constant and (**b**) a large spring constant.

The results of varying the initial energy release rate using the same spring are shown in Figure 7 for the load–displacement relationship and in Figure 8 for the crack length and energy release rate versus time. Because the spring constants were the same, the slopes of the load–displacement results were the same for different G_{ini}. In addition, G_{th} was almost the same, even when G_{ini} varied. Therefore, the threshold was considered to be independent of the spring constant and the initial load level.

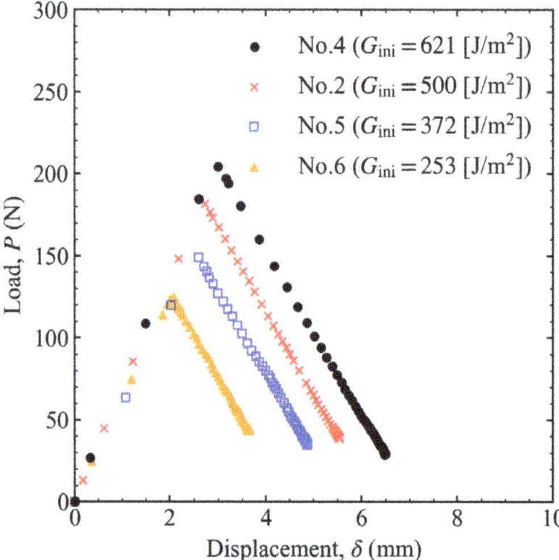

Figure 7. Relationship between the load and displacement when the initial load level was changed to 621, 500, 372, and 253 J/m^2.

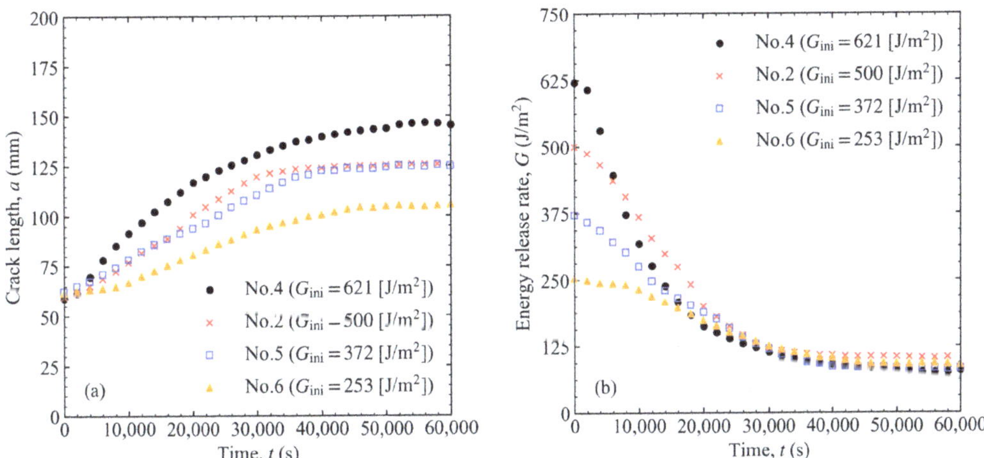

Figure 8. (**a**) Crack length and (**b**) energy release rate variation over time when the initial load level was changed to 621, 500, 372, and 253 J/m^2.

4.2. Effects of the Surface Treatment and Immersion Temperature

The changes in the crack length and energy release rate over time are shown in Figures 9 and 10 at different temperatures and surface treatments. Comparing the crack propagation at different water temperatures, it is clear that the hotter the water, the more the crack propagated. The decrease in the energy release rate also tended to increase with the increasing water temperature. Comparing the differences in the surface preparation, it can be seen that the pickling treatment suppressed crack propagation more than the sandblasting treatment.

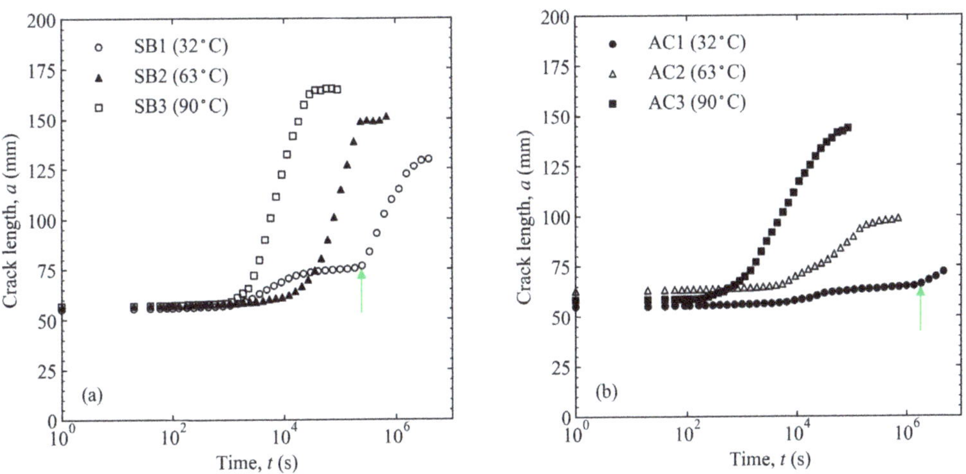

Figure 9. Crack length variation over time at different temperatures and surface treatments: (**a**) sandblasting and (**b**) pickling. The light-green arrows are the failure mode changing points, which relate to the locations of the light-green triangles in Figure 11.

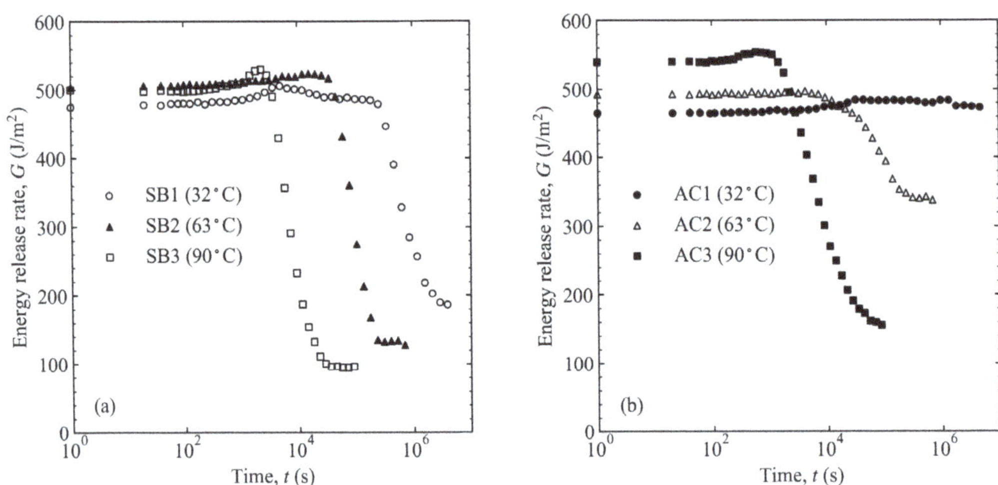

Figure 10. Energy release rate variation over time at different temperatures and surface treatments: (**a**) sandblasting and (**b**) pickling.

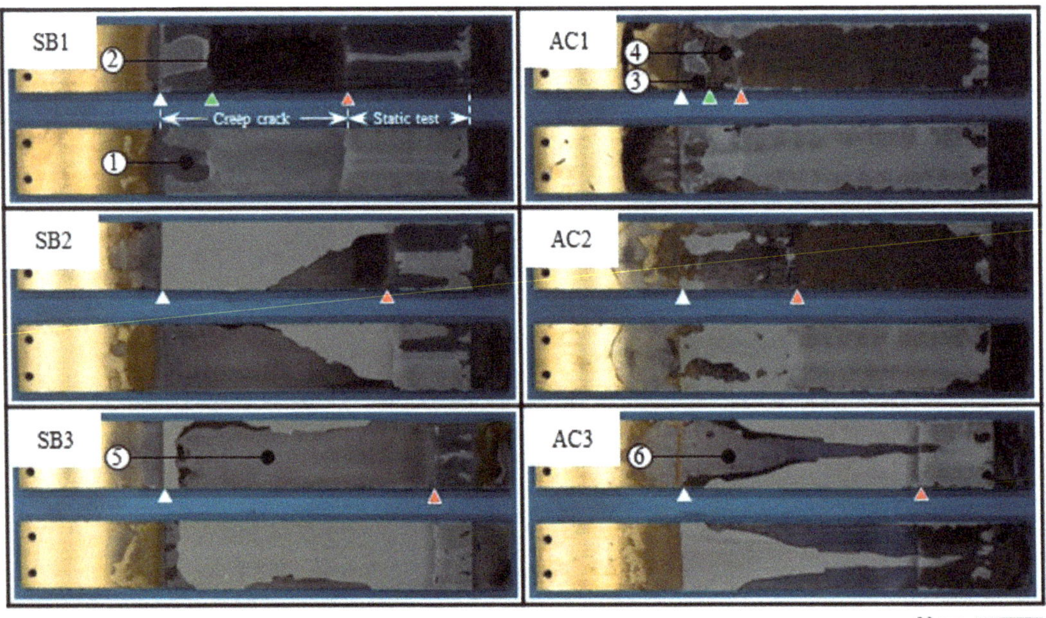

Figure 11. The macroscopic image of the fractured surfaces of the specimens after the spring-loaded DCB tests. White triangles: initial crack positions; light-green triangles: failure mode changing points; red triangles: crack positions after creep immersion tests. Numbered locations are magnified using a microscope in Figure 12.

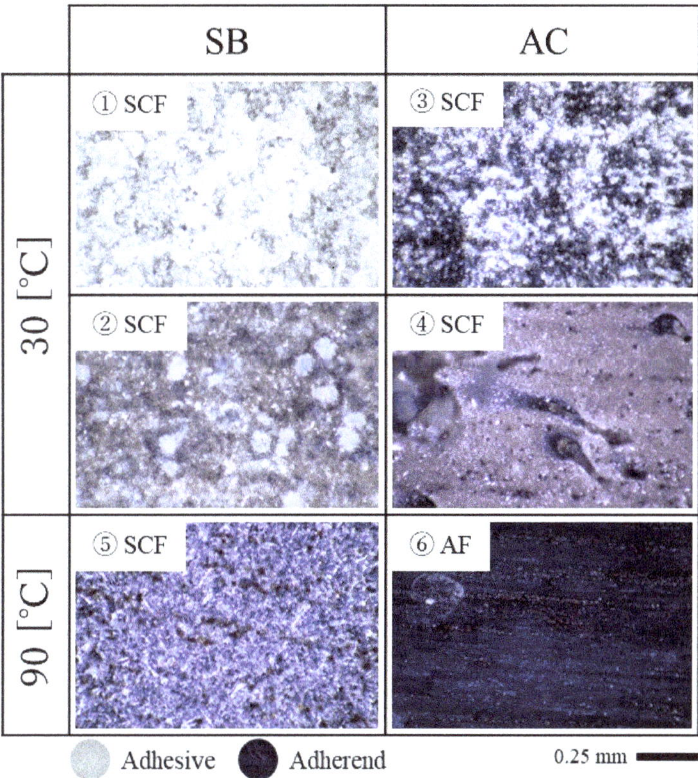

Figure 12. Magnified photograph under a microscope at selected points from Figure 11 (SCF: special cohesive failure: AF: adhesive failure).

After the tests, the specimens were completely separated, and fractured surfaces were observed, as shown in Figure 11. The initial crack positions are indicated by the white triangles, and the crack positions after the creep immersion tests are indicated by red triangles. Macroscopic observation of the surface showed that most of the epoxy remained on one side of the surface, and it appears as if the adhesive failure (AF) occurred in all tests. Here, AF is a failure mode of an adhesive layer when a failure occurs at the interface between the adhesive and either of the two adherends. Therefore, as shown in Figure 12, magnified observations were made at six specific points using a polarizing microscope (BX53P, Olympus Corp., Tokyo, Japan). Because the epoxy resin is white in color and aluminum alloy is dark in color when viewed with the polarizing microscope, it can be seen that some of the epoxy resin remained on the surfaces and that there were differences in the amount of epoxy resin remaining. In most cases, a thin layer of epoxy was seen on the surface, and the failure mode was considered substrate-near cohesive failure rather than AF. Thus, it is commonly classified as special cohesive failure (SCF). Here, cohesive failure (CF) is a failure mode of an adhesive layer when a failure occurs inside the layer and is classified as SCF, especially when a large amount of adhesive remains on either surface of the adherends. A detailed surface analysis showed that the sandblasting left more epoxy on the surface than the pickling, and AF was observed only when the pickling-treated specimen was immersed at 90 °C. In general, AF exhibits weak interfacial bonding. However, the crack propagated more in the sandblasted specimen with SCF than in the pickled specimen with AF at 90 °C immersion. In hot water, epoxy undergoes various chemical changes, and cracks can be caused by factors other than water penetration. Such complications may have reversed the trend.

It is also noteworthy that two steps of crack propagation were observed in both surface treatments with 32 °C immersion (SB1 and AC1). When comparing points 1 and 2 or 3 and 4, as shown in Figure 12, differences in the amount of epoxy remaining were observed depending on the location. The light-green triangles in Figure 11 show the transition point of failure, which was approximately 70 mm for SB1 and 65 mm for AC1. In the first step, a lot of epoxy remained on the aluminum surfaces, but with longer immersion time, the amount of epoxy remaining on the surface decreased. Along with this change, the crack propagated further, as shown by the arrows in Figure 9.

4.3. Creep Crack Growth Rate

The relationship between the crack growth rate and fracture parameters is used to discuss the crack growth behavior of the adhesive or interface under creep-loading conditions [8,42–48]. Similar to Paris' law of fatigue crack growth (FCG) behavior, a power law relationship:

$$\frac{da}{dt} = A\left(\frac{G_{\mathrm{I}}}{G_{\mathrm{IC}}}\right)^{m}$$

(7)

is obtained for the CCG behavior when the adhesive is assumed to be a viscoelastic material, where A and m are constants, and t is time [42]. In fatigue testing, data smoothing of the FCG rate is required because of data scattering, and the incremental polynomial method and power-law fitting approach are commonly used [49]. The same is true for the CCG rate. Therefore, in this study, a linear approximation was made for every five consecutive data points of crack length versus time, and the slope was used as the differential coefficient of the central point.

The CCG relationship (i.e., the CCG rate versus normalized energy release rate) is shown in Figure 13. Three characteristic trends were observed: initial stagnation, crack propagation, and threshold onset, as shown in Figure 14. In the case of immersion at 32 and 60 °C, the crack hardly grew for a while after loading, and the energy release rate was almost constant. Therefore, a vertical change was observed in the initial stage of the CCG relationship (region I). In contrast, for the 90 °C immersion, hardly any change was observed in the vertical direction, and the crack started to propagate immediately after loading, i.e., region II started without region I. Even at lower immersion temperatures, the crack started to propagate after a sufficient time had passed. In this case, the results follow the power law relationship after the CCG rate recovery, i.e., it flips at the turning point from regions I to II and after a while moves to the bottom left. The delay in the crack growth is considered, because it takes longer for water to penetrate the adhesive at lower temperatures. The penetration rate was faster at a hotter immersion, resulting in a difference in the initial trend. The arrows in Figure 13a,d refer to the light-green triangles in Figure 11, so the change in the fracture surfaces was related to the change in the CCG relationship. Pickling has a larger exponent in the power law m than sandblasting, and it is larger with a lower immersion temperature. Thus, it can be seen that the crack grew more slowly with a larger m. Comparing coefficient A, it increased with increasing temperature, but dependence on the surface treatment method was not observed. After region II, a vertical change associated with the slowing of the crack growth was observed (region III). Therefore, with the help of the CCG relationship, it is possible to clearly determine whether the results are approaching the threshold.

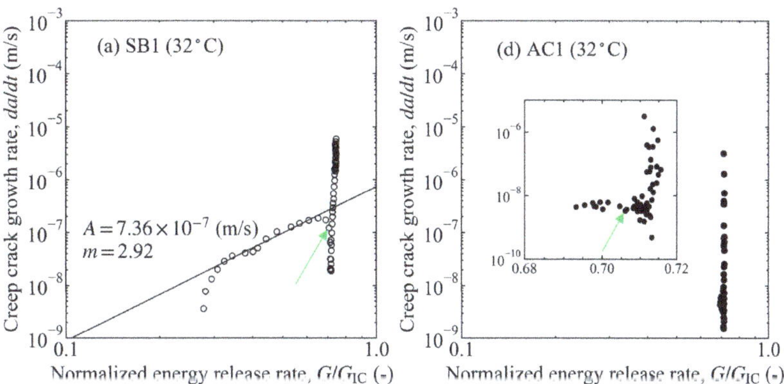

Figure 13. *Cont.*

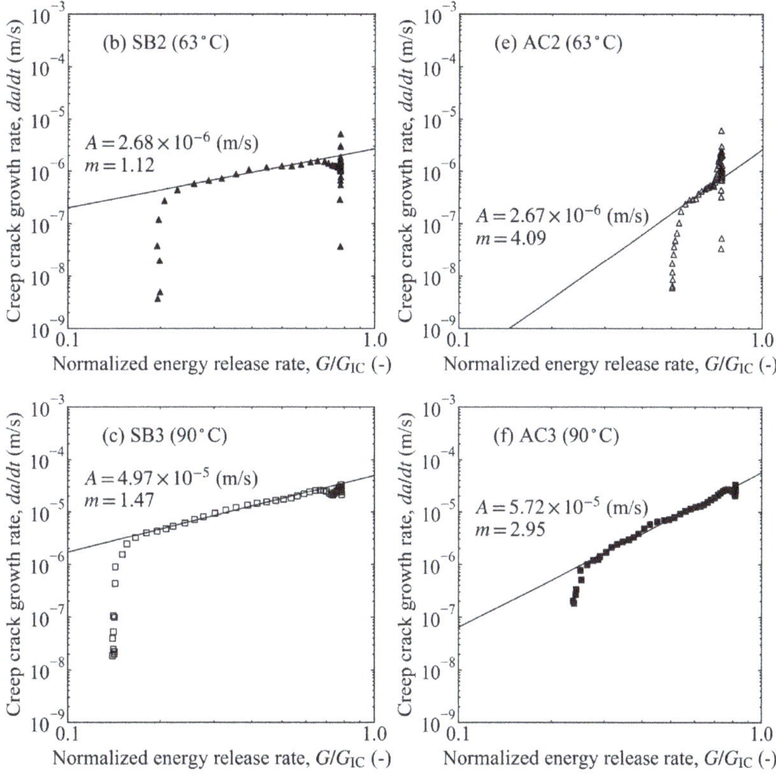

Figure 13. Creep crack growth relationship between the creep crack growth rate and normalized energy release rate at different temperatures and surface treatments: (**a–c**) sandblasting; (**d–f**) pickling at 32, 63, and 90 °C, respectively. A magnified graph is included in (**d**). The light-green arrows in (**a**) and (**d**) are the failure mode changing points, which relate to the locations of the light-green triangles in Figure 11.

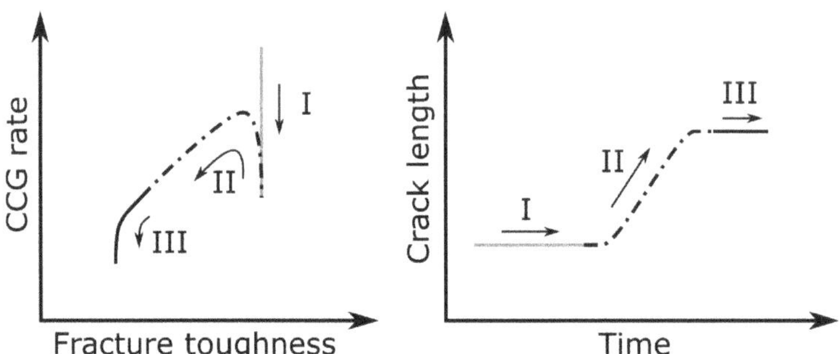

Figure 14. Schematic diagram of the state changes accompanying crack growth due to the creep immersion.

5. Conclusions

In this study, a DCB specimen consisting of aluminum plates bonded with an epoxy adhesive was set in a spring-loaded jig and subjected to a creep load. Degradation due to the creep and immersion was experimentally investigated by placing the loaded specimens in temperature-controlled water. In the original DCB test method, three parameters are needed to calculate the energy release rate: the crack length, load, and displacement. However, by introducing a data-reduction scheme using the compliance method, the energy release rate can be calculated with two parameters. Moreover, the relationship between displacement and load can be theoretically derived for the spring-loaded DCB test method when inserting a spring with a known spring constant. Therefore, another data-reduction scheme is possible, and it has been experimentally confirmed that only a single parameter, the load or displacement, is sufficient to calculate the energy release rate of the spring-loaded DCB tests. By changing the spring constant and the initial value of the load at a constant immersion temperature, it became clear that the threshold of the energy release rate depended only on the temperature. Conversely, the threshold value increased at lower immersion temperatures. Moreover, the crack growth was inhibited more by the pickling treatment than by the sandblasting treatment under a combination of creep and immersion conditions. The change in the crack growth behavior was evident when the results were plotted using the relationship between the creep crack growth rate and energy release rate. From various points of view, the accuracy of the spring-loaded DCB test was shown to be very high, and it became clear that it is a well-suited test method for evaluating hydrothermal creep.

Author Contributions: Conceptualization, H.A. and C.S.; Methodology, K.N. and Y.S.; Software, K.N.; Analysis, K.N., K.S., K.H., and H.A.; Investigation, K.N.; Writing—Original Draft Preparation, K.N.; Writing—Review and Editing, Y.S.; Funding Acquisition, C.S. All authors have read and agreed to the published version of the manuscript.

Funding: This research was funded by the New Energy and Industrial Technology Development Organization (NEDO) under project number: JPNP14014.

Institutional Review Board Statement: Not applicable.

Informed Consent Statement: Not applicable.

Data Availability Statement: The data presented in this study are available upon request from the corresponding author.

Acknowledgments: This work was based on the results obtained from a project commissioned by the New Energy and Industrial Technology Development Organization (NEDO), and we would like to thank all the project members for their contribution. We express our appreciation to Cemedine Co., Ltd., for the supply of materials.

Conflicts of Interest: The authors declare no conflict of interest.

References

1. Viana, G.; Costa, M.; Banea, M.D.; da Silva, L.F.M. A review on the temperature and moisture degradation of adhesive joints. *Proc. Inst. Mech. Eng. L* **2017**, *231*, 488–501. [CrossRef]
2. Houjou, K.; Shimamoto, K.; Akiyama, H.; Sato, C. Effect of cyclic moisture absorption/desorption on the strength of epoxy adhesive joints and moisture diffusion coefficient. *J. Adhes.* **2022**, *98*, 1535–1551. [CrossRef]
3. Han, X.; Crocombe, A.D.; Anwar, S.N.R.; Hu, P.; Li, W.D. The effect of a hot-wet environment on adhesively bonded joints under a sustained load. *J. Adhes.* **2014**, *90*, 420–436. [CrossRef]
4. Shimamoto, K.; Akiyama, H. Estimating the mechanical residual strength from IR spectra using machine learning for degraded adhesives. *J. Adhes.* **2022**, *98*, 2423–2445. [CrossRef]
5. Borges, C.S.P.; Marques, E.A.S.; Carbas, R.J.C.; Ueffing, C.; Weißgraeber, P.; da Silva, L.F.M. Review on the effect of moisture and contamination on the interfacial properties of adhesive joints. *Proc. Inst. Mech. Eng. C* **2021**, *235*, 527–549. [CrossRef]
6. Shimamoto, K.; Batorova, S.; Houjou, K.; Akiyama, H.; Sato, C. Accelerated test method for water resistance of adhesive joints by interfacial cutting of open-faced specimens. *J. Adhes.* **2021**, *97*, 1255–1270. [CrossRef]
7. Shimamoto, K.; Batorova, S.; Houjou, K.; Akiyama, H.; Sato, C. Degradation of epoxy adhesive containing dicyandiamide and carboxyl-terminated butadiene acrylonitrile rubber due to water with open-faced specimens. *J. Adhes.* **2021**, *97*, 1388–1403. [CrossRef]
8. Jhin, G.; Azari, S.; Ameli, A.; Datla, N.V.; Papini, M.; Spelt, J.K. Crack growth rate and crack path in adhesively bonded joints: Comparison of creep, fatigue and fracture. *Int. J. Adhes. Adhes.* **2013**, *46*, 74–84. [CrossRef]
9. Pascoe, J.A.; Alderliesten, R.C.; Benedictus, R. Methods for the prediction of fatigue delamination growth in composites and adhesive bonds—A critical review. *Eng. Fract. Mech.* **2013**, *112–113*, 72–96. [CrossRef]
10. Sekiguchi, Y.; Sato, C. Effect of bond-line thickness on fatigue crack growth of structural acrylic adhesive joints. *Materials* **2021**, *14*, 1723. [CrossRef]
11. Costa, M.; Viana, G.; da Silva, L.F.M.; Campilho, R.D.S.G. Environmental effect on the fatigue degradation of adhesive joints: A review. *J. Adhes.* **2017**, *93*, 127–146. [CrossRef]
12. Datla, N.V.; Ameli, A.; Azari, S.; Papini, M.; Spelt, J.K. Effects of hygrothermal aging on the fatigue behavior of two toughened epoxy adhesives. *Eng. Fract. Mech.* **2012**, *79*, 61–77. [CrossRef]
13. Wang, M.; Liu, A.; Liu, Z.; Wang, P.-C. Effect of hot humid environmental exposure on fatigue crack growth of adhesive-bonded aluminum A356 joints. *Int. J. Adhes. Adhes.* **2013**, *40*, 1–10. [CrossRef]
14. Sugiman, S.; Crocombe, A.D.; Aschroft, I.A. The fatigue response of environmentally degraded adhesively bonded aluminium structures. *Int. J. Adhes. Adhes.* **2013**, *41*, 80–91. [CrossRef]
15. Houjou, K.; Sekiguchi, Y.; Shimamoto, K.; Akiyama, H.; Sato, C. Energy release rate and crack propagation rate behaviour of moisture-deteriorated epoxy adhesives through the double cantilever beam method. *J. Adhes.* **2022**. *Online First version.* [CrossRef]
16. Han, X.; Crocombe, A.D.; Anwar, S.N.R.; Hu, P. The strength prediction of adhesive single lap joints exposed to long term loading in a hostile environment. *Int. J. Adhes. Adhes.* **2014**, *55*, 1–11. [CrossRef]
17. Manterola, J.; Zurbitu, J.; Renart, J.; Turon, A.; Urresti, I. Durability study of flexible bonded joints: The effect of sustained loads in mode I fracture tests. *Polym. Test.* **2020**, *88*, 106570. [CrossRef]
18. de Zeeuw, C.; de Freitas, S.T.; Zarouchas, D.; Schilling, M.; Fernandes, R.L.; Portella, P.D.; Niebergall, U. Creep behaviour of steel bonded joints under hygrothermal conditions. *Int. J. Adhes. Adhes.* **2019**, *91*, 54–63. [CrossRef]
19. Tan, W.; Na, J.-X.; Zhou, Z.-F. Effect of temperature and humidity on the creep and aging behavior of adhesive joints under static loads. *J. Adhes.* **2022**. *Online First version.* [CrossRef]
20. *ASTM D2919-01*; Standard Test Method for Determining Durability of Adhesive Joints Stressed in Shear by Tension Loading. ASTM International: West Conshohocken, PA, USA, 2014. [CrossRef]
21. Ebnesajjad, S. Chapter 10—Durability of adhesive bonds. In *Adhesive Technology Handbook*, 2nd ed.; William Andrew Publishing: Norwich, NY, USA, 2009. [CrossRef]
22. Hayashibara, H.; Iwata, T.; Ando, T.; Murakami, C.; Mori, E.; Kobayashi, I. Degradation of structural adhesive bonding joints on ship exposure decks. *J. Mar. Sci. Technol.* **2020**, *25*, 510–519. [CrossRef]
23. *ASTM D3433-99*; Standard Test Method for Fracture Strength in Cleavage of Adhesives in Bonded Metal Joints. ASTM International: West Conshohocken, PA, USA, 2020. [CrossRef]
24. *ISO 25217:2009*; Adhesives–Determination of the Mode I Adhesive Fracture Energy of Structural Adhesive Joints Using Double Cantilever Beam and Tapered Double Cantilever Beam Specimens. International Organization for Standardization: Geneva, Switzerland, 2009.
25. Sun, F.; Blackman, B.R.K. Using digital image correlation to automate the measurement of crack length and fracture energy in the mode I testing of structural adhesive joints. *Eng. Fract. Mech.* **2021**, *255*, 107957. [CrossRef]
26. Terasaki, N.; Fujio, Y.; Sakata, Y.; Horiuchi, S.; Akiyama, H. Visualization of crack propagation for assisting double cantilever beam test through mechanoluminescence. *J. Adhes.* **2018**, *94*, 867–879. [CrossRef]
27. de Moura, M.F.S.F.; Morais, J.J.L.; Dourado, N. A new data reduction scheme for mode I wood fracture characterization using the double cantilever beam test. *Eng. Fract. Mech.* **2008**, *75*, 3852–3865. [CrossRef]
28. *ISO 10354:1992*; Adhesives–Characterization of Durability of Structural-Adhesive-Bonded Assemblies–Wedge Rupture Test. International Organization for Standardization: Geneva, Switzerland, 1992.

29. *ASTM D3762-03*; Standard Test Method for Adhesive-Bonded Surface Durability of Aluminum (Wedge Test). ASTM International: West Conshohocken, PA, USA, 2010. [CrossRef]
30. Cognard, J. The mechanics of the wedge test. *J. Adhes.* **1986**, *20*, 1–13. [CrossRef]
31. Adams, R.D.; Cowap, J.W.; Farquharson, G.; Margary, G.M.; Vaughn, D. The relative merits of the Boeing wedge test and the double cantilever beam test for assessing the durability of adhesively bonded joints, with particular reference to the use of fracture mechanics. *Int. J. Adhes. Adhes.* **2009**, *29*, 609–620. [CrossRef]
32. Sargent, J.P. Durability studies for aerospace applications using peel and wedge tests. *Int. J. Adhes. Adhes.* **2005**, *25*, 247–256. [CrossRef]
33. Yu, S.; Tong, M.N.; Critchlow, G. Wedge test of carbon-nanotube-reinforced epoxy adhesive joints. *J. Appl. Polym. Sci.* **2009**, *111*, 2957–2962. [CrossRef]
34. Kempe, M.; Wohlgemuth, J.; Miller, D.; Postak, L.; Booth, D.; Phillips, N. Investigation of a wedge adhesion test for edge seals. In Proceedings of the Reliability of Photovoltaic Cells, Modules, Components, and Systems IX, San Diego, CA, USA, 26 September 2016; Volume 9938, p. 993803. [CrossRef]
35. van Dam, J.P.B.; Abrahami, S.T.; Yilmaz, A.; Gonzalez-Garcia, Y.; Terryn, H.; Mol, J.M.C. Effect of surface roughness and chemistry on the adhesion and durability of a steel-epoxy adhesive interface. *Int. J. Adhes. Adhes.* **2020**, *96*, 102450. [CrossRef]
36. Budzik, M.; Jumel, J.; Imielińska, K.; Shanahan, M.E.R. Accurate and continuous adhesive fracture energy determination using an instrumented wedge test. *Int. J. Adhes. Adhes.* **2009**, *29*, 694–701. [CrossRef]
37. Budzik, M.K.; Jumel, J.; Shanahan, M.E.R. An in situ technique for the assessment of adhesive properties of a joint under load. *Int. J. Fract.* **2011**, *171*, 111–124. [CrossRef]
38. Dillard, D.A.; Liechti, K.M.; Lefebvre, D.R.; Lin, C.; Thornton, J.S.; Brinson, H.F. *Development of Alternative Techniques for Measuring the Fracture Toughness of Rubber-to-Metal Bonds in Harsh Environments. Adhesively Bonded Joints: Testing, Analysis and Design*; ASTM STP; Johnson, W.S., Ed.; American Society for Testing and Materials: Philadelphia, PA, USA, 1988; Volume 981, pp. 83–97. [CrossRef]
39. Hasegawa, K.; Crocombe, A.D.; Coppuck, F.; Jewel, D.; Maher, S. Characterising bonded joints with a thick and flexible adhesive layer—Part 1: Fracture testing and behaviour. *Int. J. Adhes. Adhes.* **2015**, *63*, 124–131. [CrossRef]
40. Sekiguchi, Y.; Katano, M.; Sato, C. Experimental study of the mode I adhesive fracture energy in DCB specimens bonded with a polyurethane adhesive. *J. Adhes.* **2017**, *93*, 235–255. [CrossRef]
41. Gheibi, M.R.; Shojaeefard, M.H.; Googarchin, H.S. A comparative study on the fracture energy determination theories for an automotive structural adhesive: Experimental and numerical investigation. *Eng. Fract. Mech.* **2019**, *212*, 13–27. [CrossRef]
42. Plausinis, D.; Spelt, J.K. Designing for time-dependent crack growth in adhesive joints. *Int. J. Adhes. Adhes.* **1995**, *15*, 143–154. [CrossRef]
43. Kook, S.-Y.; Dauskardt, R.H. Moisture-assisted subcritical debonding of a polymer/metal interface. *J. Appl. Phys.* **2002**, *91*, 1293–1303. [CrossRef]
44. Lane, M.W.; Liu, X.H.; Shaw, T.M. Environmental effects on cracking and delamination of dielectric films. *IEEE Trans. Device Mater. Reliab.* **2004**, *4*, 142–147. [CrossRef]
45. Tan, T.; Rahbar, N.; Buono, A.; Wheeler, G.; Soboyejo, W. Sub-critical crack growth in adhesive/marble interfaces. *Mater. Sci. Eng. A* **2011**, *528*, 3697–3704. [CrossRef]
46. Birringer, R.P.; Shaviv, R.; Besser, P.R.; Dauskardt, R.H. Environmentally assisted debonding of copper/barrier interfaces. *Acta Mater.* **2012**, *60*, 2219–2228. [CrossRef]
47. Bosco, N.; Tracy, J.; Dauskardt, R. Environmental influence on module delamination rate. *IEEE J. Photovolt.* **2019**, *9*, 469–475. [CrossRef]
48. Springer, M.; Bosco, N. Environmental influence on cracking and debonding of electrically conductive adhesives. *Eng. Fract. Mech.* **2021**, *241*, 107398. [CrossRef]
49. Sekiguchi, Y.; Houjou, K.; Shimamoto, K.; Sato, C. Two-parameter analysis of fatigue crack growth behavior in structural acrylic adhesive joints. *Fatigue Fract. Eng. Mater. Struct.* **2022**. *Online First version.* [CrossRef]

Disclaimer/Publisher's Note: The statements, opinions and data contained in all publications are solely those of the individual author(s) and contributor(s) and not of MDPI and/or the editor(s). MDPI and/or the editor(s) disclaim responsibility for any injury to people or property resulting from any ideas, methods, instructions or products referred to in the content.

Article

A Novel Technique for Substrate Toughening in Wood Single Lap Joints Using a Zero-Thickness Bio-Adhesive

Shahin Jalali [1], Catarina da Silva Pereira Borges [1], Ricardo João Camilo Carbas [1,2,*], Eduardo André de Sousa Marques [2], Alireza Akhavan-Safar [1,*], Ana Sofia Oliveira Ferreira Barbosa [1], João Carlos Moura Bordado [3] and Lucas Filipe Martins da Silva [2]

[1] Institute of Science and Innovation in Mechanical and Industrial Engineering (INEGI), Rua Dr. Roberto Frias, 4200-465 Porto, Portugal

[2] Departamento de Engenharia Mecânica, Faculdade de Engenharia, Universidade do Porto, Rua Dr. Roberto Frias, 4200-465 Porto, Portugal; lucas@fe.up.pt (L.F.M.d.S.)

[3] Centro de Recursos Naturais E Ambiennte (CERENA), Instituto Superior Técnico, University of Lisbon, 1049-001 Lisbon, Portugal

* Correspondence: rcarbas@fe.up.pt (R.J.C.C.); aakhavan-safar@inegi.up.pt (A.A.-S.)

Abstract: In contemporary engineering practices, the utilization of sustainable materials and eco-friendly techniques has gained significant importance. Wooden joints, particularly those created with polyurethan-based bio-adhesives, have garnered significant attention owing to their intrinsic environmental advantages and desirable mechanical properties. In comparison to conventional joining methods, adhesive joints offer distinct benefits such as an enhanced load distribution, reduced stress concentration, and improved aesthetic appeal. In this study, reference and toughened single-lap joint samples were investigated experimentally and numerically under quasi-static loading conditions. The proposed research methodology involves the infusion of a bio-adhesive into the wooden substrate, reinforcing the matrix of its surfaces. This innovative approach was developed to explore a synergetic effect of both wood and bio-adhesive. The experimentally validated results showcase a significant enhancement in joint strength, demonstrating an 85% increase when compared to joints with regular pine substrates. Moreover, the increased delamination thickness observed in toughened joints was found to increase the energy absorption of the joint.

Keywords: wooden joints; toughened joint; transverse strength; delamination; polyurethan bio-adhesives

1. Introduction

The recent concerted drive towards reducing the impact of fossil-based products has spurred the development and increasingly larger adoption of bio-based materials [1–3]. This endeavor seeks to usher in a more sustainable era, where the utilization of tree and plant-derived products assumes a pivotal role. This is particularly true in the creation of eco-friendly load-bearing structures, enabled by the fabrication of composites comprising natural fibers, such as flax, jute, and palm tree. Within this context, the significance of wood as a naturally sourced structural material comes to the fore. Across the annals of history, wood has been explored for diverse daily applications due to its status as a replenishable natural material [4–6]. Noteworthy attributes such as its versatility in assuming various shapes, coupled with its impressive durability, resistance to wear and environmental factors, mechanical robustness, relatively low weight, global abundance, and economic viability, position it as a valid and sustainable material choice. Moreover, wood's adaptability to distinct environmental conditions, including temperature fluctuations, loading rates, and moisture influences, has led to its widespread use in structural applications, especially in the civil engineering field [7–9]. Wood thus stands out as a natural composite material but, and akin to most composites, design of complex wooden structures is not trivial, since it exhibits

Citation: Jalali, S.; Borges, C.d.S.P.; Carbas, R.J.C.; Marques, E.A.d.S.; Akhavan-Safar, A.; Barbosa, A.S.O.F.; Bordado, J.C.M.; da Silva, L.F.M. A Novel Technique for Substrate Toughening in Wood Single Lap Joints Using a Zero-Thickness Bio-Adhesive. *Materials* **2024**, *17*, 448. https://doi.org/10.3390/ma17020448

Academic Editors: Marco Lamberti and Aurelien Maurel-Pantel

Received: 30 November 2023
Revised: 13 January 2024
Accepted: 15 January 2024
Published: 17 January 2024

Copyright: © 2024 by the authors. Licensee MDPI, Basel, Switzerland. This article is an open access article distributed under the terms and conditions of the Creative Commons Attribution (CC BY) license (https://creativecommons.org/licenses/by/4.0/).

sensitivity to the introduction of holes and notches. Traditional joining methods like riveting and bolting are unsuitable for wood due to its composition [10]. In light of this, adhesive bonding emerges as a suitable technique for joining wood parts. This method circumvents the creation of stress concentrations and establishes a broader, more uniform bonded area. Adhesive joints are primarily designed to operate under shear forces, enhancing the material's resistance. However, in the most commonly used joint configuration, the single-lap joint (SLJ), the overlapped and offset substrates can bend to ensure an alignment of the load path. This alignment generates a flexural moment at the ends of the overlap, leading to the formation of stress concentrations at the overlap ends. Consequently, the stress distribution within the adhesive becomes a combination of both shear and tensile forces. This combination of loading modes highlights the complex interplay that adhesive joints must withstand when subjected to varying loads and conditions [11,12]. As a result, the failure load and the failure mode of the joint become contingent on not only the cohesive properties of the adhesive but also that of its substrates. In adhesive joints involving wood substrates, the contribution of peel loading is particularly concerning, since it leads to delamination between distinct grain layers and ultimately causes joint failure. To counteract this issue, multiple corrective approaches have been proposed, such as wood densification, substrate toughening, and even physical modifications of the adhesive [3]. These processes aim to tackle the challenge of delamination and bolster the integrity of the adhesive and the wood substrate.

Multiple recent research works have been devoted to improving the efficacy of wooden adhesive joints, integrating bio-adhesives and wood-based substrates in joint configurations. The failure mode and load capacity of joints are significantly influenced by the properties of both the adhesive and the substrate. As discussed above, peel stress is a critical factor that affects the strength and failure of composite materials, especially when their matrix is weak. This is because the fibers within the matrix may not reach their peak load before the matrix fails [13,14]. Therefore, any method that enhances the peel strength of wood directly contributes to the overall strength of the joints [15]. In contrast, the strength of a given bio-adhesive and its additives is still an unexplored path, since these materials are still mostly used in non-structural applications.

Baumberger et al. [16] studied films with up to 30% kraft lignin and starch, created using extrusion and thermal molding, seeking to improve the performance of these natural materials. The mechanical properties of the composite were evaluated through tensile tests at two humidity levels. The study found that lignin increased the strength and strain-to-failure of the composite at 58% relative humidity (RH) for lignin contents up to 20%. It was observed that lignin reduced the affinity of the films to water, and this behavior was attributed to a combination of the hydrophilic starch matrix and hydrophobic lignin. This aspect is of concern in adhesives with additives since the adhesive matrix is generally more hydrophilic than the reinforcement, making the adhesive/reinforcement interface a favorable pathway for water diffusion [17].

One of the methods that can improve the strength of SLJ is to use a matrix with high toughness in the transverse direction, which can enhance the resistance to peel stress. This method can be combined with other methods that aim to reduce the local stress concentration in the composite adherends [18].

Stucki et al. [19] investigated the use of bio-based bonding additives, especially kraft lignin and its derivatives, to enhance the moisture resistance and bond strength of friction-welded wood joints. Results of this work show that kraft lignin has the best performance among the tested biomolecules, and that its molecular structure and thermal behavior affect the bonding quality. The paper also discusses the possible bonding mechanisms involved in the friction welding process, and suggests further research directions to optimize the lignin-based bonding additives.

Shang et al. [14] presented a methodology aimed at mitigating delamination issues in adhesive joints featuring composite substrates. The investigation encompassed both experimental and numerical aspects, evaluating parameters such as joint strength, failure modes, and stress distribution within the SLJs. The toughened composite material yielded a substantial 22% increase in joint strength and, remarkably, a shift in failure mode from adherend delamination to cohesive failure within the adhesive. The findings of this study suggest that the toughened composite material holds the potential to effectively mitigate delamination in adhesive joints involving composite substrates while enhancing overall performance.

Conventional wood joints have several drawbacks, such as low transverse strength, poor durability, and high environmental impact of synthetic adhesives. Therefore, alternative methods are needed to improve the performance and sustainability of wood joints. to overcome these challenges a novel and effective method was proposed to enhance the strength and durability of wood joints by increasing the peel and transverse strength of the substrate using a polyurethane-based adhesive derived from natural resources. This method could improve the mechanical properties and resistance to environmental factors of wood joints, as well as reduce the carbon footprint and toxicity of the adhesive. By applying the bio-adhesive on the surface of the wood substrates, the shear and peel strength of the SLJs can increase, as well as the absorbed energy during failure. A series of experiments were conducted to compare the performance of the proposed method with that of conventional methods. The peel and transverse strength of the wood and testing SLJ joints under static loading conditions were measured. The proposed method significantly increased the peel and transverse strength of the wood joints, as well as their durability and stability under various environmental factors. The results suggest that the method can provide a viable solution to the challenges and limitations of conventional wood joints. The study contributes to the advancement of wood engineering and the development of eco-friendly adhesives. A numerical model was also developed to clarify the mechanisms behind the mechanical behaviors that were observed.

2. Material

2.1. Bio-Adhesive

A polyurethane-based bio-adhesive, derived from 70% renewable biomass sources, such as vegetable oils, according to the ASTM D6866 standard [20], was used, which exhibits excellent affinity for wood and holds significant potential as a viable alternative to synthetic adhesives. This is a novel monocomponent system with a 100% solids content, which means it does not contain any solvents or low molecular weight oligomers that could compromise its performance. It has an average molecular weight of 5200 and three terminal aliphatic isocyanate groups (-NCO) that react with moisture and active hydrogen atoms on the surfaces of the substrates, forming strong covalent bonds. The adhesive layer thickness is a key parameter for the lap joint performance, and it was optimized for our experimental setup. The adhesive also has a unique composition that replaces the conventional propylene oxide (PO) based polyhydroxyl alcanoates with ethylene oxide (EO) based ones in the "3 legs" of the structure. This results in aliphatic polar bio-derived structures that enhance the wetting and physical adhesion to polar substrates. The nominal viscosity of the adhesive at 25 °C is 8 Poise, which facilitates its application and curing.

The bio-adhesive was developed by the team of Professor João Bordado at Instituto Superior Técnico. The bio-adhesive is not yet commercially available, but it shows consistent performance with typical moisture-curing polyurethane adhesives. Furthermore, due to the zero-thickness curing process of the adhesive, it lends itself well to be used in many wooden applications. The bio-adhesive is produced in an irreversible reaction, without humidity, in a reactor under a nitrogen atmosphere, and heating is achieved with a thermal oil coil. The bio-adhesive uses an aliphatic isocyanate as a basis, which contains 70% plant matter, as these are more easily biodegradable. Manufacturing the bio-adhesive is estimated to consume 15–20% less energy than those derived from petroleum. The bio-adhesive consists of pentamethylene diisocyanate and polyisocyanate, which form strong chemical bonds with the wood substrate by reacting with its hydroxyl (OH) groups. The curing process of the bio-adhesive is accelerated by increasing the humidity of the substrates. To achieve uniform curing of the bio-adhesive in the joints, all samples were maintained at consistent moisture levels. The bio-adhesive offers multiple benefits due to the synergistic interplay between its components and the wood's OH groups, as well as the influence of humidity. The bio-adhesive not only enhances the mechanical interlocking and the physical and mechanical bonds within the joints, which are typical for normal adhesives, but also provides superior chemical bonding effects. The combination of these factors contributes to the overall strength of the joints (see Figure 1).

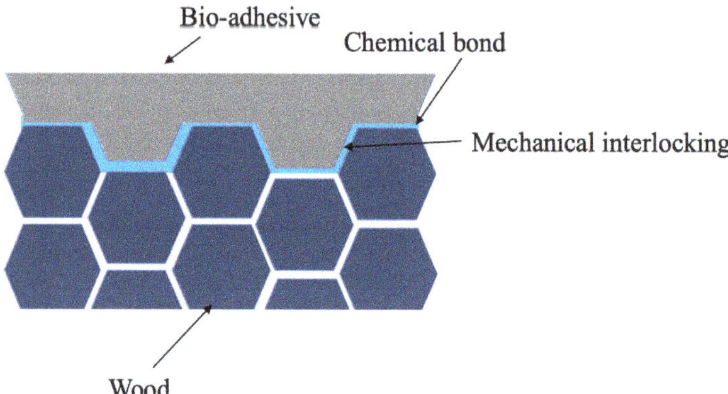

Figure 1. Schematic molecular bonds of moisture curing adhesive.

This bio-adhesive requires a curing process at 100 °C for a duration of 8 h, followed by a post-curing period of 48 h at room temperature. This consistent curing procedure was applied to all tested samples, to ensure comparability. The mechanical properties of the adhesive were previously characterized under quasi-static conditions (1 mm/min) by Jalali et al. [21], and can be found in Table 1.

Table 1. Bio-adhesive properties [21].

Young's Modulus (MPa)	Tensile Strength (MPa)	Mode I Fracture Energy (N/mm)	Mode II Fracture Energy (N/mm)
197.09 ± 9.76	3.27 ± 0.14	0.33 ± 0.03	1.27 ± 0.1

2.2. Substrate

The main substrate under investigation consisted of pine timber obtained from Portuguese sources. Initially, the timber was sourced in beams with a length of 1000 mm, which was subsequently cut into smaller pieces with dimensions suitable for the joint and bulk sample configurations. The thickness of the wood was standardized at 6 mm, while the width was set at 25 mm.

The selection of pine wood as the primary material was based on various factors, including its widespread availability, cost-effectiveness, and desirable mechanical properties, such as strength, stiffness, and durability. It is important to note that wood, being a complex and heterogeneous material, exhibits properties that can vary due to factors such as species, growth conditions, grain slope and size, defects, knots, shakes, and age. The substrates, selected from the beams that had the most symmetrical grains and no noticeable damage, had a moisture content ranging between 12% and 18%, which was guaranteed by the supplier. The moisture content of the wood was within the standard range of 8–25% by weight for different types and uses of wood. The wood was considered dry, as it had a moisture content of 19% or less, which is the maximum value for sawn lumber design. The moisture content of the wood was an important factor for its properties and performance, as it affected its dimensional stability, strength, durability, and biological resistance.

To further emphasize the significance of the chosen substrate, Oliveira et al. [22] conducted extensive testing on pine wood and compiled the results, the elastic properties Young's modulus (*E*), Poisson's ratio (*v*), and shear modulus (*G*) are summarized in Table 2. All the wood directions are shown in Figure 2.

Table 2. Nine elastic properties of the pine wood along longitudinal (*L*), radial (*R*) and tangential (*T*) directions [22].

E_L (GPa)	E_R (GPa)	E_T (GPa)	v_{LT}	v_{LR}	v_{TR}	G_{LR} (GPa)	G_{LT} (GPa)	G_{TR} (GPa)
12.0	1.9	1.0	0.5	0.4	0.3	1.1	1.0	0.3

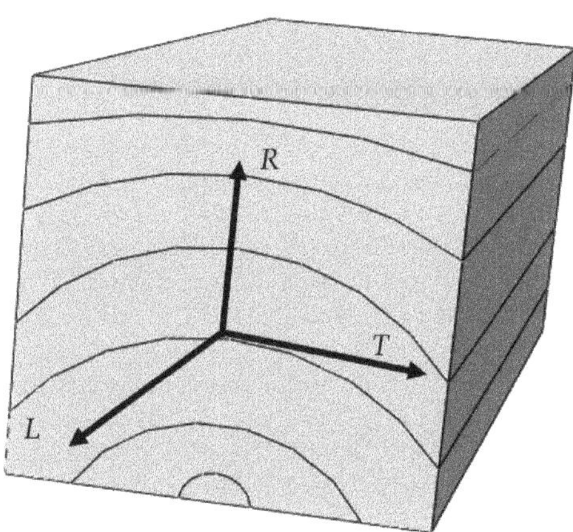

Figure 2. Wood beam directions.

2.3. Toughened Substrate

The substrate edges were toughened by infusing the bio-adhesive in the pine wood substrates. This method was chosen due to the higher strength exhibited by the bio-adhesive in comparison to the wood matrix (lignin).

The adhesive penetration process in the wood involved two main steps. In the first step, all the surfaces of the substrate were thoroughly wetted with the bio-adhesive using a brush. This ensured that the adhesive was evenly distributed and absorbed by the entire surface area of the wood. The purpose of this step was to facilitate effective bonding between the adhesive and the wood.

Following the initial wetting step, the substrate was then soaked in the adhesive pot for 1 h. This extended soaking allowed for optimal penetration and interaction between the bio-adhesive and the wood substrate, further strengthening the bond. Figure 3 presents a schematic diagram, illustrating the step-by-step procedure of wood toughening through compression.

Also, the density of the toughened samples was measured after the curing process of the bio-adhesive. The results indicated that the average density of the reference wood was 0.56 ± 0.03 g/cm^3 and that the bio-adhesive increased the wood density by 4%.

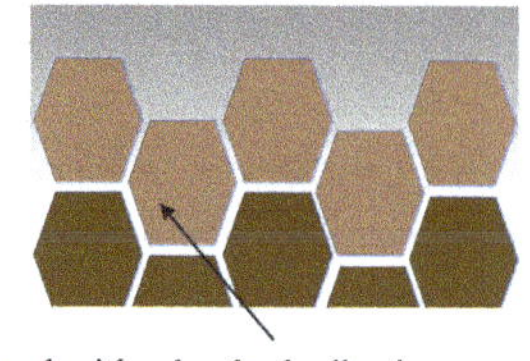
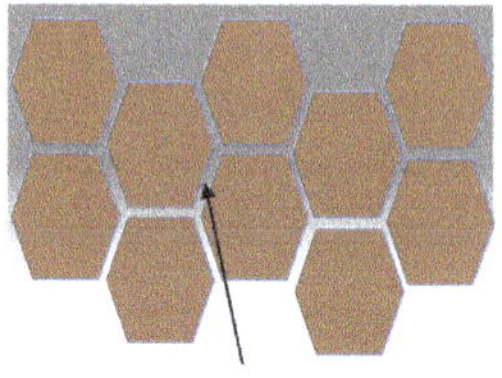

Figure 3. Toughening procedures.

3. Experimental Procedure

3.1. Bulk Testing

The mechanical behavior of the reference, non-toughened, pine wood and toughened pine wood was assessed by conducting traction tests on 1 mm thick pine wood samples in both the fiber direction and matrix direction. The reference pine wood samples underwent a simple sanding process to ensure smooth surfaces and minimize the presence of microcracks and surface stress concentrations.

The toughened samples were prepared by applying the previously mentioned toughening procedures on a 1 mm thick wood plate. Wooden end tabs were then bonded to the ends of the samples, as depicted in Figure 4a. The geometry and dimensions of the bulk samples can be observed in Figure 4b. Through this testing approach, a comprehensive comparative analysis was conducted between the toughened and reference samples, thus providing valuable insights into the mechanical performance of the toughened layer.

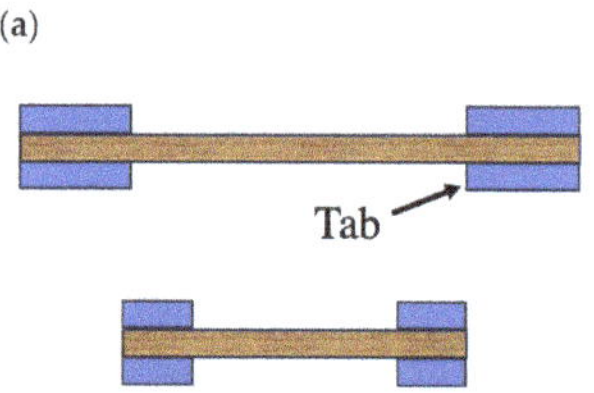
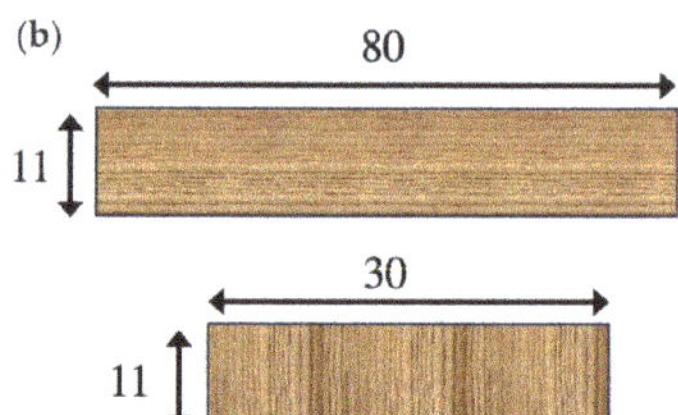

Figure 4. Pine wood bulk sample, tested sample (**a**), geometry and dimension of the tested wood (**b**). (Dimensions are in mm.).

3.2. Joint Testing

SLJ samples were prepared with an overlap length of 25 mm, ensuring direct bonding of the substrates to meet the zero-thickness requirement of the bio-adhesive. To achieve uniform pressure distribution along bonding area, clamps were used to apply a constant level of pressure to the joints. In order to protect the substrates from damage caused by the applied pressure, 2 mm thick aluminum plates were employed, as illustrated in Figure 5. To facilitate the easy debonding of the joint from the aluminum plates and clamps, a mold-release agent was applied the aluminum surfaces.

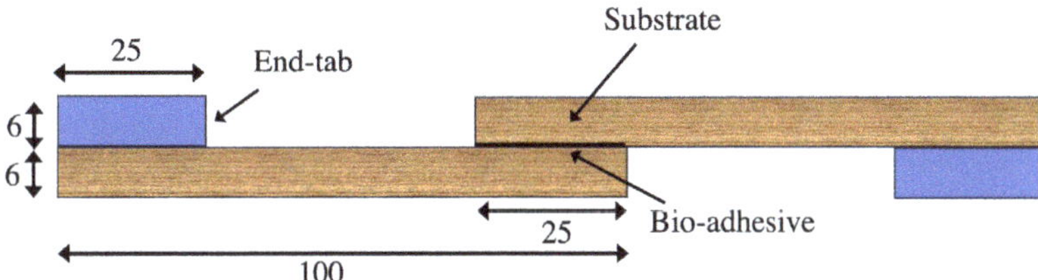

Figure 5. SLJ geometry and dimensions. (Dimensions are in mm.)

Figure 5 shows the joint geometries and dimensions of the SLJs. It is worth noting that, assuming no edge effects, the stress and strain conditions along the width direction should remain consistent for SLJs with different widths. As a result, the failure load of the SLJs is expected to be directly proportional to the width. For the purpose of maintaining consistency and facilitating comparison, a sample width of 25 mm was selected in this study. Previous research has already investigated the impact of overlap length, with several authors conducting tests and comparisons using different values [23,24].

Surface Preparation

The successful adhesion of the pine wood substrate relies on a correct preparation of the bonding surfaces [4]. To achieve the desired zero-thickness bond with the adhesive, the surfaces underwent a sanding process, employing 400-grade sandpaper. This step ensured that the surfaces were smooth and uniform, providing an optimal foundation for bonding.

Following the sanding procedure, compressed air was employed to thoroughly clean the wood surfaces, effectively removing any loose particles that could impede proper adhesion. The presence of dust particles on the surface creates a barrier, hindering the adhesive's ability to form a robust bond and potentially resulting in weak or non-bonded joints.

To further ensure the absence of contaminants on the wood surfaces, an additional cleaning step was carried out, using a solvent (acetone) to remove impurities such as oils waxes and other organic substances from the surface. Once the wood surfaces were properly prepared, a layer of adhesive was applied to the bonding area of each substrate before they were assembled. A total of ten SLJs were manufactured simultaneously. The entire setup was then transferred to an oven, where it was subjected to an appropriate cycle to cure the adhesive.

After the curing process, excess adhesive was removed and fixing tabs at the ends of the substrates were bonded using Araldite AV138 (Duxford, UK), a two-component epoxy adhesive that cures at room temperature for 24 h.

3.3. Testing Condition

In real-world applications, adhesive joints experience various types of loads, including quasi-static conditions. For the quasi-static testing phase, an INSTRON 3367(Norwood, MA, USA) universal testing machine equipped with a 30 kN load cell was utilized at ambient room temperature. A constant cross head displacement rate of 1 mm/min was maintained throughout the testing process. For each specific joint the configuration was tested, and a load-displacement curve was generated by the machine. To ensure the statistical validity of the results, four repetitions were performed for each configuration under analysis. At least three samples were tested for each configuration.

4. Experimental Results and Discussion
4.1. Bulk Testing

The effect of toughening on the mechanical properties of pine wood was investigated along both the fiber and matrix directions. The results revealed that adhesive penetration increased the strength of the toughened wood by 10% when tested along the fiber direction. However, adhesive penetration did not have a significant effect on the elastic modulus of the wood along the fiber direction, as shown in Figure 6a. On the other hand, the strain at failure decreased by 60% in these samples, and the adhesive made the wood significantly more brittle along this direction. In contrast, adhesive penetration notably enhanced the wood strength along the matrix direction. The superior strength of the adhesive, compared to the wood matrix, resulted in penetration between the wood cells and the formation of chemical bonds. This led to a remarkable 180% increase in both the failure strength and the failure strain of the wood along the matrix direction. Moreover, the failure strain and the absorbed energy of the toughened wood increased by two and seven times, respectively, compared to the reference wood samples (as shown in Figure 6b). The testing results of the toughened ply showed significant improvements in its strength compared to the reference wood. The elastic properties of the samples are presented in Table 3.

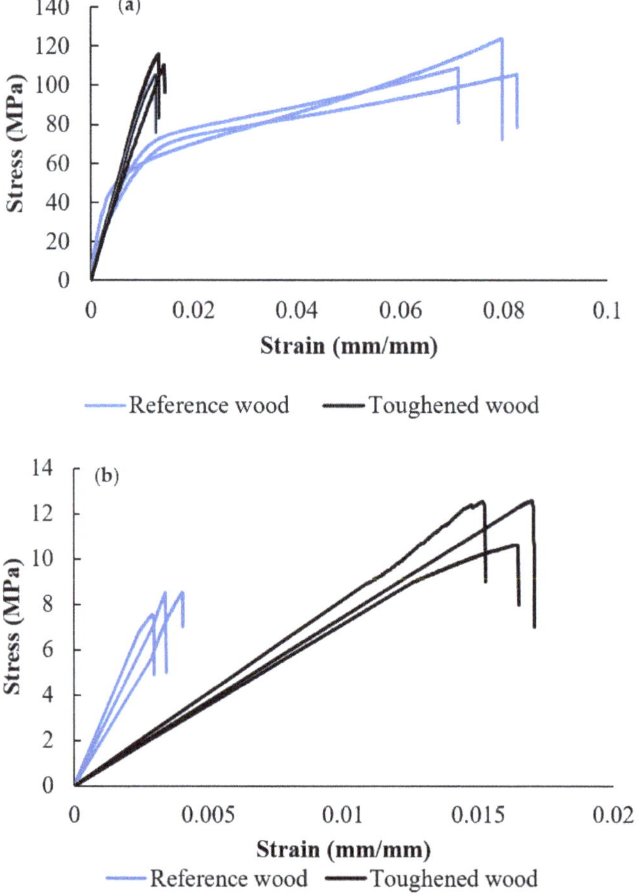

Figure 6. Typical stress-strain behavior of bulk samples. Fiber direction (**a**), matrix direction (**b**).

Table 3. Elastic behavior of bulk samples.

		Young's Modulus (MPa)	Tensile Strength (MPa)	Strain at Failure (%)
Fiber direction	Reference	12.3 ± 1.2	93.2 ± 4.2	7.6 ± 0.3
	Toughened	11.9 ± 1.8	102.3 ± 8.2	1.3 ± 0.1
Matrix direction	Reference	2.1 ± 0.1	8.1 ± 0.3	0.4 ± 0.1
	Toughened	0.8 ± 0.1	11.8 ± 0.8	1.6 ± 0.1

4.2. Joint Testing

The experimental results depicted in Figure 7 display the load-displacement behavior of the joints under quasi-static conditions. It is observed that the toughened joints exhibited a higher failure load, with an increase of approximately 20%, compared to the reference joints. Furthermore, the stiffening effect on the joints was notable, as indicated by an increase in stiffness of approximately 38%. Of particular interest is the behavior of the toughened joints at the point of failure. When the toughened joints reached the failure load of the reference joints, the toughening process induced plastic deformation within the joint structure. As a result, the displacement at failure increased by approximately 85%, indicating a greater capacity for energy absorption before failure. This increase in absorbed energy was substantial, amounting to approximately 170% compared to the reference joints. The improved energy absorption capability of the toughened joints highlights their enhanced resilience and ability to dissipate energy effectively.

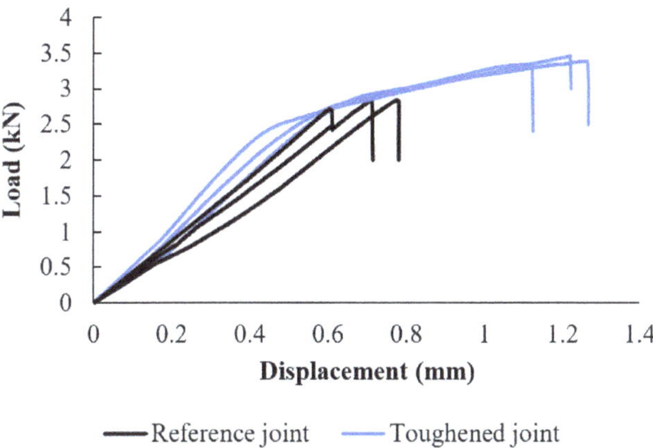

Figure 7. Typical load-displacement of SLJ.

The nonlinear behavior of the joints results in a significantly larger area underneath the curve leading to substantial enhancement in the quantity of energy absorbed during testing. Additionally, it is noteworthy that the process of toughening has a more pronounced effect on the absorbed energy of the joint than on its failure load, as the amount of energy absorbed has increased by around 230%.

4.2.1. Fracture Surfaces

Figure 8a provides digital images of the fracture surfaces that were analyzed to gain insight into the behavior of the joints. The examination of these images revealed that delamination of the wood was the primary failure mode observed in both the toughened and reference joints. This finding indicates that the adhesive achieved good adhesion and could be cured effectively.

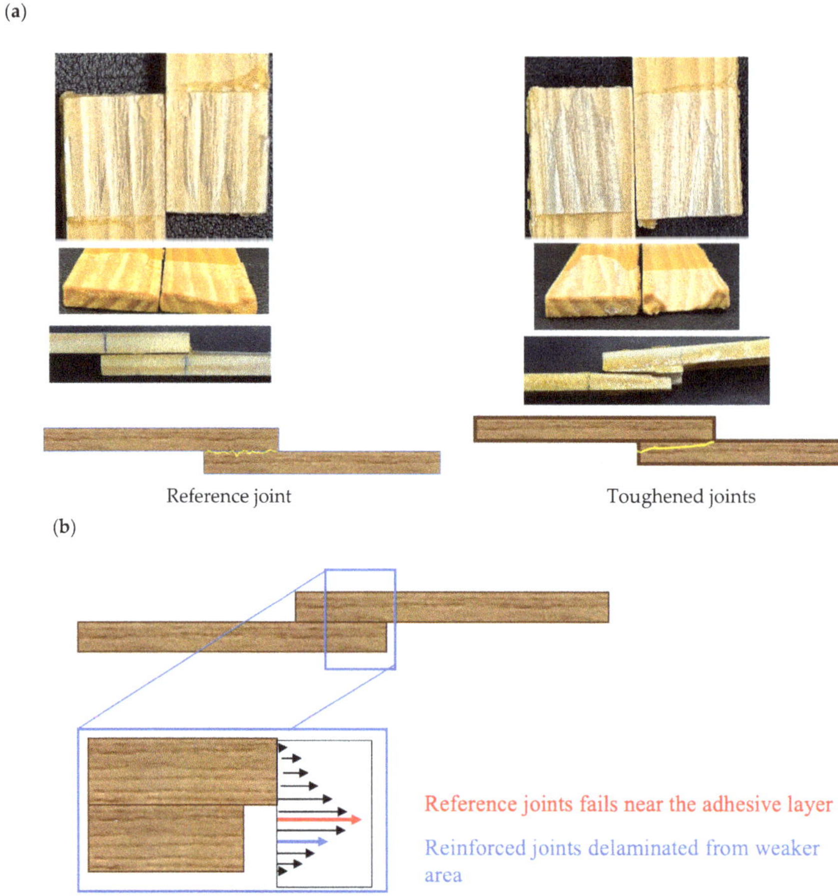

Figure 8. Fracture surface of SLJ (**a**), fracture behavior explanation (**b**).

However, a notable difference was observed in the toughened joints, indicating a significant change from the behavior seen in the reference joints. In the reference joints, delamination typically occurred at a relatively shallow depth within the wood substrate. This meant that the failure was concentrated in the surface layers of the wood. In contrast, the toughened joints revealed a notably different failure pattern. Within the toughened joints, an intriguing phenomenon unfolded. The failure predominantly transpired within the non-toughened region of the wood substrate. It is worth emphasizing the significance of this observation as while the non-toughened core of the wood bore the brunt of the failure, the outer plies, which had experienced enhanced adhesion due to adhesive penetration, were found to be more resilient. This particular behavior was instrumental in driving up the failure load of the toughened joints. The key to this phenomenon lies in the stress distribution along the overlap. At the ends of the overlap, where stresses tend to concentrate, stress levels approached the failure threshold of the wood. This stress concentration effect was instrumental in the change in failure mode observed.

Figure 8a presents the difference fracture behavior of the joints. In the toughened joint, the crack path assumed a more complex and notably deeper trajectory. This complexity led to a substantial increase in the amount of energy required for the joint to reach its failure point. On the flip side, in the reference joint, the delamination path remained relatively uniform, with the crack predominantly propagating horizontally along the overlap. However, in the toughened joint, the crack's propagation followed a more intricate path. Not only did it traverse the overlap horizontally, but it also exhibited a tendency to change direction vertically, interacting with different grain orientations, and even propagating vertically through the thickness of the wood substrate.

Figure 8b provides a schematic explanation of this failure mechanism. It visually demonstrates how the toughening process leads to a redistribution of the failure zone, with the failure occurring in the non-toughened area while the toughened outer plies retained their integrity. This behavior highlights the positive effect of the toughening technique, which enhanced the load-bearing capacity and resistance to failure of the joints.

4.2.2. Scanning Electron Microscopy (SEM) Analysis

To gain further insight into the behavior of the joints, SEM analysis was conducted to examine the adhesive's penetration and its interaction with the wood substrates. SEM analyses were performed using a JEOL JSM 6301F/Oxford INCA Energy 350/Gatan Alto 2500 microscope (Tokyo, Japan) at CEMUP (University of Porto, Portugal). The microscope was operated at an accelerating voltage of 20 kV and a working distance of 10 mm. A secondary electron detector was used to image the adhesive composition and the wood structure. Since the analyzed samples were not conductive, they were coated with a thin film of gold (Au)/palladium (Pd) alloy, by sputtering, using the SPI module sputter coater equipment. The sputtering time was 120 s and the current was 15 mA. As depicted in Figure 9, reference substrate and edges of toughened substrate were compared. The SEM images revealed the effective penetration of the adhesive into the wood. This analysis confirmed that not only the wood matrix was reinforced, but also that the adhesive penetrated the wood fibers. The SEM analysis provided visual evidence of the adhesive's ability to penetrate and interact with the wood fibers, enhancing the overall strength and toughness of the joint. By saturating the outer layers of the wood fibers, the adhesive created a reinforced zone that contributed to the joint's improved performance. This reinforcement allowed the wood fibers to undergo plastic deformations, enhancing the joint's ability to withstand applied loads and reducing the likelihood of premature failure.

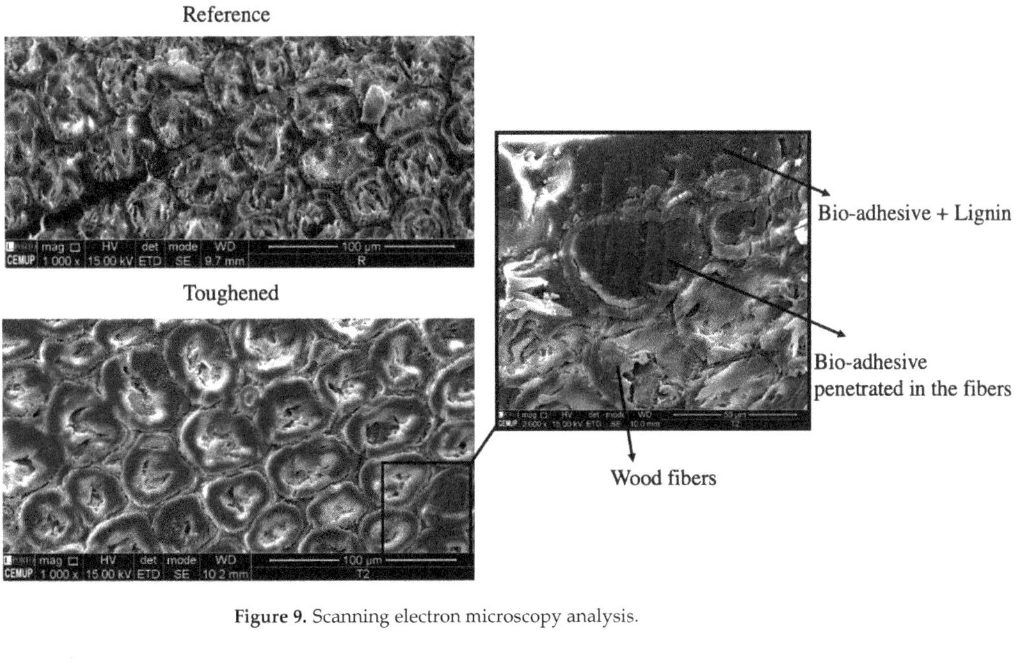

Figure 9. Scanning electron microscopy analysis.

5. Numerical Analysis

5.1. Numerical Details

To validate the experimental results and to support the investigation into the failure mechanism, a static finite element analysis was conducted using ABAQUS v2017 software. In order to simplify the analysis and reduce computational time, a two-dimensional (2D) model was employed, assuming the stress distribution along the width of the SLJ to be uniform.

The primary objective of this analysis was to predict the load-displacement behavior and failure modes of the joints. The numerical results were evaluated by considering a representative experimental curve that could capture the average response of the samples.

The wood samples in all tested cases exhibited elastic deformation, and thus, linear elastic behavior was assumed for the substrates that the model was meshed using quadrilateral elements with four nodes and plane stress conditions. Also, to predict the initiation and propagation of potential damages and delamination of the substrates, a cohesive zone model (CZM) was employed. Four-node cohesive quadrilateral elements were used to model the behavior of the substrates, employing cohesive elements with a triangular (bilinear) traction-separation law based on the mechanical and cohesive properties of pine wood Tables 2 and 4. The CZM elements were introduced to both the reference and toughened substrates, by considering the delamination thickness observed in the experiments, in order to effectively simulate the delamination phenomenon.

Table 4. Strength (σ) properties of pine wood [22].

σ_L (MPa)	σ_R (MPa)	σ_T (MPa)	σ_{LR} (MPa)	σ_{LT} (MPa)	σ_{RT} (MPa)
97.5	7.9	4.2	16.0	16.0	4.5

Cohesive elements were placed at a distance of 0.1 mm from the substrate interface for the reference joints, and for toughened joints the cohesive elements were placed in thicknesses of 0.1 and 1.3 mm from the interface of the substrates in order to predict any possible delamination in the toughened wood or core wood. The thickness of cohesive elements was measured from the average thickness of delamination obtained in the experiments. The adhesive behavior was simulated by using cohesive contact between two substrates assuming the bio-adhesive properties. In regions with significant stress gradients, the element sizes were refined by reducing the mesh size. The mesh was distributed along the substrates and bondline in the x-direction using single and double biasing, where element sizes of 0.01 mm and 0.05 mm were employed. Along the end tabs (see Figure 10a,b) a uniform mesh distribution was applied. The assumed boundary condition is illustrated in Figure 10c where the left side of SLJ was fixed and the right side subjected to a displacement of 2 mm.

5.2. Numerical Results

In Figure 11, a comparison between the numerical and experimental load-displacement data for the reference joint is presented. The predicted failure load from the numerical model closely matched the results of the experiments. This congruence serves as a validation of the practicality and accuracy of the mechanical properties attributed to the materials within the simulation. Just as in the experiments, the cohesive zones situated beneath the initial wood plies displayed a notable and extensive level of degradation. In stark contrast, the adhesive layer exhibited minimal signs of degradation. The cracks started where the first layer of wood meets the second layer along the overlap's edge. This spot is weaker because of stress considerations at the overlap ends. So, this is where failure often begins in these joints. This localized vulnerability played a pivotal role in the initiation of the failure process, shedding light on a critical aspect of the joint's mechanical behavior.

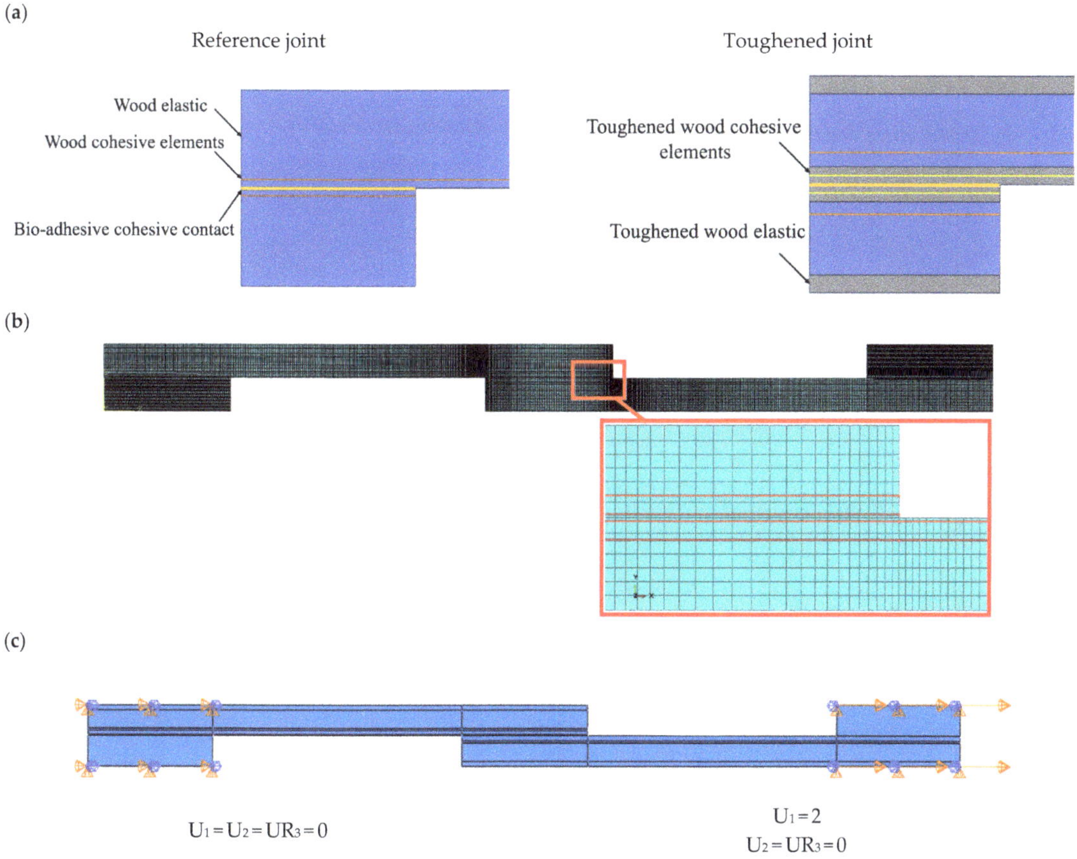

(**a**)

Reference joint

Toughened joint

(**b**)

(**c**)

Figure 10. (**a**) Partition, (**b**) mesh used in SLJ simulation, (**c**) boundary condition.

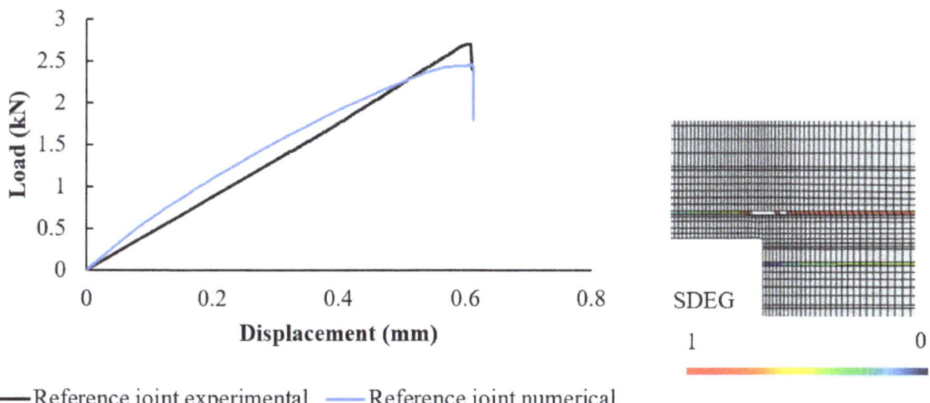

Figure 11. Numerical analysis of reference joint. The image on the right shows the damage distribution, where the SDEG is the damage index, which varies between 0 (no damage) and 1 (complete failure).

As illustrated in Figure 12, the numerically derived failure mechanism and load-displacement curve for toughened joints display a good agreement with the experimental results at the point of failure load. This close correlation between the numerical models and the experimental data highlights the reliability of the simulation. The numerical simulation has proven its efficacy in accurately predicting the experimental failure mechanism, as can be observed in Figure 8. Notably, the observed failure mechanism in toughened joints involves the cohesive failure of the wood's cohesive layer.

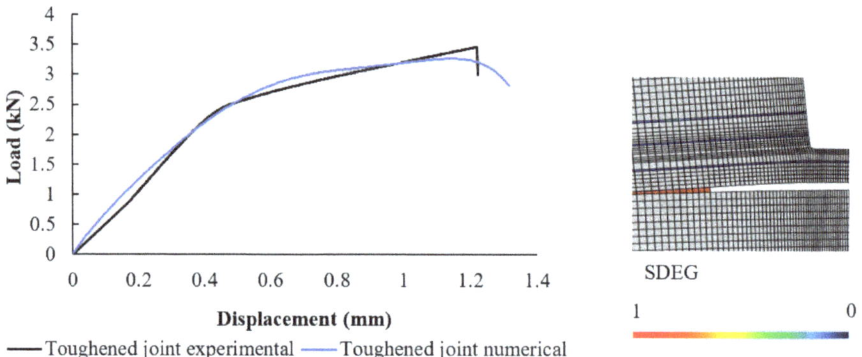

Figure 12. Toughened joint numerical analysis. The image on the right shows the damage distribution, where the SDEG is the damage index, which varies between 0 (no damage) and 1 (complete failure).

Both experimental and numerical findings lead to the conclusion that the reinforcement of the substrate through the application of bio-adhesive leads to a noticeable augmentation in the failure load. Furthermore, a significant increase in delamination thickness is observed in cases where the substrate has been toughened. This dual enhancement process, which results in an elevated failure load and larger delamination thickness, shows the effectiveness and future potential of the process of reinforcing the substrate with bio-adhesive.

6. Conclusions

This study has successfully introduced a novel and practical approach to enhance the loading capacity of joints with wood substrates and bonded with bio-adhesives. The reinforcement strategy involves infusing bio-adhesive into the wooden substrate, reinforcing the matrix near the surface. The method for producing the toughened substrate demonstrated consistent and repeatable results, establishing its feasibility for real-world applications.

The experimental assessment of SLJs utilizing the toughened substrate yielded promising outcomes. Notably, the substrate's delamination thickness experienced a significant increase, showing an 85% rise in joint strength compared to conventional SLJ samples with regular pine substrates.

The exploration of failure behavior, augmented failure load, and improved absorbed energy was supported by the development of a finite element method model. This modeling approach effectively minimized the stress concentration at the overlap edges and efficiently distributed stresses over a wider area. As a result, the joints exhibited a significant increase in failure load.

This comprehensive study not only presents a practical technique for enhancing wooden joint performance but also offers insights into the underlying mechanics of the observed improvements. The combination of bio-adhesive infusion, experimental validation, and advanced modeling techniques marks a significant stride toward enhancing the robustness and reliability of wooden joints in structural applications.

The main findings of this study were:

- The toughened substrates increased the delamination thickness and the joint strength by 85% compared to regular pine substrates.

- The failure behavior, failure load, and absorbed energy of the joints were explored and supported by a finite element method model.
- The model reduced the stress concentration at the overlap edges and increased the stress distribution over a wider area.
- The study presented a practical technique for enhancing wooden joint performance using bio-adhesive infusion.
- The study offered insights into the underlying mechanics of the observed improvements.

Author Contributions: S.J.: Conceptualization, investigation, simulation, resources, validation, writing original draft and editing. C.d.S.P.B.: Validation, investigation, review, recourses, data correction. R.J.C.C.: Validation, formal analysis, investigation, simulation, data curation, supervision, project administration. E.A.d.S.M.; Methodology, formal analysis, data curation, review, supervision. A.A.-S.: Formal analysis, review, and editing. A.S.O.F.B.: Validation, formal analysis, data curation. J.C.M.B.: Methodology, validation, supervision. L.F.M.d.S.: Methodology, formal analysis, investigation, supervision, project administration, funding acquisition. All authors have read and agreed to the published version of the manuscript.

Funding: This research was funded by the Project No. PTDC/EME-EME/6442/2020 "A smart and eco-friendly adhesively bonded structure for the next generation mobility platforms", and the in-dividual grants 2022.12426.BD and CEECIND/03276/2018, funded by national funds through the Portuguese Foundation for Science and Technology (FCT).

Institutional Review Board Statement: Not applicable.

Informed Consent Statement: Not applicable.

Data Availability Statement: Data are contained within the article.

Conflicts of Interest: The authors declare no conflict of interest.

References

1. Gajula, S.; Antonyraj, C.A.; Odaneth, A.A.; Srinivasan, K. A consolidated road map for economically gainful efficient utilization of agro-wastes for eco-friendly products. *Biofuels Bioprod. Biorefining* **2019**, *13*, 899–911. [CrossRef]
2. Zhang, Y.; Duan, C.; Bokka, S.K.; He, Z.; Ni, Y. Molded fiber and pulp products as green and sustainable alternatives to plastics: A mini review. *J. Bioresour. Bioprod.* **2022**, *7*, 14–25. [CrossRef]
3. Borges, C.S.P.; Jalali, S.; Tsokanas, P.; Marques, E.A.S.; Carbas, R.J.C.; da Silva, L.F.M. Sustainable Development Approaches through Wooden Adhesive Joints Design. *Polymers* **2022**, *15*, 89. [CrossRef] [PubMed]
4. Hu, M.; Duan, Z.; Zhou, X.; Du, G.; Li, T. Effects of surface characteristics of wood on bonding performance of low-molar ratio urea–formaldehyde resin. *J. Adhes.* **2022**, *99*, 803–816. [CrossRef]
5. Oliveira, P.R.; May, M.; Panzera, T.H.; Scarpa, F.; Hiermaier, S. Reinforced biobased adhesive for eco-friendly sandwich panels. *Int. J. Adhes. Adhes.* **2020**, *98*, 102550. [CrossRef]
6. Bogue, R. Recent developments in adhesive technology: A review. *Assem. Autom.* **2011**, *31*, 207–211. [CrossRef]
7. Gurr, J.; Barbu, M.C.; Frühwald, A.; Chaowana, P. The bond strength development of coconut wood in relation to its density variations. *J. Adhes.* **2022**, *98*, 1520–1533. [CrossRef]
8. de Queiroz, H.F.M.; Banea, M.D.; Cavalcanti, D.K.K. Adhesively bonded joints of jute, glass and hybrid jute/glass fibre-reinforced polymer composites for automotive industry. *Appl. Adhes. Sci.* **2021**, *9*, 1–14. [CrossRef]
9. Ammann, S.; Niemz, P. Specific fracture energy at glue joints in European beech wood. *Int. J. Adhes. Adhes.* **2015**, *60*, 47–53. [CrossRef]
10. Tserpes, K.; Barroso-Caro, A.; Carraro, P.A.; Beber, V.C.; Floros, I.; Gamon, W.; Kozłowski, M.; Santandrea, F.; Shahverdi, M.; Skejić, D.; et al. A review on failure theories and simulation models for adhesive joints. *J. Adhes.* **2021**, *98*, 1855–1915. [CrossRef]
11. Gooch, J.W. ASTM D638. In *Encyclopedic Dictionary of Polymers*; Springer: New York, NY, USA, 2011; p. 51.
12. Faidzi, M.K.; Abdullah, S.; Abdullah, M.F.; Singh, S.S.K.; Azman, A.H. Evaluating an adhesive effect on core surface configuration for sandwich panel with peel simulation approach. *J. Mech. Sci. Technol.* **2021**, *35*, 2431–2439. [CrossRef]
13. Santoni, I.; Pizzo, B. Evaluation of alternative vegetable proteins as wood adhesives. *Ind. Crop. Prod.* **2013**, *45*, 148–154. [CrossRef]
14. Shang, X.; Marques, E.; Machado, J.; Carbas, R.; Jiang, D.; da Silva, L. A strategy to reduce delamination of adhesive joints with composite substrates. *Proc. Inst. Mech. Eng. Part L J. Mater. Des. Appl.* **2018**, *233*, 521–530. [CrossRef]
15. Wiktor, W.; Pała, T. Selected aspects of cohesive zone modeling in fracture mechanics. *Metals* **2021**, *11*, 302.
16. Baumberger, S.; Lapierre, C.; Monties, B.; Della Valle, G. Use of kraft lignin as filler for starch films. *Polym. Degrad. Stab.* **1998**, *59*, 273–277. [CrossRef]

17. Azari, S.; Papini, M.; Spelt, J.K. Effect of Surface Roughness on the Performance of Adhesive Joints Under Static and Cyclic Loading. *J. Adhes.* **2010**, *86*, 742–764. [CrossRef]
18. Rangaswamy, H.; Sogalad, I.; Basavarajappa, S.; Acharya, S.; Patel, G.C.M. Experimental analysis and prediction of strength of adhesive-bonded single-lap composite joints: Taguchi and artificial neural network approaches. *SN Appl. Sci.* **2020**, *2*, 1–15. [CrossRef]
19. Stucki, S.; Lange, H.; Dreimol, C.H.; Weinand, Y.; Burgert, I. The influence of wood surface treatments with different biomolecules on dry and wet strength of linear friction welded joints. *J. Adhes. Sci. Technol.* **2023**, *37*, 1–20. [CrossRef]
20. *ASTM D6866-22*; Standard Test Methods for Determining the Biobased Content of Solid, Liquid, and Gaseous Samples Using Radiocarbon Analysis. ASTM: West Conshohocken, PA, USA, 2022.
21. Jalali, S.; Borges, C.d.S.P.; Carbas, R.J.C.; Marques, E.A.d.S.; Bordado, J.C.M.; da Silva, L.F.M. Characterization of Densified Pine Wood and a Zero-Thickness Bio-Based Adhesive for Eco-Friendly Structural Applications. *Materials* **2023**, *16*, 7147. [CrossRef]
22. Oliveira, J.; Demoura, M.; Silva, M.; Morais, J. Numerical analysis of the MMB test for mixed-mode I/II wood fracture. *Compos. Sci. Technol.* **2007**, *67*, 1764–1771. [CrossRef]
23. Gültekin, K.; Akpinar, S.; Özel, A. The effect of the adherend width on the strength of adhesively bonded single-lap joint: Experimental and numerical analysis. Compos. *Part B Eng.* **2014**, *60*, 736–745. [CrossRef]
24. Pisharody, A.P.; Blandford, B.; Smith, D.E.; Jack, D.A. An experimental investigation on the effect of adhesive distribution on strength of bonded joints. *Appl. Adhes. Sci.* **2019**, *7*, 1–12. [CrossRef]

Disclaimer/Publisher's Note: The statements, opinions and data contained in all publications are solely those of the individual author(s) and contributor(s) and not of MDPI and/or the editor(s). MDPI and/or the editor(s) disclaim responsibility for any injury to people or property resulting from any ideas, methods, instructions or products referred to in the content.

Article

Analytical Calculation of Relationship Temperature and Fatigue and Creep Strength Based on Thermal Activation

Keiji Houjou [1,*], Kazumasa Shimamoto [1], Haruhisa Akiyama [1], Yu Sekiguchi [2] and Chiaki Sato [2]

[1] Nanomaterials Research Institute, The National Institute of Advanced Industrial Science and Technology, Tsukuba 305-8564, Japan; kazumasa.shimamoto@aist.go.jp (K.S.); h.akiyama@aist.go.jp (H.A.)

[2] Precision and Intelligence Laboratory, Tokyo Institute of Technology, Yokohama 226-8501, Japan; sekiguchi.y.aa@m.titech.ac.jp (Y.S.); csato@pi.titech.ac.jp (C.S.)

* Correspondence: zz-houjou@aist.go.jp

Abstract: The purpose of this study was to formulate a mathematical expression for the temperature dependence of adhesive strength using various parameters. Adhesive structures are typically exposed to a broad temperature range, spanning from low to high temperatures; therefore, understanding how their strength depends on temperature is crucial. The strength was measured through tensile, fatigue, and creep tests at temperatures ranging from −60 °C to 135 °C. The properties of these test types were thoroughly investigated by analyzing the strength of the test results from a thermal activity perspective. The results demonstrate that there is a clear relationship between temperature and strength. The intensity decreased with temperature according to the exponential function and could be accurately represented using the parameters of thermal activity. The temperature at which the strength begins to decrease in the fatigue test was higher than in the static tests. Consequently, we were able to accurately express the relationship between the temperature and intensity using certain parameters. Few studies successfully developed a precise nonlinear relationship between temperature and intensity using approximate expressions.

Keywords: epoxy adhesive; lap joint; fatigue strength; creep strength; temperature dependence

Citation: Houjou, K.; Shimamoto, K.; Akiyama, H.; Sekiguchi, Y.; Sato, C. Analytical Calculation of Relationship Temperature and Fatigue and Creep Strength Based on Thermal Activation. *Materials* **2024**, *17*, 3055. https://doi.org/10.3390/ma17133055

Academic Editors: Marco Lamberti, Aurelien Maurel-Pantel and Aniello Riccio

Received: 24 April 2024
Revised: 11 June 2024
Accepted: 19 June 2024
Published: 21 June 2024

Copyright: © 2024 by the authors. Licensee MDPI, Basel, Switzerland. This article is an open access article distributed under the terms and conditions of the Creative Commons Attribution (CC BY) license (https://creativecommons.org/licenses/by/4.0/).

1. Introduction

In recent years, the demand for lightweight transportation equipment has significantly increased, driven by concerns related to environmental protection and energy conservation. One of the most effective methods for achieving lightweighting is adhesive bonding, which enables the fabrication of dissimilar material joints. Automobiles are produced in a variety of ways to reduce energy consumption [1]. In particular, weight reduction of bodies is essential to reduce energy consumption. For example, structural design innovations and improvements in steel materials were made. However, the use of lightweight materials is the most effective way to significantly reduce weight, and the development of adhesive bonding technology is rapidly advancing for this purpose [2]. To apply adhesive bonding technology to the structural components of transportation equipment, ensuring durability and reliability against environmental loads is essential. In addition to weather resistance against factors such as temperature, humidity, ultraviolet rays, salt, and acid rain, load durability must be studied simultaneously [3–9]. In this study, we focused on the relationship between environmental temperature and adhesive strength, which is a key factor affecting adhesive joints. The purpose of this study was to elucidate and mathematically express the relationship between environmental temperature and adhesive joint strength, thus allowing the temperature dependence of strength to be expressed objectively. Banea MD et al. [10] experimented in detail on the relationship among temperature, strength, and fracture toughness of epoxy adhesives, which was based on mode1. Tests were conducted at temperatures from $R.T.$ to 200 °C. The strength at temperatures above the T_g (glass transition temperature) point was revealed. JINGXIN NA et al. [11] measured the static strength of polyurethane from −40 °C to 90 °C using butt joints and a single lap join and obtained an approximate linear equation between temperature and strength. However, the relationship among fatigue, creep strength, and temperature was not experimented with. Moreover, they did not generalize the formula. Therefore, we conducted a study to accurately represent the nonlinear relationship between temperature and intensity. Various studies have explored the temperature dependence of material strength and adhesive strength [10–15], demonstrating that the relationship between the two can be expressed as a formula [16–18]. Williams et al. [19] proposed a method (WLF, Eq.) to express the relationship between the temperature and time of strain in polymer materials. Since then, numerous research examples using the time–temperature conversion rule have been published [20–22]. However, in this study, for adhesive joints, the strength was tested together with the metal or resin adherend. Consequently, creating a master curve based on the strain during the test was not feasible.

In this study, the definition of the relationship between temperature and strength was investigated based on the hypothesis that strength reduction depends on the increase in the number of defects due to heat. This study aimed to establish the relationship between temperature and intensity.

2. Specimen and Experimental Procedure

2.1. Adhesive

A thermoset epoxy adhesive supplied by Cemedine Co., Ltd., Koga, Japan, was selected for this study. Its chemical composition is listed in Table 1. Carboxyl-terminated butadiene-nitrile rubber (CTBN) was added to the adhesive to improve toughness and elongation at fracture.

Table 1. Chemical composition of adhesive (mass%).

Material	Mass %
Bisphenol A epoxy resin	24
CTBN-modified epoxy resin (elastomer 40%) CTBN; carboxyl-terminated butadiene acrylonitrile rubber	39
Fumed silica	3
Filler (CaCO$_3$)	26
CaO	2
Dicyane diamide	5
3-(3,4-dichlorophenyl)-1,1'-dimethylurea	1

2.2. Test Specimen

Figure 1 shows the test specimens used in this study, consisting of A6061-T6 aluminum plates with a width of 25 mm and thickness of 3 mm, which served as adherends. Prior to assembly, the adherends underwent a surface treatment process, including degreasing with acetone, followed by alkaline and acid cleaning in a hot bath at 60 °C for 30 s. Afterward, the adhesive was applied to the aluminum adherends and cured at 180 °C for 1 h, ensuring a controlled adhesive layer thickness of 0.3 mm.

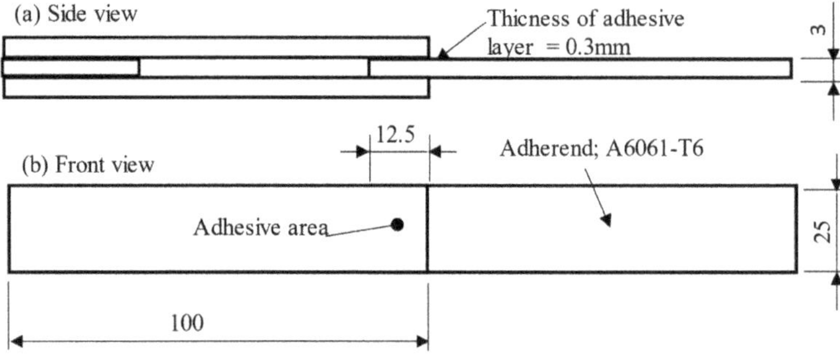

Figure 1. Schematic and dimensions of test specimen used in this study: (a) side view; (b) front view.

2.3. Experimental Procedure

The experiments included tensile shear, fatigue, and creep tests. The atmospheric temperature was measured stepwise from $-60\ °C$ to $135\ °C$ in an electric furnace. However, humidity was not controlled. The test conditions were as follows: the tensile test rate was 1 mm/min, the fatigue test was conducted at a frequency f = 10 Hz, with a stress ratio R = -1, and the creep test was conducted under load control. If the specimen did not fracture within a certain time or number of times, the maximum stress was defined as the creep or fatigue limit. Tensile and creep tests were performed using a tensile testing machine (RTF-1350 Tensilon, A&D Corporation, Kitamoto, Japan) with an electric furnace and a maximum load of 50 KN. The control temperature of the electric furnace ranged from $-45\ °C$ to $210\ °C$ in air. A hydraulic-type fatigue testing machine (EHF-E50KN Servo Pulser, Shimadzu Corporation, Kyoto, Japan) with an electric furnace was used. The control temperature of the electric furnace ranged from $-60\ °C$ to $250\ °C$ in air. The clamps were hydraulically operated, and a maximum cyclic load of 50 KN was applied. The fracture surfaces of the specimens were observed using an optical microscope, and the chemical structure of the adhesive was investigated using Fourier transform infrared (FTIR) spectroscopy.

3. Results and Discussion

3.1. Relationship between Test Temperature and Shear Strength

Figure 2a shows the relationship between test temperatures ranging from $-60\ °C$ to $135\ °C$ and the shear strength of adhesive joints. The strength decreased as the test temperature increased and improved as the test temperature decreased; however, it did not exhibit further increases below a certain temperature. In all the cases, cohesive failure was observed as the failure mode. However, the fracture surface tested below $0\ °C$ was a brittle cleavage fracture surface accompanied by thin-layer cohesive failure. As the temperature increased, the fracture changed to ductile fracture, accompanied by elongation. The gray shaded area in Figure 2a represents the T_g of the adhesive. Table 2 lists the bulk mechanical properties and T_g points measured in previous studies. The Young's modulus, Poisson's ratio, and tensile strength of the bulk specimens were 1100 MPa, 0.41, and 27 MPa at $R.\ T.$, respectively. T_g was 110–125 $°C$ [7,9]. The adhesive strength gradually decreased up to 135 $°C$, and a significant drop in strength was not observed even above T_g.

Table 2. Mechanical properties of cured adhesive.

Bulk Mechanical Properties [7,9]	
Tensile strength at R.T. (MPa)	30
Young's modulus at R.T. (MPa)	1100
Poisson's ratio at R.T.	0.41
T_g point ($°C$)	110–125

In prior research, the authors of the current study developed a mathematical relationship between test temperature T and shear strength τ_B within a range from $-20\ °C$ to $135\ °C$, as shown in Equation (1) [15]. Figure 2b shows the relationship between temperature and strength reduction schematically. The proposal suggested that the strength τ_B at the test temperature T could be calculated by subtracting the amount of strength reduction τ_T owing to the test temperature from the strength C [15]. However, various studies have demonstrated the existence of a maximum temperature T_0 at which the strength does not decrease [12,16,18,23–25]. Therefore, the equation was modified; when $T < T_0$, Equation (2) was used in this study.

$$T \geq T_0 \quad \tau_B = C - a \cdot exp\left\{ -\frac{H}{T - T_0} \right\} \tag{1}$$

$$T_0 > T \qquad \tau_B = C \tag{2}$$

where C represents the original strength of the material before it decreases with temperature, a represents a proportionality constant that converts the probability of the existence of defects into material strength, and H includes the Boltzmann constant and activation energy. The constants are discussed later in Sections 3 and 4. However, at low temperatures, it is influenced by the difference in the linear expansion coefficient between the adherend and adhesive [25]. Therefore, the applicable lower-limit temperature of Equation (2) is unknown. The dashed line in Figure 2a represents the curve generated by substituting the intensity data obtained in the experiment into Equations (1) and (2) and determining the constants C, a, H, and T_0 using the least-squares method.

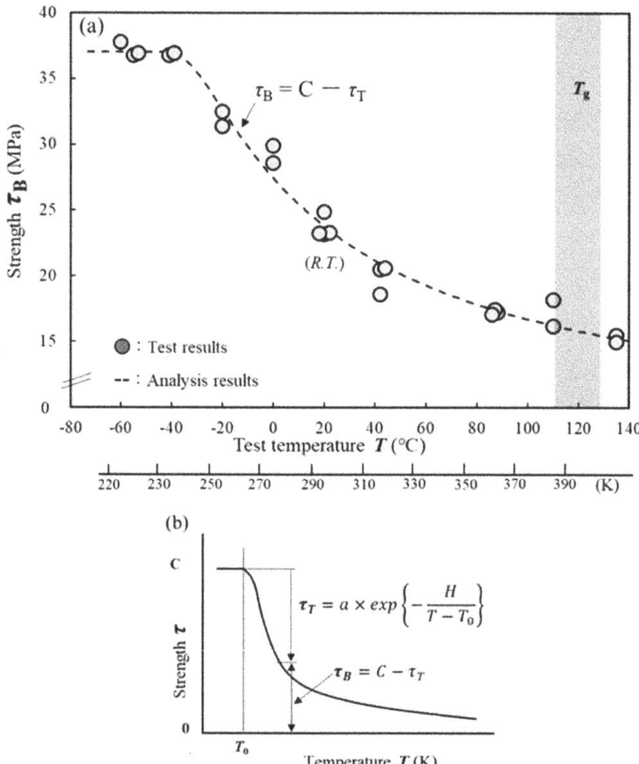

Figure 2. Effect of temperature on shear strength. (**a**) Relationship between test temperature T and shear strength τ_B. (**b**) Schematic showing the relationship between temperature and strength.

3.2. Relationship between Test Temperature and Fatigue Strength

We then studied whether Equations (1) and (2) could be applied to the relationship between the test temperature and fatigue limit. Figure 3a shows the fatigue test results from -55 °C to 135 °C. The frequency and stress ratios were $f = 10$ Hz and $R = -1$. The tests were terminated when the specimens did not fail until $N = 5 \times 10^6$. The maximum amplitude stress τ_a was defined as the fatigue limit τ_w. The arrows indicate tests that were terminated without failure. The fatigue limit improved significantly as the test temperature T decreased. Figure 3b represents the relationship between test temperature T and fatigue limit τ_w. The dashed line is an approximation line calculated by substituting the test results into Equations (1) and (2).

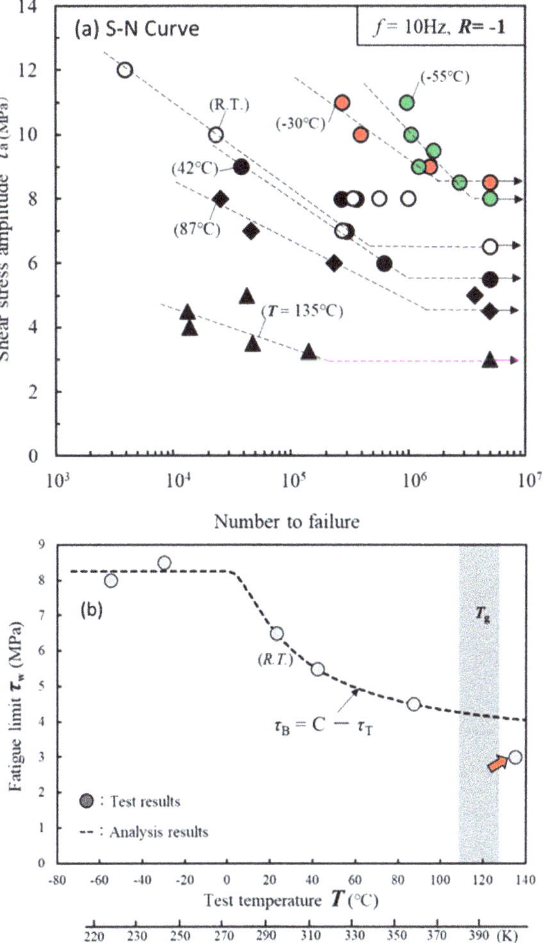

Figure 3. Effect of test temperature on fatigue strength. (**a**) S-N curve from −55 °C to 135 °C. (**b**) Relationship between test temperature T and fatigue limit τ_w.

This approximation line closely matches the experimental results. However, the fatigue limit of 135 °C (red arrow) is out of the approximate line. As explained later in Figure 4, at 135 °C, Equation (1) was not able to approximate the experimental results because the oxidation or hydrolysis of the adhesive contributed to reducing the strength, in addition to τ_T. Figure 4 shows photographs of the fracture surface at each temperature. All fracture surfaces exhibited cohesive failure. Extremely thin adhesive layers remained on the opposite adherent surfaces. The test temperatures in Figure 4a,b are −55 °C and −30 °C, respectively, and fatigue fracture surfaces, indicated by the arrows, can be observed. Figure 4h shows a detailed view of the fatigue-fractured surface with wave patterns perpendicular to the crack propagation direction. In contrast, in Figure 4g, a brittle fracture surface can be observed, indicating an unstable fracture. Figure 4c–e show ductile cohesive failures of the adhesive. In particular, Figure 4e shows that failure occurred deep within the adhesive layer. A discolored oxidized area with reduced strength is visible within a width of 3 mm from the edge.

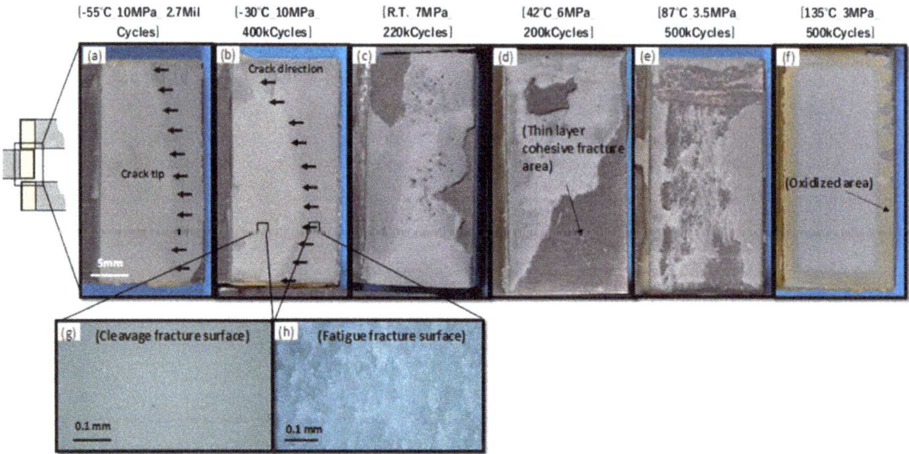

Figure 4. Photographs of fractured surfaces during fatigue tests at various temperatures. (**a–f**) Fracture surface under each test condition. (**g,h**) Details of fatigue fracture surface.

3.3. Relationship between Test Temperature and Creep Strength

We then studied whether Equations (1) and (2) could be applied to the relationship between the test temperature and creep limit. Figure 5a shows the creep test results from $T = -45\,°C$ to $135\,°C$. The creep limit τ_w was defined as the maximum stress τ that did not fail for over 400,000 s. The arrows in the figure indicate that the tests were terminated without fracture. The range of fatigue limit τ_w was 5.5 MPa (8.5 MPa to 3 MPa, Figure 3a); however, that of creep limit τ_w was 27 MPa (28 MPa to 1 MPa, Figure 5a). The creep strength was sensitive to the test temperature.

Figure 5b shows the relationship between test temperature T and creep limit τ_w. The dashed line is an approximation calculated by substituting the test results into Equations (1) and (2). This approximation line closely matches the experimental results. For the same reason as observed in the fatigue test, the creep limits of $87\,°C$ and $135\,°C$ (red arrows) are also out of the approximate line. Because the exposure time of the creep tests were longer than that of the fatigue tests, the deterioration of the adhesive progressed significantly during the high-temperature test, resulting in a decrease in strength.

Figure 6 shows photographs of the fractured surfaces during testing. These fracture surfaces exhibited cohesive or thin-layer cohesive failures. As shown in Figure 6a,b, the adhesive exhibits brittle fracture behavior at low temperatures. The arrows indicate the crack tip at the time of fracture. When the test temperature exceeded *R.T.* (controlled to $23–28\,°C$), cohesive failure transitioned to ductile failure, as shown in Figure 6c–f. Cohesive failure occurred deep within the adhesive as the test temperature increased. In the test at $135\,°C$, as shown in Figure 5f, the adhesive color changed, indicating deterioration due to oxidation. Figure 5b shows that the strengths at $87\,°C$ and $135\,°C$ are significantly lower than the values calculated from Equations (1) and (2) (dashed line).

FTIR measurements were performed to investigate the deterioration of the adhesive. Specimens that endured for 400,000 s were then subjected to fracture testing. Subsequently, FTIR measurements were taken at a location 2–3 mm from the edge of the adhesive. Figure 7a shows the FTIR profiles of the adhesive that underwent the creep test from $-45\,°C$ to $135\,°C$ for 400,000 s. The peak at 1740 cm^{-1} (*Io*) indicates a C=O bond. *Io* is the peak generated by the oxidation or hydrolysis of adhesives in a high-temperature atmosphere and can be considered a barometer for adhesive deterioration [26]. The *Io* peak at $135\,°C$ clearly increases. The *Io* peaks were extracted from the profile and normalized using the intensity (*Is*) of the fingerprint area peak at 1505 cm^{-1}, and the (*Io*/*Is*) is shown in Figure 7b. The C=O peak did not change significantly from $-45\,°C$ to $55\,°C$

but exhibited a notable increase above 87 °C. Consequently, the creep strength decreased in Figure 5b. Therefore, in conjunction with Figure 5b, these results suggest that, if oxidation and hydrolysis do not progress significantly, the creep strength can be explained by Equations (1) and (2).

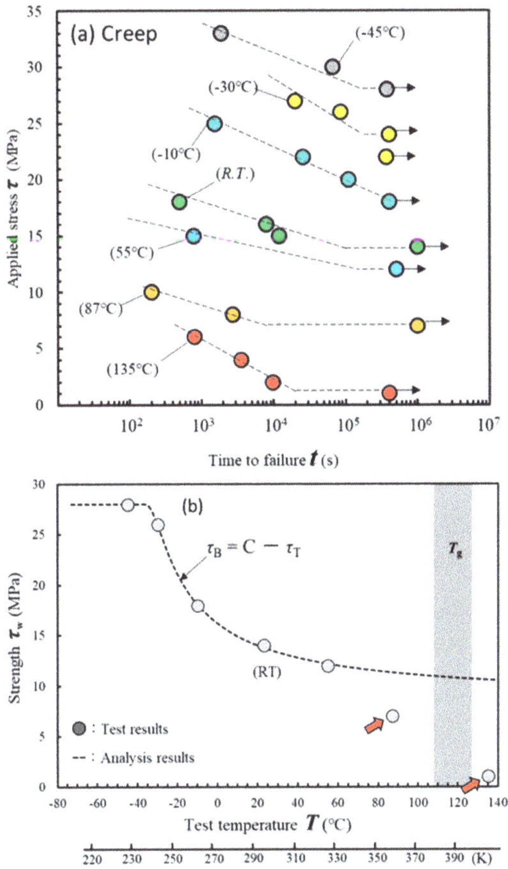

Figure 5. Effect of test temperature on creep strength. (**a**) Relationship between test temperature T and creep limit τ_w. (**b**) Schematic showing the relationship between temperature and strength.

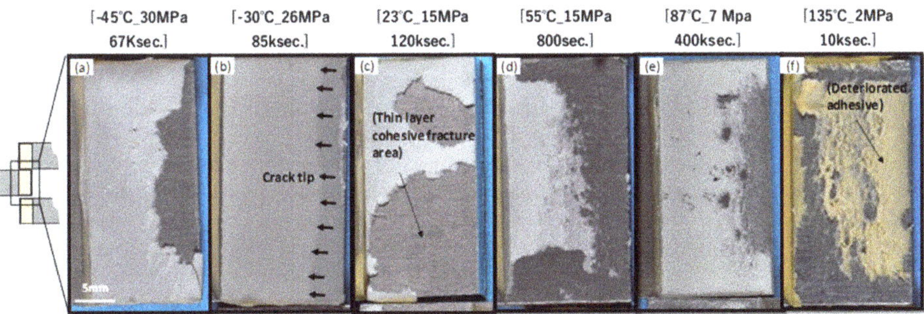

Figure 6. Photographs of fractured surfaces during creep tests at various temperatures. (**a**–**f**) Fracture surface under each test condition.

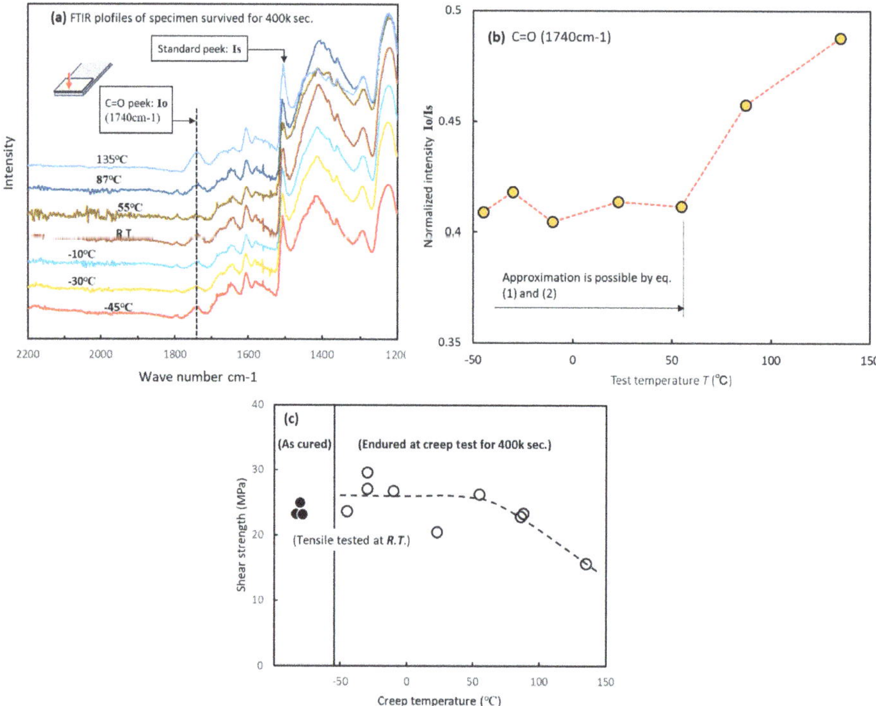

Figure 7. Degradation behavior of adhesives during creep tests. (**a**) FTIR profiles of specimens that survived for four million seconds. (**b**) Relationship between test temperature and intensity of C=O peaks. (**c**) Relationship between test temperature and shear strength of specimens that endured creep tests for 400,000 s.

Figure 7c shows the results of measuring the *R.T.* shear strength of the specimens that did not fail during creep tests. A clear decrease in shear strength was not observed from $-55\ ^\circ$C to $54\ ^\circ$C, but the shear strength decreased above $87\ ^\circ$C. These findings alongside the results in Figure 7b suggest that the adhesive maintained sound strength and chemical structure at test temperatures up to $55\ ^\circ$C.

3.4. Temperature Dependence of LJ Strength

The temperature-dependent properties for each test type were studied based on the approximation lines of the tensile, fatigue, and creep tests. Figure 8 shows the approximation lines for the three tests used in this study. Some of these trends are described below.

T_0 of the fatigue test was considerably higher than those of the tensile and creep tests. In Equation (1), the assumption is that the strength decrease depends on the increase in defects owing to temperature (thermal activation and thermal vibration). However, the temperature at which defects begin to be produced owing to thermal vibration is not necessarily at the same temperature. Therefore, the strength did not decrease until the temperature reached the point at which the heat began to exert an influence in the case of creep and tensile tests. However, in the fatigue tests, cyclic strain was added to the effects of thermal vibration. Consequently, the specimens were in a high-energy state, causing the strength to begin decreasing at higher temperatures compared to the creep and tensile tests. As shown in Table 3, the fatigue T_0 is 40 to $50\ ^\circ$C higher than the T_0 for the other tests. Another characteristic of Figure 8 is the high sensitivity of the creep strength to the test temperature. The triangle mark in the figure indicates that the slope of the strength in the creep test was steeper than that in the other tests. As indicated in Table 3, constant *H*, which depends on activation energy, was the lowest during the creep test. This corresponds directly to the shape of the curve shown in Figure 8. Based on the aforementioned results, we developed a method to describe the temperature dependence of the strength using the parameters T_0 and H.

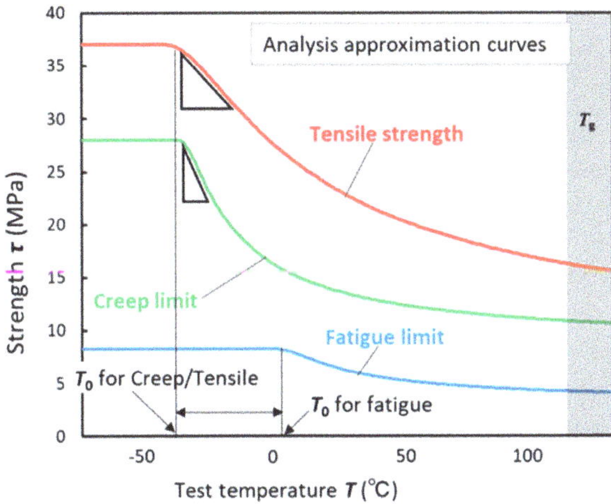

Figure 8. Comparison of approximate lines for tensile, fatigue, and creep tests.

Table 3. Constants T_0 and H calculated from experimental results using the least-squares method.

Test Type	T_0 (°C)	H (Q/B) ($\times 10^3$ K/mol)
Tensile	−49	54
Fatigue	−3	28
Creep	−39	19

4. Conclusions

A detailed investigation of tensile, fatigue, and creep strengths in epoxy adhesive joints across a range of temperatures was conducted. Consequently, we reached the following conclusions:

(1) Tensile strength, fatigue limit, and creep limit improved with lower test temperatures; however, no improvement occurred below a certain temperature.
(2) The relationship between test temperature and strength can be approximated using a thermal activation equation.
(3) In long-term high-temperature tests, such as 135 °C fatigue tests, the adhesive deteriorates and does not follow the established curve.
(4) Observations of the fracture surface in the fatigue test revealed that brittle fracture occurred when the test temperature was low, whereas ductile fracture occurred when it was high.
(5) The T_0 point for fatigue was higher than those of the other tests, and the temperature sensitivity of strength, represented by H, was highest in the creep test.
(6) A method to express the temperature dependence of adhesive strength using constants T_0 and H was proposed.

In this study, we conducted experiments using assembled lap joints. However, it is desirable to conduct clear research using only adhesive materials. Furthermore, adaptation of Equations (1) and (2) to metallic and inorganic materials as well as appreciation of statistical analysis for fatigue and creep test results are a future study.

Author Contributions: Conceptualization, K.H.; Methodology, K.H.; Software, Investigation, K.S.; Investigation, K.H.; Data curation, Investigation, Y.S.; Writing—original draft, K.H.; Funding acquisition, Investigation, H.A. and C.S. All authors have read and agreed to the published version of the manuscript.

Funding: This paper is based on results from a project, JPNP20005, commissioned by the New Energy and Industrial Technology Development Organization (NEDO; funding number 2020005008480). This research did not receive any specific grant from funding agencies in the public, commercial, or not-for-profit sectors.

Institutional Review Board Statement: Not applicable.

Informed Consent Statement: Not applicable.

Data Availability Statement: Data are contained within the article.

Conflicts of Interest: The authors declare no conflict of interest.

References

1. Joost, W.J. Reducing Vehicle Weight and Improving U.S. Energy Efficiency Using Integrated Computational Materials Engineering. *JOM* **2012**, *64*, 1032–1038. [CrossRef]
2. Miller, W.S.; Zhuang, L.; Bottema, J.; Wittebrood, A.J.; De Smet, P.; Haszler, A.; Vieregge, A. Recent development in aluminum alloys for the automotive industry. *Mater. Sci. Eng. A* **2000**, *280*, 37–49. [CrossRef]
3. Ferreira, J.A.M.; Reis, P.N.; Costa, J.D.M.; Richardson, M.O.W. Fatigue behavior of composite adhesive lap joints. *Compos. Sci. Technol.* **2002**, *62*, 1373–1379. [CrossRef]
4. Da Silva, L.F.M.; Sato, C. *Design of Adhesive Joints under Humid Condition*; Springer: Berlin/Heidelberg, Germany, 2013. [CrossRef]
5. Puigvert, F.; Crocombe, A.D.; Gil, L. Fatigue and creep analyses of adhesively bonded anchorages for CFRP tendons. *Int. J. Adhes. Adhes.* **2014**, *54*, 143–154. [CrossRef]
6. Savvilotidou, M.; Keller, T.; Vassilopoulos, A.P. Fatigue performance of a cold-curing structural epoxy adhesive subjected to moist environments. *Int. J. Fatigue* **2017**, *103*, 405–414. [CrossRef]
7. Houjou, K.; Shimamoto, K.; Akiyama, H.; Sato, C. Dependence of fatigue limit on stress ratio and influence of cyclic stress on shear strength for an adhesive lap joint. *J. Adhes.* **2021**, *97*, 1153–1165. [CrossRef]
8. Houjou, K.; Shimamoto, K.; Akiyama, H.; Sato, C. Effect of cyclic moisture absorption/desorption on the strength of epoxy adhesive joints and moisture diffusion coefficient. *J. Adhes.* **2022**, *98*, 1535–1551. [CrossRef]
9. Houjou, K.; Sekiguchi, Y.; Shimamoto, K.; Akiyama, H.; Sato, C. Energy release rate and crack propagation rate behavior of moisture-deteriorated epoxy adhesives through the double cantilever beam method. *J. Adhes.* **2023**, *99*, 1016–1030. [CrossRef]
10. Banea, M.D.; da Silva, L.F.M.; Campilho, R.D.S.G.; de Jesus, A.M.P. Characterization of aluminum single-lap joints for high temperature applications. *Mater. Sci.* **2013**, *730–732*, 721–726. [CrossRef]
11. Jingxin, N.A.; Liu, Y.; Wang, Y.; Pan, L.; Yan, Y. Effect of Temperature on the Joints Strength of an Automotive Polyurethane Adhesive. *J. Adhes.* **2016**, *92*, 52–64.
12. Adams, R.D.; Coppendale, J.; Mallick, V.; Al-Hamdan, H. The effect of temperature on the strength of adhesive joints. *Int. J. Adhes. Adhes.* **1992**, *12*, 185–190. [CrossRef]
13. Wrosch, M.; Xian, G.; Karbhari, V.M. Moisture absorption and desorption in a UV cured urethane acrylate adhesive based on radiation source. *J. Appl. Polym.* **2008**, *107*, 3654–3662. [CrossRef]
14. da Silva, L.F.M.; Adams, R.D. Adhesive joints at high and low temperatures using similar and dissimilar adherends and dual adhesives. *Int. J. Adhes. Adhes.* **2007**, *27*, 216–226. [CrossRef]
15. Houjou, K.; Shimamoto, K.; Akiyama, H.; Sato, C. Effect of test temperature on the shear and fatigue strengths of epoxy adhesive joints. *J. Adhes.* **2022**, *98*, 2599–2617. [CrossRef]
16. Argon, A.S. A theory for the low-temperature plastic deformation of glassy polymers. *Philos. Mag.* **1973**, *28*, 839–865. [CrossRef]
17. Bascom, W.D.; Cottington, R.L. Effect of temperature on the adhesive fracture behavior of an elastomer-epoxy resin. *J. Adhes.* **1976**, *7*, 333–346. [CrossRef]
18. Li, W.; Zhang, X.; Kou, H.; Wang, R.; Fang, D. Theoretical prediction of temperature dependent yield strength for metallic materials. *Int. J. Mech. Sci.* **2016**, *105*, 273–278. [CrossRef]
19. Williams, M.L.; Landel, R.F.; Ferry, J.D. Temperature dependence of relaxation mechanism in amorphous polymers and other glass-forming liquids. *J. Am. Chem. Soc.* **1955**, *77*, 3701–3707. [CrossRef]
20. Gent, A.N.; Lai, S.-M. Interfacial bonding, energy dissipation, and adhesion. *J. Polym. Sci. B Polym. Phys.* **1994**, *32*, 1543–1555. [CrossRef]
21. Shanahan, M.E.R.; Garré, A. Viscoelastic dissipation in wetting and adhesion phenomena. *J. Am. Chem. Soc.* **1995**, *11*, 1396–1402. [CrossRef]
22. Shioya, M.; Kuroyanagi, Y.; Ryu, M.; Morikawa, J. Analysis of the adhesive properties of carbon nanotube- and graphene oxide nanoribbon-dispersed aliphatic epoxy resins based on the Maxwell model. *Int. J. Adhes. Adhes.* **2018**, *84*, 27–36. [CrossRef]

23. Li, X.; Schnecken, S.; Simon, E.; Bergqvist, L.; Zhang, H.; Szunyogh, L. Tensile strain-induced softening of iron at high temperature. *Sci. Rep.* **2015**, *5*, 16654. [CrossRef] [PubMed]
24. Maraveas, C.; Fasoulakis, Z.C.; Tsavdaridis, K.D. Mechanical properties of High and Very High Steel at elevated temperatures and after cooling down. *Fire Sci. Rev.* **2017**, *6*, 3. [CrossRef]
25. Kumazawa, H.; Nakagawa, H.; Saito, T.; Ogawa, T. Effect of cryogenic environment on strength properties of double-lap composite joints. *J. Compos. Mater.* **2020**, *46-4*, 150–161. (In Japanese) [CrossRef]
26. Shimamoto, K.; Batorova, S.; Houjou, K.; Akiyama, H.; Sato, C. Degradation of epoxy adhesive containing dicyandiamide and carboxyl-terminated butadiene acrylonitrile rubber due to water with open-faced specimens. *J. Adhes.* **2021**, *97*, 1388–1403. [CrossRef]

Disclaimer/Publisher's Note: The statements, opinions and data contained in all publications are solely those of the individual author(s) and contributor(s) and not of MDPI and/or the editor(s). MDPI and/or the editor(s) disclaim responsibility for any injury to people or property resulting from any ideas, methods, instructions or products referred to in the content.

Article

Properties and Performance of Epoxy Resin/Boron Acid Composites

Anna Rudawska [1], Mariaenrica Frigione [2,*], Antonella Sarcinella [2], Valentina Brunella [3], Ludovica Di Lorenzo [3] and Ewa Olewnik-Kruszkowska [4]

[1]   Faculty of Mechanical Engineering, Lublin University of Technology, Nadbystrzycka 36, 20-618 Lublin, Poland; a.rudawska@pollub.pl
[2]   Department of Innovation Engineering, University of Salento, Via Arnesano, 73100 Lecce, Italy; antonella.sarcinella@unisalento.it
[3]   Department of Chemistry, University of Torino, Via P. Giuria 7, 10125 Torino, Italy; valentina.brunella@unito.it (V.B.); ludovica.dilorenzo@unito.it (L.D.L.)
[4]   Faculty of Chemistry, Nicolaus Copernicus University in Toruń, Gagarin 7 Street, 87-100 Toruń, Poland; olewnik@umk.pl
*   Correspondence: mariaenrica.frigione@unisalento.it

Abstract: This research study focused on the effect of adding boric acid to epoxy resin in order to obtain a composite material with improved properties and performance. To this end, a fine powder of boric acid (H_3BO_3) was introduced into epoxy resin in different amounts, i.e., 0.5 g, 1.0 g, and 1.5 g. As the matrix of the epoxy composites, styrene-modified epoxy resin based on bisphenol A (BPA) (Epidian 53) was used. It was cross-linked with two types of curing agents, i.e., an amine (ET) and a polyamide (PAC). The mechanical properties of the obtained epoxy composites (in terms of compressive strength, compressive modulus, and compressive strain) were determined at room temperature in order to assess the effect of the addition of boron acid and of the type of curing agent employed to cure the epoxy on these characteristics. Calorimetric measurements were made to highlight any changes in the glass transition temperature (Tg) as a result of the addition of boric acid to epoxy resin. Finally, flammability tests were performed on both Epidian 53/PAC and Epidian 53/ET epoxy composites to analyze their fire behavior and consequently establish the effectiveness of the selected additive as a flame retardant.

Keywords: boron acid; curing agents; epoxy resin composite; mechanical properties; thermal properties; flammability tests

Citation: Rudawska, A.; Frigione, M.; Sarcinella, A.; Brunella, V.; Di Lorenzo, L.; Olewnik-Kruszkowska, E. Properties and Performance of Epoxy Resin/Boron Acid Composites. *Materials* **2024**, *17*, 2092. https://doi.org/10.3390/ma17092092

Academic Editors: Marco Lamberti and Aurelien Maurel-Pantel

Received: 4 April 2024
Revised: 16 April 2024
Accepted: 20 April 2024
Published: 29 April 2024

Copyright: © 2024 by the authors. Licensee MDPI, Basel, Switzerland. This article is an open access article distributed under the terms and conditions of the Creative Commons Attribution (CC BY) license (https://creativecommons.org/licenses/by/4.0/).

1. Introduction

Epoxy resins are thermosetting polymers [1–4] whose molecules contain at least two epoxy groups. They are extensively used in various fields, including adhesives, coatings, sealing materials, matrices for composite materials reinforced with different fibers/fillers, structural elements, laminates, and dielectric materials [5–8]. The most commonly used epoxy resins are those based on diglycidyl ether of bisphenol A (DGEBA) [4,9–12]. In order to obtain specific functional properties, epoxy resin is subjected to a curing (i.e., cross-linking) process in the presence of a suitable curing agent [6,13–16]. The selection of the most appropriate curing agent for the curing of an epoxy resin is one of the most important elements in the development of an epoxy material with the required functional characteristics [1,6,17,18]. As Gotto [19] emphasizes, "the correct choice of curing agent can dramatically improve the properties of the formulation such as heat resistance and flexibility while also allowing curing at lower temperatures for example". It should be mentioned that the process of cross-linking of epoxy resins with various curing agents is widely discussed in the literature [17,20,21], but many authors emphasize that the mechanism of epoxy curing is complex [13,19].

In order to modify epoxy resins and create composites with optimal properties for a specific application, different compounds can also be chosen; they are added into the resin before the curing process takes place [3]. The epoxy resin and the selected curing agent thus form the matrix for the final composite [3,5]. An appropriate selection of components allows for improvements in terms of (i) mechanical, thermal, electrical, and dielectric properties [13,22–24], (ii) technological properties of materials that facilitate the processing, or (iii) providing special functional properties, e.g., low flammability, durability, low coefficient of friction, or resistance to dirt [25–28].

Among the properties mentioned, flame retardant properties are very important [4,29–33], and this applies to all polymeric materials, especially if used in particular sectors, such as in electrical works or in the construction field [26,34–36]. The presence of fragments derived from epichlorohydrin in the epoxy chain results in epoxy resins being characterized by a certain reduction in flammability compared to other halogen-free polymer materials. Unfortunately, this level of flame retardancy is insufficient in industrial practice, and these resins do not meet the increasingly stringent safety regulations [37].

In order to obtain better flame retardant properties, additives—antypirenes—with various chemical bases are used [4,30,35,38]. These include boron acid [25,39–42], phosphorus compounds [9,43], halogenated flame retardants (bromine and chlorine) [39], and halogen-free retardant [8,44], among others. Kandola et al. [44] reviewed the different types of flame retardants required to achieve a specific level of flame retardancy. Hamciuc et al. [25] presented a study focused on epoxy-based composites with improved flame-resistant properties, which were obtained by adding two flame retardant additives to the epoxy resin, one of them being boric acid. These authors noticed, inter alia, that the higher charring efficiency of epoxy-based composites containing H_3BO_3 was favorable for improving flame resistance. Nazarenko et al. [39] demonstrated that the incorporation of boric acid into the polymer matrix increases the thermal stability of epoxy composites and leads to a 2–2.7 times reduction in toxic gaseous products. Visakh et al. [42] showed that the thermal properties of the tested epoxy composites depend on the filler content. The results showed that the addition of 10% wt. fillers of both boric acid and natural zeolite significantly improved the thermal properties of the obtained composites. Flame retardants for epoxy resins can also contain combinations of elements that allow synergistic and cooperative effects to be achieved. For example, phosphorus–nitrogen-based modifiers and flame retardants of epoxy resins are presented by, e.g., Konstantinova et al. [45], Orlov et al. [46], Terekhov et al. [47], or Benin et al. [48].

In most of the cited works [25,32,36,42], however, the research was mainly focused on determining flame retardant properties [8], which are an extremely important aspect, especially when epoxies are used in building applications as resins for injection or for wall or floor coatings. Boric acid is a common, fairly economic additive that can act as an effective flame retardant for epoxy resins in such applications; it was, therefore, selected in this study. However, in the present work, attention has also been paid to mechanical and thermal properties. When designing structures containing composite materials, mechanical properties and thermal characteristics are as important as behavior in the presence of fire. In fact, both specific functional and mechanical properties [26,33,49] should be taken into account when selecting the most appropriate material for a specific application. Thermal properties, especially the glass transition temperature, determine the temperature range of use of resins and influence their durability. Due to the large number of epoxy resin types, curing agents, and modifying additives [3], as well as the numerous possibilities of mixing them in different quantitative compositions, investigations on these materials are still an interesting research field. In this work, therefore, the results of a research study focused on determining the effect of the addition of boric acid on the mechanical and thermal characteristics of an epoxy resin are presented. Fine-grained boric acid (H_3BO_3) powder was introduced into epoxy composites. A styrene-modified epoxy resin based on bisphenol A (BPA) was used as the matrix of the epoxy composites. The epoxy matrix was cross-linked with two types of curing agents, i.e., an amine and a polyamide. The effect of high temperatures as a function of the formulation of the epoxy composites was assessed by thermogravimetric analysis. To the best of the authors' knowledge, the thermal and mechanical properties of epoxy composites containing boric acid have never been contemporarily analyzed in the literature. Finally, flammability tests were performed on unmodified epoxies and on those modified with boric acid.

2. Materials and Methods

2.1. Boron Acid/Epoxy Resin Composites

Boron acid/epoxy resin composites were produced using an epoxy resin based on bisphenol A (BPA), which is a mixture of a resin with an epoxy number of 0.48–0.51 mol/100 g with styrene solvent. Its scheme is presented in Figure 1a. The epoxy value of this epoxy resin (trade name Epidian 53, Sarzyna Resins, Nowa Sarzyna, Poland) is min. 0.41 mol/100 g; the viscosity range is 900–1500 m·Pas; the density at a temperature of 25 °C is 1.15 g/cm^3; and the average molecular weight is $\leq$700. Amine (whose chemical structure is illustrated in Figure 1b) and polyamide curing agents were used to cure the epoxy resin. The properties of these curing agents are presented in Table 1 [50,51]. These agents are, in fact, often used to cure epoxy systems used as adhesives, coating materials, or other structural elements. In particular, the amine curing agent is mainly used for epoxies applied as a floor coating. In the case of structural adhesive applications, a polyamide is often used as the epoxy resin curing agent.

Table 1. Physicochemical properties of the curing agents used (Sarzyna Resins, Nowa Sarzyna, Poland).

	Curing Agent Type	
	Amine	Polyamide
Properties	**Adduct of Aliphatic Amine (Triethylenetetramine) and Aromatic Glycidyl Ether**	**Polyaminoamide**
	Trade Name	
	ET	PAC
Amine number [mg KOH/g]	700–900	290–360
Viscosity 25 °C [m·Pas]	200–300	10,000–25,000
Density at 20 °C [g/cm^3]	1.02–1.05	1.10–1.20
Stoichiometric ratio: epoxy resin/curing agent	100:18	100:80

The epoxy resin/curing agent ratio resulted from the stoichiometric ratios of the epoxy resin and the curing agents, as specified in Table 1. The epoxy composites were produced by adding fine powder of boric acid (H$_3$BO$_3$, Chempur Company, Piekary Śląskie, Poland) to the resin. This compound has a wide range of applications [31,36,52–54], one of which is to improve the flame retardancy of polymers [31,39]. The ortho form of boric acid was used, as shown in Figure 1c [52]. Three amounts of boric acid were used, i.e., 0.5%, 1.0%, and 1.5% (per 100 g resin). In the available literature [33,42], epoxy composites have various contents of boric acid, often combined with a different filler. The choice of these quantities was justified by the work presented in [42]. On the other hand, the experiments presented in [32] contain an overview of information regarding the amounts of various antipyrene substances, including antiperspirants (phosphorus), which were used in amounts ranging from 0 to 1.4%. In the study presented by Demirham et al. [36], various amounts of boric acid (i.e., 1.25, 2.5, 3.75, and 5.0% by weight) were used as flame retardant additives. Visakh et al. [42] studied epoxy composites filled with boric acid and natural zeolite with different contents (1, 5, and 10% by weight). Murat Unlu et al. [55] used boron compounds in three concentrations: 1, 3, and 5% by weight.

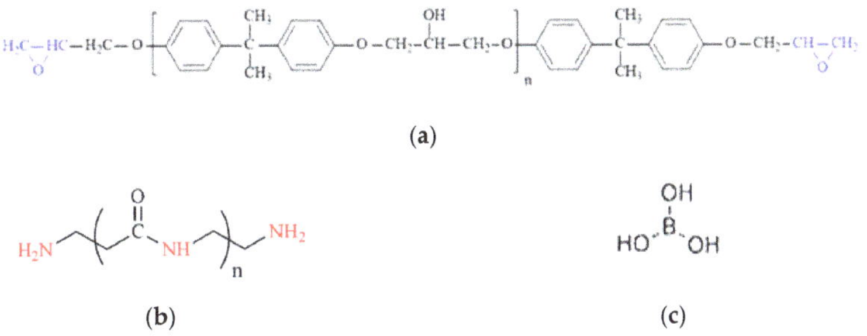

(a)

(b)

(c)

Figure 1. Schemes of (**a**) bisphenolic epoxy resin [56]; (**b**) polyaminoamide curing agent [56]; (**c**) ortho form of boric acid [48].

The composition and designation of boron acid/epoxy composites are presented in Table 2. For comparative purposes, neat epoxy compounds without the addition of a modifier were also produced.

Table 2. Boron acid/epoxy composites and relative neat epoxy resin investigated in the present study.

Resin	Curing Agent	Boric Acid (H_3BO_3) Content (%/per 100 g Resin)	Denotation
		0.5	E53/ET/H_3BO_3/0.5
	Amine	1.0	E53/ET/H_3BO_3/1.0
Epoxy resin	(ET)	1.5	E53/ET/H_3BO_3/1.5
average molecular		0.0	E53/ET
weight $\leq$ 700		0.5	E53/PAC/H_3BO_3/0.5
(Epidian 53)	Polyamide	1.0	E53/PAC/H_3BO_3/1.0
	(PAC)	1.5	E53/PAC/H_3BO_3/1.5
		0.0	E53/PAC

2.2. Preparation of Boron Acid/Epoxy Composites and Their Mechanical and Thermal Characterization

The preparation of the composite samples for mechanical tests and microscopic observations took place in several stages:

1.	Preparation of cylindrical polymeric molds;
2.	Weighing the ingredients;
3.	Homogenization of boric acid in the epoxy resin;
4.	Introduction of the curing agent;
5.	Dispersing the curing agent in the modified epoxy resin;
6.	Curing and post-curing processes;
7.	Finishing the cured epoxy composite samples and conditioning them before any tests.

To prepare the composite samples, two types of cylindrical molds were prepared. A silicone-based release agent Soudal (Soudal, Pionki, Poland) was applied to the inner surface of the mold using a spraying method from a distance of approximately 80–100 mm. The components of the composites were weighed using a laboratory balance (TM-1,5TN, FAWAG S.A, Lublin, Poland) with an accuracy of 0.1 g. The epoxy resin was weighed in a beaker with a capacity of 500 cm^3 and the appropriate amount of boric acid was then added. The mixture underwent mechanical homogenization (mechanical disc mixer) for 3 min at a speed of 144 rpm. After this process, the appropriate amount of curing agent was added to the epoxy mixture. Each epoxy compound with the respective curing agent was mixed with a low-speed laboratory mixer for 3 min, deaerated, and poured into cylindrical molds (shown in Figure 2) with dimensions of 13.0 mm × 26.0 mm (Type 1—for mechanical tests) and 10.0 mm × 10.0 mm (Type 2—for microscopic observations), covered with a release agent. The composites were cured at laboratory temperature (21 ± 1 °C) and humidity (21 ± 2% R.H.) for 168 h, and then finished by milling. This treatment aimed at obtaining accurate dimensions (in the case of the mechanical test, 12.7 mm and 25.4 mm, according to the ASTM D695 standard (ASTM D695; Standard Test Method for Compressive Properties of Rigid Plastics. ASTM International, West Conshohocken, PA, USA, 2023) and perpendicularity deviations from the axis. The specimens were then conditioned at laboratory temperature (21 ± 1 °C) and humidity (21 ± 2% R.H.) before being tested. For comparison purposes, some epoxy specimens not containing boric acid were also produced.

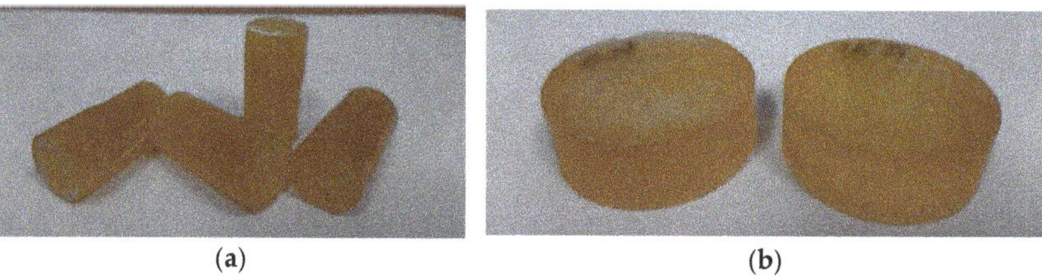

(a)   (b)

Figure 2. Samples of boron acid/epoxy resin composites: (**a**) type 1; (**b**) type 2.

A total of 48 composite samples were prepared for mechanical tests (6 samples for each composite formulation), and a total of 8 samples were prepared for thermal analyses (1 sample for each composite formulation).

After conditioning the epoxy composite samples, mechanical tests in compression mode in accordance with the ISO 604 standard (ISO 604; Plastics. Determination of compressive properties. International Organization for Standardization, Geneva, Switzerland, 2002) were carried out using a Zwick/Roell Z150 testing machine(Zwick/Roell GmbH&Co. KG, Ulm, Germany). The mechanical tests were executed using the following parameters: initial force of 200 N, test speed of 10 mm/min. The results of the mechanical tests were statistically analyzed using correlation and regression. The Pearson linear correlation coefficient r (X,Y) was adopted as a measure of the correlation of one variable (X) with the other variable (Y). Assumptions were made for the statistical analysis, and the corresponding statistical tests were used [57]. Statistical analysis was performed using the Statistica 13.1 program.

Thermal properties, in terms of glass transition temperature (Tg), were evaluated employing a differential scanning calorimetry (DSC1 Stare System) produced by Mettler Todedo (Columbus, OH, USA). The tests were carried out by scanning samples weighing approximately 10–20 mg and sealed in aluminum pans in an inert atmosphere (flow rate of nitrogen: 80 mL/min) between −15 °C and 150 °C. A constant heating rate of 10 °C/min was employed in each test. The average of the results of at least three experiments performed on each sample of epoxy composites was calculated.

The thermal degradation process as a function of the boric acid content in the epoxies was evaluated by thermogravimetric (TGA) analysis employing a thermogravimetric analyzer (TGA55) purchased by TA Instruments Company (Waters™, New Castle, DE, USA). This analysis was performed in the range of 25 °C–1000 °C, using a heating rate of 10 °C/min and nitrogen atmosphere (flow rate: 25 mL/min). The samples, weighing between 10 and 15 mg, were placed in an alumina crucible and tested. Three samples of each composition were analyzed and the results were averaged.

2.3. Flammability Tests

Flammability tests were performed based on the guidelines contained in the PN-EN 60695-11-10 standard (PN-EN 60695-11-10. Fire hazard testing—Part 11-10: Test flames—50 W horizontal and vertical flame test methods. Polish Committee for Standardization, Warsaw, Poland, 2014), Part 11-10. To this end, samples were placed horizontally on a universal stand used for testing the flammability of materials. The ignition source was a 50 W gas burner powered by methane. Comparative studies of changes in the temperature field in the area of the burning sample were carried out using a Flir A70 thermal imaging camera (FLIR Systems, Inc., Meer, Belgium). Thermal images were recorded from a distance of 700 mm with a thermal image recording frequency of 15 Hz. The recorded temperature ranged from 175 to 1000 °C. Images were recorded immediately after the ignition source was removed and after times t of 10, 20, 30, 40, 50, and 60 s. The analysis of the recorded thermal images (i.e., thermograms) was carried out using Flir Research Studio Player software v.1.7.

In the flammability test performed using method A, samples were mounted horizontally in the holder. Keeping the central axis of the burner tube inclined at an angle of 45° to the horizontal, the burner flame was applied to the lower edge of the free end of each sample in such a way that the flame covered it for a length of approximately 6 mm. The test flame was held for 10 ± 1 s without changing its position. After moving the burner away, the time meter was started and the time after which the flame front reached the marking made at a distance of 10 mm from the ignited end of the sample was recorded. The flame was extinguished when the time stopped.

The material was classified according to the criteria specified in the recommendations of the PN-EN 60695-11-10 standard.

3. Results

3.1. Mechanical Tests

3.1.1. Compressive Strength

The compressive strength values (average values from six samples) of the boron acid/epoxy resin composites are presented in Figure 3.

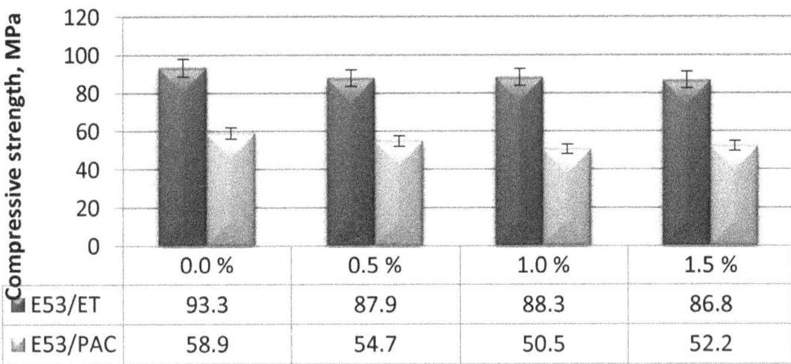

Figure 3. Compressive strengths calculated on the boron acid/epoxy resin composites.

The compressive strengths of the boron acid/epoxy resin composites hardened by the amine curing agent (i.e., E53/ET specimens) were only slightly affected by the content of boric acid and ranged from 86.8 to 87.9 MPa (Figure 3). On the other hand, the highest compressive strength (93.3 MPa) was achieved by the neat (reference) epoxy samples (E53/ET), suggesting a limited negative effect of the addition of boric acid to this resin. In fact, the differences in compressive strength measured on the boron acid/epoxy resin composites and the reference epoxy compound are equal to the following:

- −6.8% for the composite containing 0.5 g boric acid (E53/ET/H$_3$BO$_3$/0.5);
- −6.3% for the composite with 1.0 g boric acid (E53/ET/H$_3$BO$_3$/1.0);
- −8.2% for the composite with 1.5 g boric acid (E53/ET/H$_3$BO$_3$/1.5).

When analyzing the compressive strength results calculated for the boron acid/epoxy resin composites hardened by the polyamide curing agent, again reported in Figure 3, it was found that in this case, the highest compressive strength (equal to 58.9 MPa) was also achieved by the neat (reference) epoxy resin (i.e., E53/PAC). This value is 16.6% higher than the lowest compressive strength measured on the boron acid/epoxy resin composite containing 1.0 g of boric acid (i.e., E53/PAC/H_3BO_3/1.0 system), equal to 50.5 MPa. On the other hand, the highest compressive strength value (54.7 MPa) was achieved by the composite containing 0.5 g of boric acid (i.e., the E53/PAC/H_3BO_3/0.5 compound), this value being 7.1% lower than that measured on the neat epoxy resin. Therefore, even for this epoxy system, the addition of small quantities of boric acid has a limited effect on compressive strength.

In order to determine the relationship between the amount of boric acid and the compressive strength of the epoxy compounds, the method of linear correlation between two variables was used. The results are summarized in Table 3.

Table 3. Correlation between the amount of boric acid and the compressive strength measured on the boron acid/epoxy resin composites.

X Variable	Amount of Boric Acid	
	Compressive Strength of Boron Acid/Epoxy Resin Composites	
Y Variable	**Cured Using Amine Curing Agent (Base: E53/ET)**	**Cured Using Polyamide Curing Agent (Base: E53/PAC)**
r (X, Y)	−0.854	−0.860
r^2	0.729	0.740
t	−2.321	−2.385
p	0.014	0.014
Regression coefficient X to Y	−3.820	−5.460
Regression coefficient Y to X	−0.191	−0.136

The correlation between the amount of boric acid and the compressive strength of boron acid/epoxy resin composites was determined using Pearson's linear correlation coefficient r (X, Y), whose results are reported in Table 3 and Figure 4. If the value of the r coefficient is close to 1, then the examined variables X and Y are linearly related to each other. The other symbols reported in Table 3 represent the following: r^2 represents the coefficient of determination; t represents the value of the t-statistic examining the significance of the correlation coefficient; and p calculates the significance level for the *t*-test.

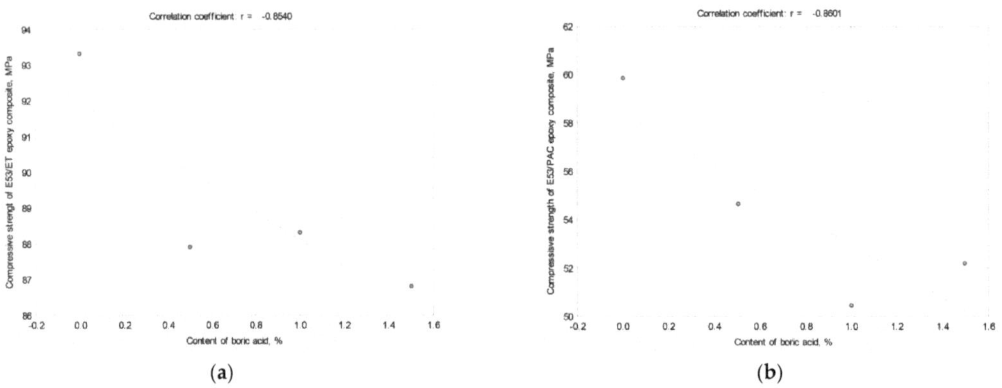

(a)　　　　(b)

Figure 4. Correlation coefficient r (X, Y) of (**a**) E53/ET boron acid/epoxy resin composites; (**b**) E53/PAC boron acid/epoxy resin composites (results are reported in Table 3).

Based on the results of the statistical analysis (reported in Table 3 and Figure 4) and the results of the relationship between compressive strength and the boric acid content in the epoxy composite, the following conclusions can be made:

(i)   For epoxy composites cured using the amine curing agent:

-   The correlation coefficient (r) was found to be −0.854, which proves a strong linear relationship between the compressive strength of epoxy composites and the boric acid content in the composites;
-   The correlation coefficient is negative, which means that as the amount of boric acid in the composite increases, the strength decreases;
-   The coefficient of determination (r^2) is 0.729, which means that almost 73% of the variation in the compressive strength can be attributed to the boric acid content in the epoxy composite.

(ii) For epoxy composites cured using the polyamide curing agent:

-   The correlation coefficient (r) was found to be −0.860, which again proves a strong linear relationship between the compressive strength of these epoxy composites and boric acid content;
-   The correlation coefficient is negative, which indicates that the strength decreases as the amount of boric acid in the composite increases;
-   There was a similar relationship as that in the case of the boron acid/epoxy resin composite hardened by amine curing agent.

It can therefore be concluded that the correlation between compressive strength and the content of boric acid in epoxy composites is not influenced by the type of curing agent employed in the resin, but rather by the amount of boron acid added.

3.1.2. Compressive Modulus

The compressive modulus values of the boron acid/epoxy resin composites are summarized in Figure 5.

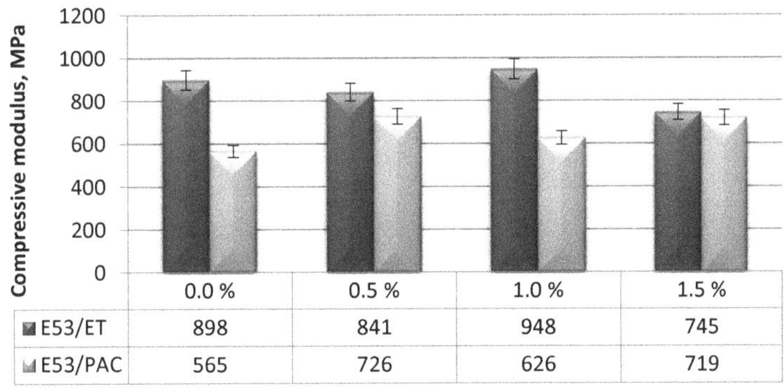

Figure 5. Compressive modulus measured on the boron acid/epoxy resin composites.

The compressive modulus measured on the boron acid/epoxy resin composites cured using the amine curing agent (i.e., E53/ET samples) ranged between 745 MPa and 948 MPa, as reported in Figure 5. The highest compressive modulus (i.e., 948 MPa) was measured for the boron acid/epoxy resin composites containing 1.0 g boric acid (i.e., the E53/ET/H_3BO_3/1.0 formulation). This value is 5.5% greater than that measured on the reference sample (i.e., the neat epoxy resin). The lowest compressive modulus value was measured on the samples of the boron acid/epoxy resin composite containing 1.5 g boric acid (i.e., E53/ET/H₃BO₃/1.5) and it was 20.5% lower than that of the reference epoxy compound (i.e., E53/ET). Taking into account the ranges of variation in these results, it can be concluded that the addition of boric acid had a certain negative influence only in the case of the highest content of this additive.

Moving on to the analysis of the boron acid/epoxy resin composites cured using the polyamide curing agent, the lowest compressive modulus, equal to 565 MPa, was found for the neat epoxy resin (i.e., the E53/PAC compound). This value was 28.5% lower than the highest compressive modulus (i.e., 726 MPa) measured on the boron acid/epoxy resin composite containing 0.5 g of boric acid (i.e., the E53/PAC/H₃BO₃/0.5 formulation). For these epoxy composites, the values of compressive modulus increased as the boric acid content increased: the modulus of the epoxy composite containing 1.0 g of boric acid (i.e., E53/PAC/H₃BO₃/1.0) was 10.8% higher than that of the neat epoxy samples and that of the epoxy composite including 1.5 g of boric acid (i.e., E53/PAC/H₃BO₃/1.5) was 27.2% higher than that of the near epoxy. We conclude that, for this epoxy system, the inclusion of boric acid was beneficial to the compression modulus.

The compressive modulus results were also subjected to basic statistical analysis using the previously introduced coefficients, i.e., correlation and regression (reported in Table 4 and Figure 6).

Table 4. Correlation between the amount of boric acid and the compressive modulus measured on the boron acid/epoxy resin composites.

X Variable	Amount of Boric Acid	
Y Variable	Compressive Modulus of Boron Acid/Epoxy Resin Composites	
	Cured Using Amine Curing Agent (Base: E5/ET)	Cured Using Polyamide Curing Agent (Base: E5/PDA)
R (X, Y)	−0.533	0.603
r²	0.284	0.364
t	−0.890	1.069
p	0.046	0.039
Regression coefficient X to Y	7.160	7.240
Regression coefficient Y to X	−0.004	0.005

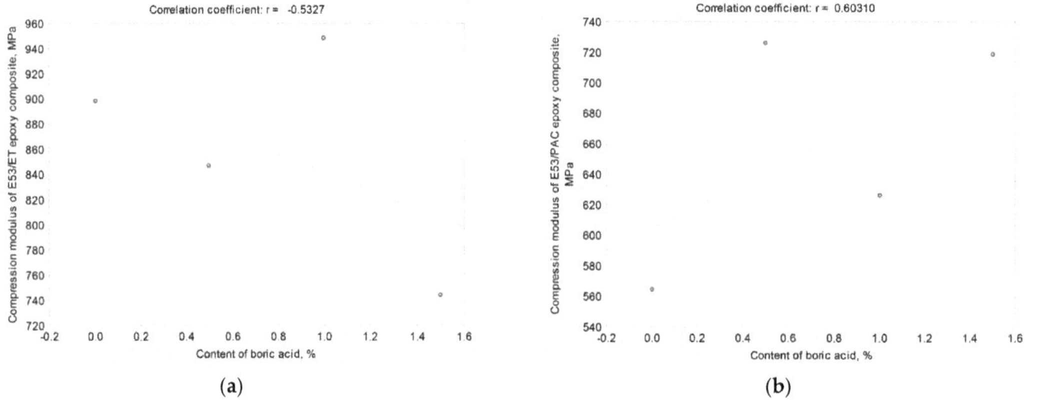

Figure 6. Correlation coefficient r (X, Y) of (**a**) E53/ET boron acid/epoxy resin composites; (**b**) E53/PAC boron acid/epoxy resin composites (results are reported in Table 4).

Based on the results of the statistical analysis (reported in Table 4 and Figure 6) and the results of the relationship between the compressive modulus and boric acid content in the epoxy composite, the following observations can be made:

(i) For epoxy composites cured using the amine curing agent:

- A correlation coefficient (r) equal to -0.533 proves a moderate linear relationship between the compressive modulus of the epoxy composites and boric acid content;
- The correlation coefficient is found to be negative, indicating that the strength decreases as the amount of boric acid in the composite increases.

(ii) For epoxy composites cured using the polyamide curing agent:

- A correlation coefficient (r) of 0.603 again indicates a moderate linear relationship between the compressive strength of this type of epoxy composite and the boric acid content;
- In this case, the correlation coefficient is positive, which means that upon increasing the amount of boric acid in the composite, its compressive strength also increases;
- There is an inverse relationship with respect to the boron acid/epoxy resin composites cured using the amine curing agent.

Based on the results of the performed statistical analysis, it can be concluded that the both the type of the curing agent and the amount of boric acid influence only to some extent the compressive modulus measured on the epoxy composites; in addition, the influence is not always detrimental.

3.1.3. Compressive Strain

The strain values calculated on epoxy composites with different boric acid contents in compression mode are shown in Figure 7. For comparison purposes, the compressive strain of the reference epoxy resin is reported in the same Figure.

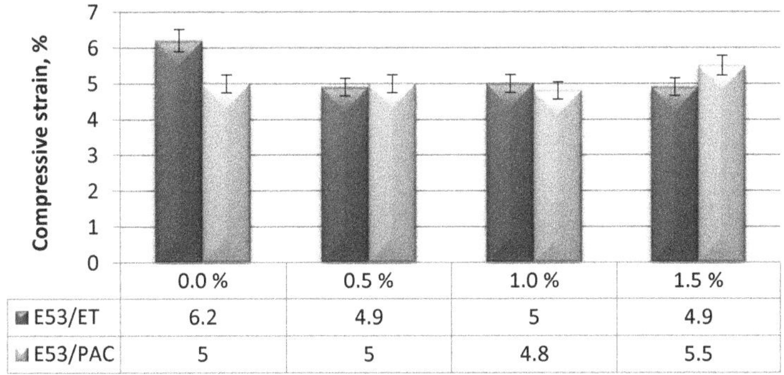

Figure 7. Compressive strain calculated on the boron acid/epoxy resin composites.

The compressive strain values of the boron acid/epoxy resin composites cured using the amine curing agent (i.e., E53/ET systems) were quite unaffected by the content of boric acid, ranging around 5% in all cases. The reference epoxy system (i.e., the E53/ET formulation) exhibits the highest compressive strain value, i.e., 6.2%, which is around 25% greater than that relative to all the boron acid/epoxy resin composites.

Analyzing the results of the boron acid/epoxy resin composites hardened using the polyamide curing agent, the lowest compressive strain value (i.e., 4.8%) was measured on the composite containing 1.0 g boric acid (i.e., E53/PAC/H_3BO_3/1.0). On the other hand, the highest value of this characteristic, namely 5.5%, was measured on the epoxy composite containing 1.5 g boric acid (i.e., E53/PAC/H$_3$BO$_3$/1.5 compound). The compressive strain of both the reference epoxy resin and the boron acid/epoxy resin composite containing 0.5 g boric acid, i.e., the E53/PAC and E53/PAC/H$_3$BO$_3$/0.5 systems, respectively, was found to be 5%. The observed results suggest that there is no precise relationship between the compressive strain value and the boric acid content, and that the addition of boric acid has little influence on this characteristic. However, in order to confirm this hypothesis, the correlation and regression coefficients of these results were also determined; they are reported in Table 5 and Figure 8.

Table 5. Correlation between the amount of boric acid and the compressive strain measured on boron acid/epoxy resin composites.

X Variable	Amount of Boric Acid	
Y Variable	**Compressive Strain of Boron Acid/Epoxy Resin Composites**	
	Cured Using Amine Curing Agent (Base: E5/ET)	**Cured Using Polyamide Curing Agent (Base: E5/IDA)**
r (X, Y)	−0.772	0.562
r^2	0.597	0.316
t	−1.720	0.961
p	0.046	0.043
Regression coefficient X to Y	−0.760	0.260
Regression coefficient Y to X	−0.785	1.215

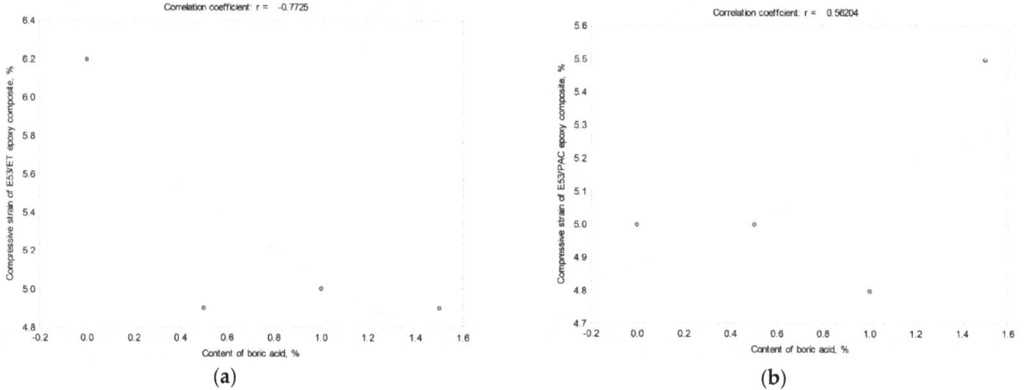

Figure 8. Correlation coefficient r (X, Y) of (**a**) E53/ET boron acid/epoxy resin composites; (**b**) E53/PAC boron acid/epoxy resin composites (Table 5).

When analyzing the results of the basic statistical analysis presented in Table 5 and Figure 8, it is possible to observe that the relationship between compressive strain and the boric acid content in the epoxy composites is similar to what was previously observed for the modulus. In fact, the compressive strain measured on the epoxy composites cured using the amine curing agent decreases as the content of boric acid increases. On the other hand, the same property increases with increasing content of boric acid in the epoxy composites hardened by the polyamide curing agent. Therefore, a limited influence of the type of curing agent on the compressive characteristics of boric acid/epoxy resin composites is confirmed.

3.2. Thermal Analyses

3.2.1. Glass Transition Temperature

Table 6 reports the average values of the glass transition temperature (Tg) measured on the epoxy composites that are the subject of the present study. The variation range of the experimental values was always less than 4%.

Table 6. Average values of the glass transition temperature (Tg) calculated on boron acid/epoxy composites and relative neat epoxy compounds.

System	Glass Transition Temperature (Tg) [°C]
E53/ET	60.8
E53/ET/H_3BO_3/0.5	58.6
E53/ET/H_3BO_3/1.0	59.8
E53/ET/H_3BO_3/1.5	60.6
E53/PAC E53/PAC/H_3BO_3/0.5	38.6
E53/PAC/H_3BO_3/1.0	39.9
E53/PAC/H_3BO_3/1.5	38.5
	40.7

From the observation of the Tg values reported in Table 6, it can be concluded that the addition of boric acid to the epoxy systems did not significantly influence this temperature. Thus, boric acid does not act as a "plasticizer" for epoxy resin, regardless of the type of curing agent used. In fact, the differences in Tg observed in Table 6 are always lower than the experimental error measured in the calculations (i.e., less than 4%). Furthermore, these values do not have a precise trend as the additive content increases. For example, the highest Tg value found for E53/ET composites was measured for the system with the highest boric acid content (i.e., E53/ET/H_3BO_3/1.5) and it was almost equal to the Tg value measured for the neat resin cured with the same hardener, i.e., the E53/ET system. On the other hand, analyzing the compounds based on epoxy cured with polyaminoamide, the greatest (very similar) Tg values were measured for the composites containing 0.5 and 1.0 gr. of boric acid and the lowest for the system with the highest boric acid content (i.e., E53/PAC/H_3BO_3/1.5). As already underlined, however, these Tg variations fall within the range of variation of the results.

Finally, a comparison of the results reported in Table 6 allows us to conclude that using an amine-based hardener, it is possible to reach a significantly higher glass transition temperature value than the Tg achieved by the same styrene-modified epoxy resin cured with a polyamide, using the same curing cycle (in terms of temperature and time) for both systems. Therefore, the choice of the most appropriate curing agent for an epoxy resin must also be made based on the (maximum) service temperature expected for that specific application.

3.2.2. Thermogravimetric Analysis

The thermal resistance of boric acid/epoxy composites in a non-oxidative atmosphere was analyzed using thermogravimetric (TGA) tests. In fact, from this analysis, it is possible to obtain useful indications on the effect of boric acid on the resistance of epoxy resin cured with two different hardeners to very high temperatures in the absence of oxygen. This information is very important for applications in which resin can reach high temperatures, even due to accidental causes. Thus, TGA was employed to evaluate the degradation range of temperatures during heating up to 1000 °C. The results of the TGA analysis are summarized in Table 7. In this table, the initial and final temperatures of the degradation process are indicated as "Onset Temperature" and "Endset Temperature", respectively. Again, a very small range of variation of the experimental values (not exceeding 4%) was calculated.

Table 7. Average results of thermogravimetric analysis (TGA) performed on boron acid/epoxy composites and relative neat epoxy compounds.

System	Onset Temperature [°C]	Endset Temperature [°C]
E53/ET	347.8	408.8
E53/ET/H_3BO_3/0.5	349.3	412.9
E53/ET/H_3BO_3/1.0	349.7	410.5
E53/ET/H_3BO_3/1.5	341.3	407.8
E53/PAC	351.7	453.2
E53/PAC/H_3BO_3/0.5	348.6	452.0
E53/PAC/H_3BO_3/1.0	349.2	452.8
E53/PAC/H_3BO_3/1.5	347.5	452.3

No clear trend can be seen in the TGA results, regardless of the type of curing agent used in the styrene-modified epoxy resin systems. Indeed, the effect of boric acid is not very evident using this test: the beginning of the thermal degradation process, in the absence of oxygen, occurs more or less starting from around 350 °C. On the other hand, the temperature at the end of the degradation process seems to depend on the type of curing agent but not on the presence, and content, of boric acid, keeping in mind that the small temperature differences measured for the different compositions are still within the experimental error.

Not having obtained conclusive results on the effect of boric acid on the thermal resistance of styrene-modified epoxy resin from this test, we analyzed the flame behavior of epoxy systems depending on the boric acid content and the type of curing agent.

3.3. Flammability Tests

Table 8 and Figure 9 present the results of flammability tests performed in a horizontal arrangement using two series of four samples, each with different epoxy resin composites. Figure 10a shows an example of the appearance of a E53/ET/H_3BO_3/1.0 sample immediately after the removal of the flame source and the appearance of the same sample after combustion (Figure 10b).

Table 8. Results of flammability tests in a horizontal arrangement.

Sample Designation of Epoxy Resin Composites	Length of Sample Failure L [mm]	Burning Time of the Measurement Section t_1 [s]	Linear Burning Rate v [mm/min]	Flammability Class According to Horizontal Test
		E53/ET		
E53/ET/H_3BO_3/0.5	11.61	103	6.76	HB
E53/ET/H_3BO_3/1.0	9.04	92	5.90	HB
E53/ET/H_3BO_3/1.5	8.13	89	5.48	HB
E53/ET	8.53	123	4.16	HB
		E53/PAC		
E53/PAC/H_3BO_3/0.5	12.95	52	14.94	HB
E53/PAC/H_3BO_3/1.0	9.79	51	11.52	HB
E53/PAC/H_3BO_3/1.5	9.44	70	8.09	HB
E53/PAC	9.21	53	10.43	HB

In the case of the E53/ET epoxy resin composites, the introduction of an additional component caused a decrease in the linear burning rate of approximately 38% compared to the base value obtained for the unmodified material, which is a positive and significant effect. In general, the values of the linear burning rate obtained for the E53/PAC epoxy resin composites were much, even more than twice, higher than for the E53/ET epoxy resin composites, which indicates a deterioration in flammability caused by the base material or modifying agent used in this case. In the case of the E53/PAC epoxy resin composites, a significant reduction in the linear burning rate was achieved with the addition of 1.0% of the modifying agent, up to 8.09 mm/min, which indicates a reduction of 6.85 mm/min (approx. 69%) in the linear burning rate compared to the unmodified sample. With a modifier content of 1.5%, an increase in the linear burning rate to 10.43 mm/min was observed, the reason for which, however, is difficult to interpret unambiguously due to the small number of samples tested. Taking into account the values of the linear combustion rate obtained and the deviations from the standard requirements, all tested samples can be conventionally classified into the HB flammability class.

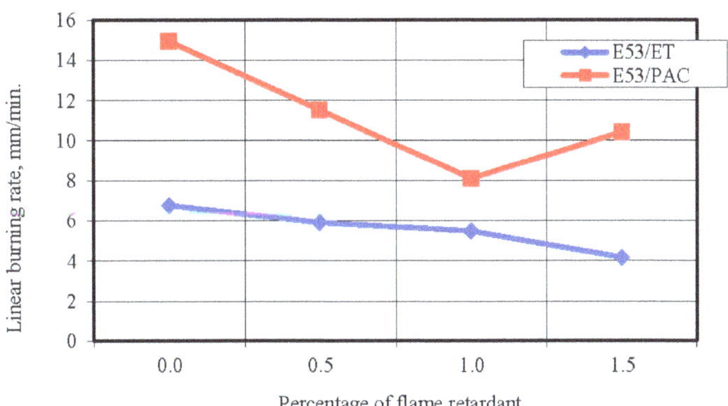

Figure 9. Linear burning rate depending on the content of the flame retardant (boron acid).

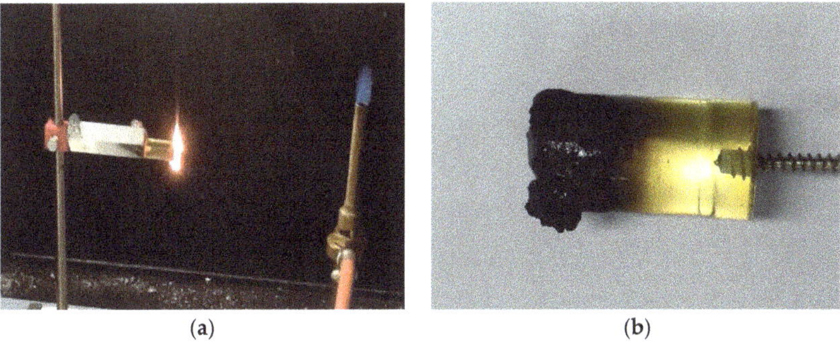

(a)   (b)

Figure 10. Views of E53/ET/H$_3$BO$_3$/1.0 sample of epoxy resin composites: (**a**) immediately after removing the flame source, (**b**) after combustion.

The analysis of the temperature field in the burning area of the samples in the horizontal burning test included determining the maximum temperature changes in the burning area, marked as "rectangle 1" in the thermal images, and the temperature along the burning sample in the area marked as "line 1". The results of changes in the maximum temperature in the burning area of samples in series 1 are presented in Table 9 and Figure 11.

Table 9. Results of maximum temperature tests in the burning area for E53/ET epoxy resin composites.

Burning Time t [s]	Maximum Temperature in the Burning Area T [°C]			
	E53/ET/H$_3$BO$_3$/0.5	E53/ET/H$_3$BO$_3$/1.0	E53/ET/H$_3$BO$_3$/1.5	E53/ET
0	467.17	565.26	471.70	432.87
10	554.30	636.54	624.01	538.27
20	597.03	608.45	647.21	643.74
30	646.99	644.10	647.18	689.55
40	658.57	674.56	652.48	724.78
50	647.21	664.77	631.39	720.64
60	630.03	685.31	698.60	601.24

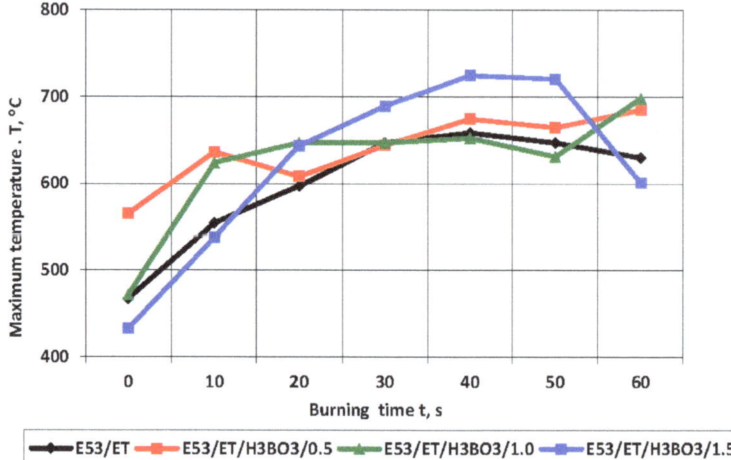

Figure 11. Maximum temperature in the burning area of E53/ET epoxy resin composite samples depending on burning time.

The maximum temperature in the burning area of E53/ET epoxy resin composite samples generally increases with the burning time, which indicates the spread of the flame. In the case of sample E53/ET, the maximum temperature is slightly reduced after burning for more than 40 s, while for sample E53/ET/H_3BO_3/1.5, its reduction was observed after burning for 60 s. However, this may result from temporary disturbances in the combustion process and would require additional research. The highest temperature value in the burning area immediately after removing the ignition source (time 0) was recorded for the sample with 0.5 modifying agent content and was 565.26 °C; it was slightly lower at 1% modifying agent content (471.7 °C), amounted to 467.17 °C in the sample with no modification, and was the lowest for the sample with a modifying agent content of 1.5% (432.87 °C). In order to facilitate the evaluation of the obtained data, they are represented graphically in Figure 11. Figure 12 presents selected thermal images of the tested samples during burning, recorded immediately after subtracting the ignition source, for comparison purposes. The visible differences in the size of the area covered by combustion in individual thermal images correspond to the linear combustion rate decreasing with increasing amounts of the modifying agent; accordingly, this area was the largest in the sample without the modifier (Figure 12a). Similar relationships can be seen in Figure 13, which presents images recorded after 60 s of burning in the same order.

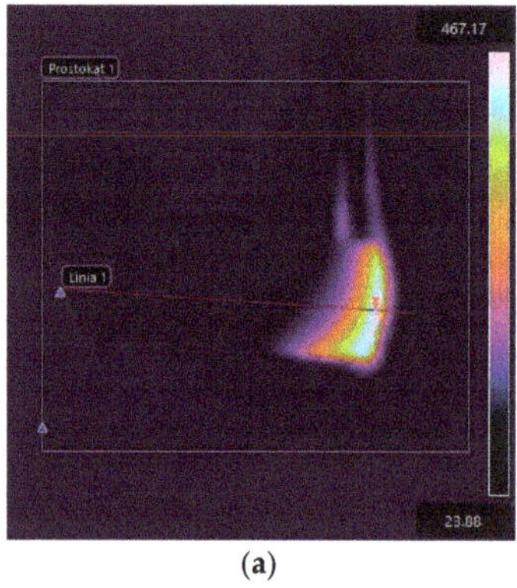

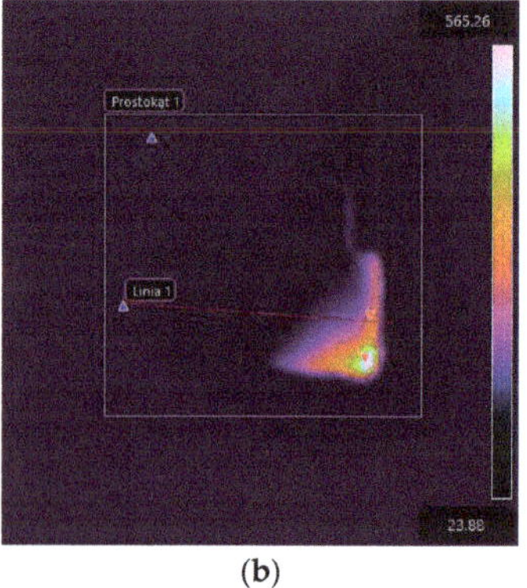

(a)   (b)

Figure 12. *Cont.*

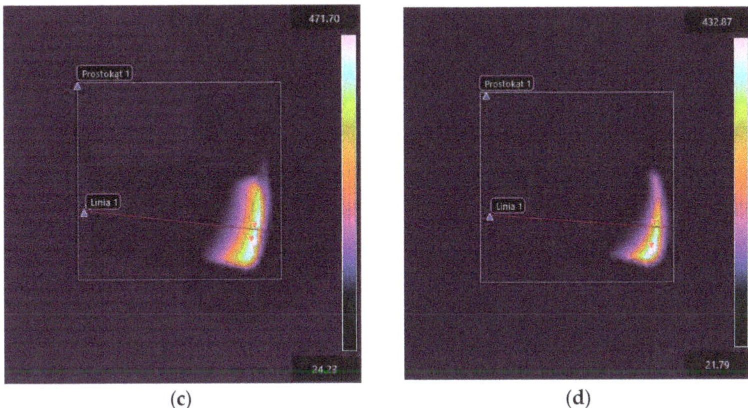

Figure 12. Thermal images of E53/ET epoxy resin composite samples recorded immediately after subtraction of the ignition source (time 0): (**a**) E53/ET; (**b**) E53/ET/H$_3$BO$_3$/0.5; (**c**) E53/ET/H$_3$BO$_3$/1.0; (**d**) E53/ET/H$_3$BO$_3$/1.5.

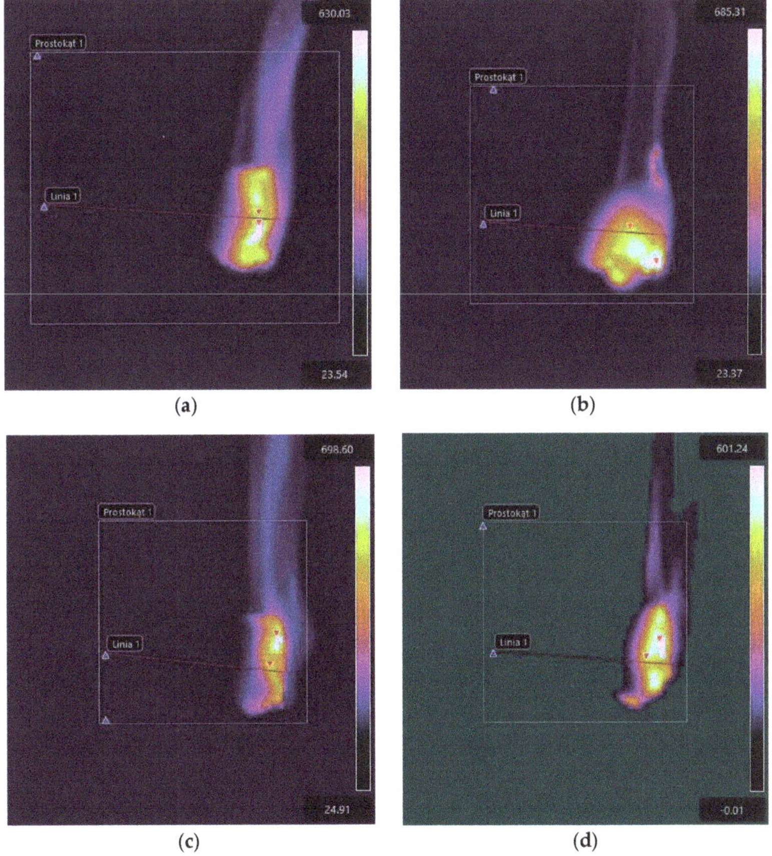

Figure 13. Thermal images of E53/ET epoxy resin composite samples recorded 60 s after ignition source subtraction: (**a**) E53/ET; (**b**) E53/ET/H$_3$BO$_3$/0.5; (**c**) E53/ET/H$_3$BO$_3$/1.0; (**d**) E53/ET/H$_3$BO$_3$/1.5.

Analogously to E53/ET epoxy resin composite samples, an analysis of the temperature field in the burning area in the horizontal burning test was carried out for E53/PAC epoxy resin composites. Due to the much higher burning rate in the case of samples of E53/ET epoxy resin composites, thermal images were only recorded up to a burning time of 40 s. The results of changes in the maximum temperature in the burning area of samples of this series are presented in Table 10 and Figure 14.

Table 10. Results of maximum temperature tests in the burning area for E53/PAC epoxy resin composites.

Burning Time t [s]	Maximum Temperature in the Burning Area T [°C]			
	E53/PAC/H$_3$BO$_3$/0.5	E53/PAC/H$_3$BO$_3$/1.0	E53/PAC/H$_3$BO$_3$/1.5	E53/PAC
0	437.53	447.89	414.97	408.71
10	474.46	491.19	464.18	446.00
20	489.77	530.04	517.38	478.10
30	510.90	536.67	586.11	492.86
40	510.81	528.86	605.18	505.39
50	437.53	447.89	414.97	408.71
60	474.46	491.19	464.18	446.00

The temperature of E53/PAC epoxy resin composites also increased with the passage of burning time, which, similarly to E53/ET epoxy resin composites, indicated the spread of the flame. The combustion temperature in the case of samples of E53/PAC epoxy resin composites was generally lower than that of E53/ET epoxy resin composites, which is a favorable phenomenon for fire conditions. The highest temperature value in the burning area immediately after removing the ignition source (time 0) was also recorded in the case of the sample with 0.5% modifying agent content and was 447.89 °C; the lowest temperature was observed with 1.5% modifying agent content (408, 71 °C). In order to facilitate the evaluation of the obtained data, they are presented graphically in Figure 14. The maximum temperature in the burning area increases to a value slightly above 500 °C and then stabilizes with increasing burning time. Only in the case of the sample containing 1% of the modifying agent does the temperature rise above 600 °C.

For comparison purposes, Figure 15 shows thermal images of the tested E53/PAC epoxy resin composite samples recorded immediately after subtracting the ignition source. The higher linear burning rate of samples of E53/PAC epoxy resin composites compared to E53/ET epoxy resin composites is reflected in the greater extent of the high-temperature area, especially on the lower surface of the samples. Combustion also occurs at a slightly lower temperature, which can be observed both in the thermograms shown in Figure 15 and those recorded after 40 s and shown in Figure 16.

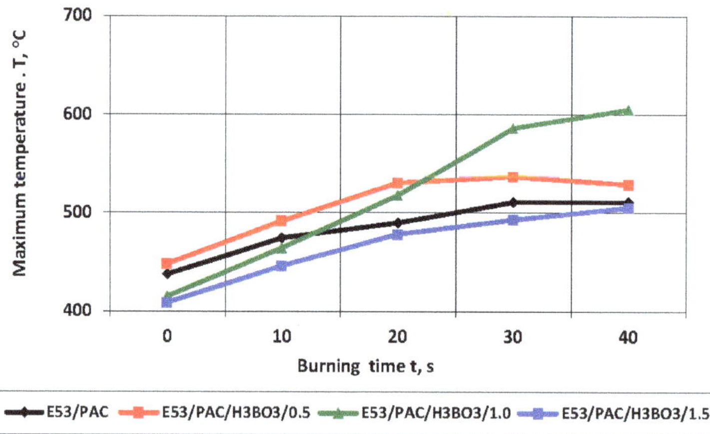

Figure 14. Maximum temperature in the burning area of E53/PAC epoxy resin composite samples depending on burning time.

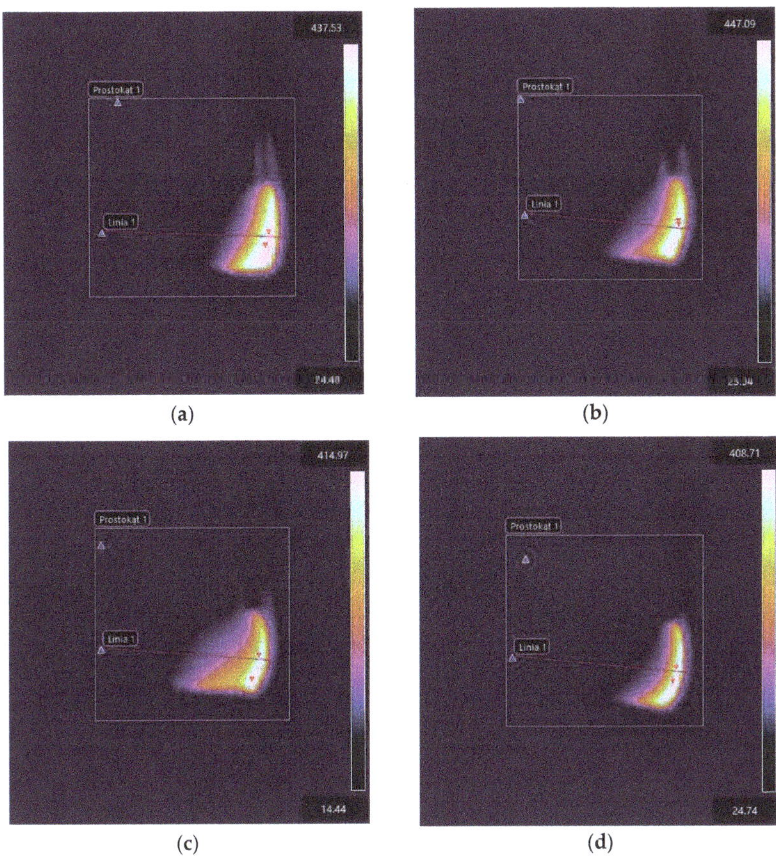

Figure 15. Thermal images of E53/PAC epoxy resin composite samples recorded immediately after the subtraction of the ignition source (time 0): (**a**) E53/PAC; (**b**) E53/PAC/H$_3$BO$_3$/0.5; (**c**) E53/PAC/H$_3$BO$_3$/1.0; (**d**) E53/PAC/H$_3$BO$_3$/1.5.

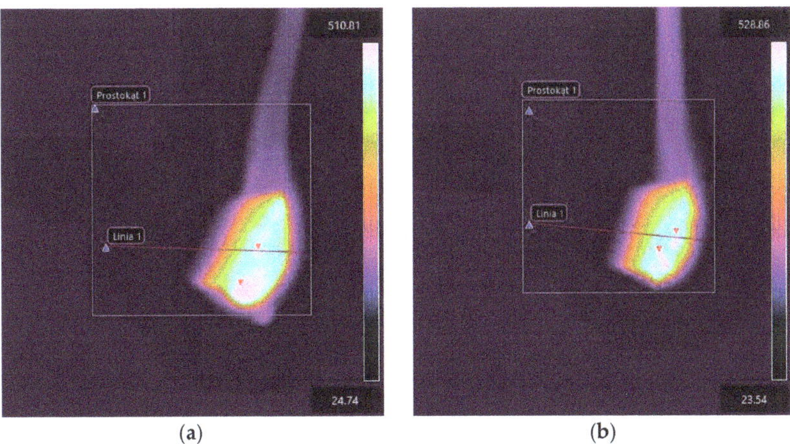

Figure 16. *Cont.*

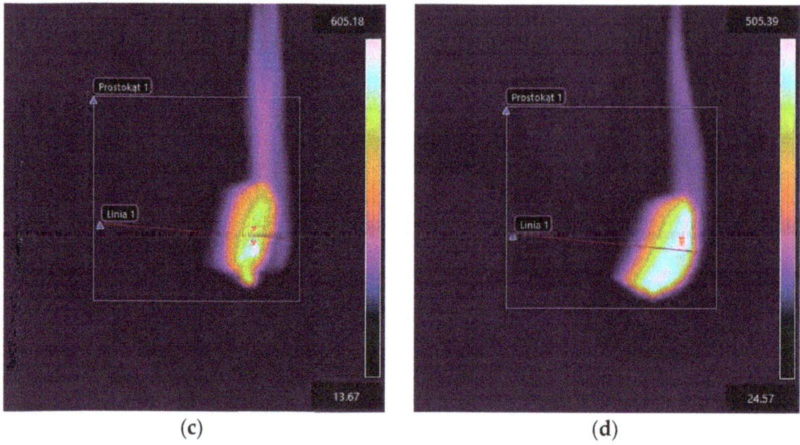

(c)   (d)

Figure 16. Thermal images of E53/PAC epoxy resin composite samples recorded 40 s after ignition source subtraction: (**a**) E53/PAC; (**b**) E53/PAC/H$_3$BO$_3$/0.5; (**c**) E53/PAC/H$_3$BO$_3$/1.0; (**d**) E53/PAC/H$_3$BO$_3$/1.5.

4. Discussion

A discussion of the results of this study is presented below, while also taking into account the findings of a previous work, reported in [53], which discusses a solvent-free resin with an epoxy number of 0.48–0.51 mol/100 g. A comparison of the average compressive strength of epoxy composites containing different amounts of boric acid and based on an unmodified solvent-free (E5) resin or the solvent-modified (E53) resin, analyzed in this study, is presented in Figure 17. The same cross-linking agents were used to cure both epoxy resins.

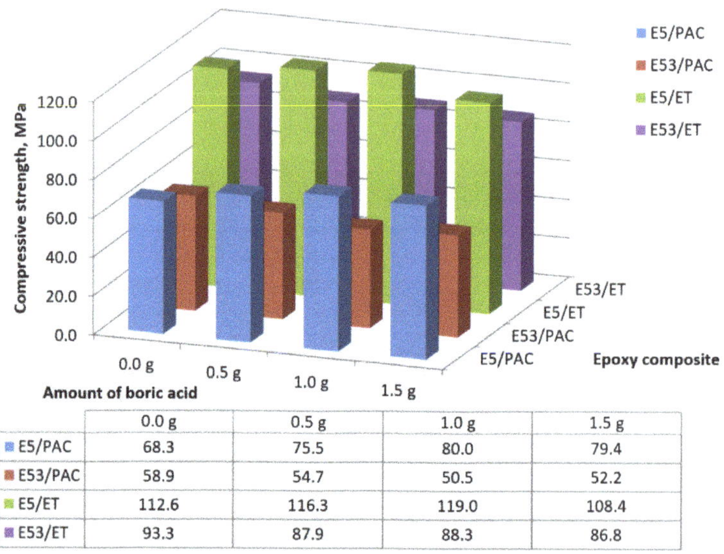

	0.0 g	0.5 g	1.0 g	1.5 g
E5/PAC	68.3	75.5	80.0	79.4
E53/PAC	58.9	54.7	50.5	52.2
E5/ET	112.6	116.3	119.0	108.4
E53/ET	93.3	87.9	88.3	86.8

Figure 17. Average compressive strength calculated on boron acid/epoxy composites containing solvent-free resin (i.e., Epidian 5) and solvent-modified resin (i.e., Epidian 53), cured using amine or polyamide curing agents.

The purpose of the comparison presented in Figure 17 is to highlight the effect of the type of epoxy resin constituting the matrix of the epoxy composites containing boron acid additive in different amounts. The average strength values are reported, not including the standard deviation of results to make the figures more readable. These data can be found in Section 3.1.1 and in the mentioned publication [58].

From the analysis of the results presented in Figure 18, it is possible to conclude that epoxy compounds based on solvent-free epoxy resin (i.e., Epidian 5) are generally characterized by a higher compressive strength with respect to epoxy compounds based on solvent-modified epoxy resin (i.e., Epidian 53), irrespective of composition. Furthermore, boron acid/epoxy resin composites based on solvent-free epoxy resin display greater compressive strength than neat epoxy, while, in the case of composites produced with epoxy modified with styrene, the opposite trend is observed. The results of the calorimetric analysis carried out in the present study demonstrated, however, that boric acid does not act as a "plasticizer" for epoxy resin, since glass transition temperatures were not significantly modified in the presence of this additive. It must be underlined that the influence of an additive on the mechanical characteristics of an epoxy compound depends on many factors, starting from the type of matrix resin. As an example, Avada et al. [52] used boric acid to increase the mechanical properties of a composite based on polyvinyl alcohol (PVOH) and cellulose fibers; their results confirmed that an increase in mechanical strength was obtained in composites containing boric acid compared to those without this additive. On the other hand, Demirhan et al. [36] found a slight decrease in tensile strength and an increase in bending strength in polypropylene–MMT (i.e., montmorillonite) composites containing various amounts of boric acid (from 1.25 to 5.0 wt.%).

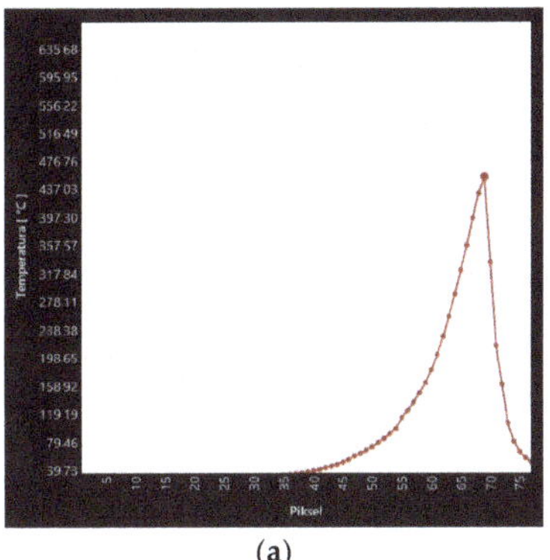
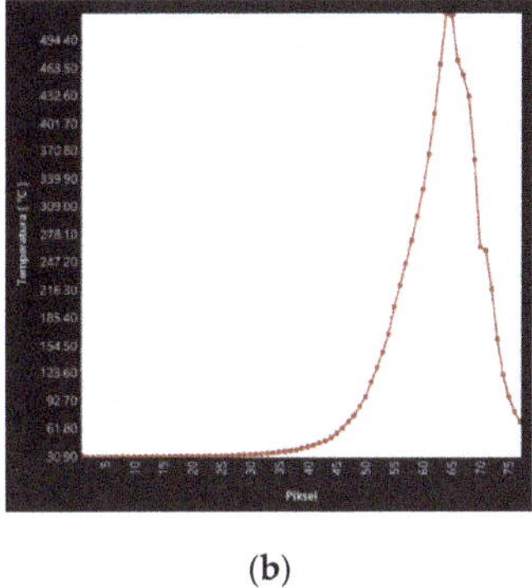

(a) (b)

Figure 18. Temperature recorded along the axis of the E53/ET epoxy resin composite samples: (**a**) immediately after removing the ignition source; (**b**) 60 s after removing the ignition source.

Referring to the type of the curing agent used to cure the epoxies, the composites cured using the amine curing agent display higher compressive strength values than those cured using the polyamide, irrespective of the kind of epoxy resin. Moreover, based on the data reported in Figure 19, it can be concluded that, in most cases, the greatest differences in compressive strength values are observed for epoxy composites cured using a polyamide curing agent with an amine number of 290–360 mg KOH/g with respect to composites hardened by an amine curing agent with an amine number of 700–900 mg KOH/g. In the case of boron acid/epoxy resin composites (matrix: Epidian 53 epoxy resin), the differences in compressive strength, depending on the type of the curing agent, were about 40% (37.8–42.8%), as reported in Figure 18. On the other hand, a comparison of the Tg values measured on the different epoxy systems analyzed in the present study (reported in Table 6) demonstrated the superiority of styrene-modified epoxy resin cured with an amine-based curing agent over that hardened using a polyamide.

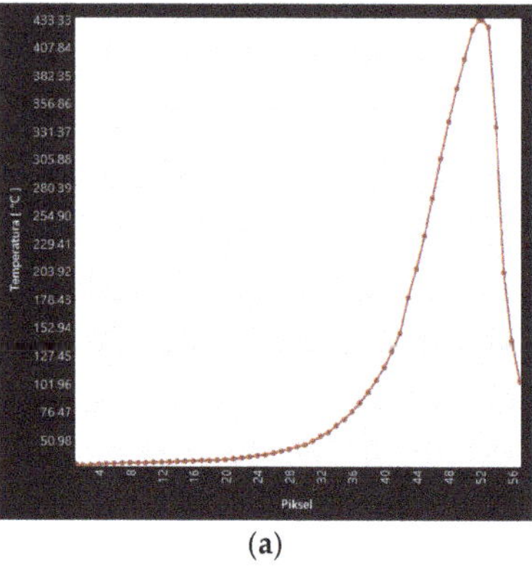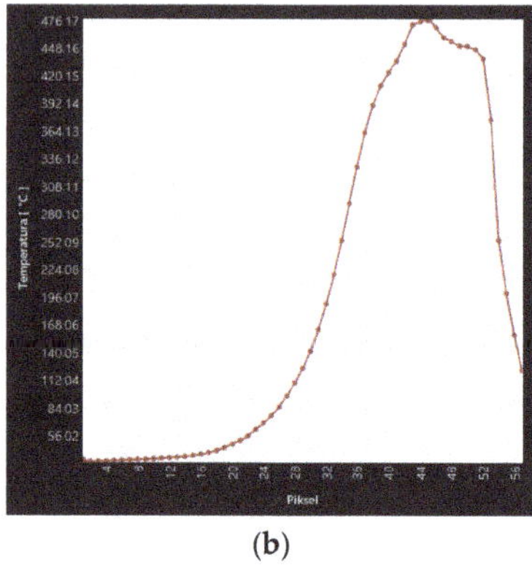

(a)   (b)

Figure 19. Temperature recorded along the axis of the E53/PAC epoxy resin composite sample: (**a**) immediately after removing the ignition source; (**b**) 40 s after removing the ignition source.

Based on the obtained results, it is confirmed that the selection of the curing agent is very important, especially when choosing the formulation of an epoxy composite to be applied in the construction field. The type of curing agent can influence the properties and characteristics of epoxy composites due to its chemical composition and structure, which are capable of giving rise, for example, to more or less rigid cross-linked structures. Furthermore, the choice of the most suitable curing agent also depends on the curing temperature. For example, aliphatic amines are suitable for curing at room temperature, while aromatic ones require high curing temperatures. Finally, the choice of hardener is also a function of the final application intended for the resin. In this regard, Saeedi et al. [13] investigated the effect of curing mechanisms on selected electrical parameters of a commercial DGEBA resin (diglycidyl ether of bisphenol-A based epoxy) cured with amine and anhydride curing agents. These authors reported that both the formation of a more or less dense network and the reactivity of the DGEBA resin depend on the curing agent used. Ignatenko et al. [21] studied the curing reactions of a DGEBA with a mixture of two curing agents. They confirmed that the properties of cured resins depend on both the curing agent employed and on the curing conditions. In relation to the results observed in the present study, it is known that polyamide curing agents, compared to amines, provide greater flexibility to cured epoxy resins at the expense of other mechanical characteristics [59].

Both epoxy resins can be modified by introducing various types of additives, including other polymers [60,61], and both resins are intended, among others, for coating applications and as sealing adhesives; however, they have different viscosities, which may be important in some applications. Generally, the use of Epidian 5 epoxy resin allows for the creation of high-viscosity compositions, while the use of Epidian 53 epoxy resin creates low-viscosity compositions (although this also depends on the type of curing agent). An unmodified, solvent-free resin with an epoxy number of 0.48–0.51 mol/100 g has a viscosity in the range of 2000–3000 mPa·s, while the viscosity of a solvent-modified epoxy resin is in the range of 900–1500 mPa·s.

The advantage of the amine-cured epoxy resin composition used is that it can be used as a polymer coating. Nowadays, such coatings are increasingly used on floors and walls in various buildings. The advantage of the composition of epoxy resins with a polyamide curing agent is that they are used to join elements exposed to deformation, because this curing agent increases the elasticity and impact strength of the composition. At the same time, a negative feature of these compositions is that, depending on the amount of the polyamide curing agent, they are less hard and less resistant to increased temperatures.

When analyzing the results obtained in the flammability test, the temperature along the sample axis was compared for both analyzed groups of epoxy composites (Figures 18 and 19).

The changes in temperature measured along the longitudinal axis of the sample (Figure 18) confirm the increase in combustion temperature and the expansion of the high-temperature area, which proves that the sample surface covered by the flame increases with the passage of combustion time.

In order to compare the temperature changes in unmodified E53/PAC epoxy resin composites (Figure 19) with those in the analogous samples of E53/ET epoxy resin composites (Figure 18) measured on sample z along its longitudinal axis, the appropriate charts are presented in Figure 19. In these charts, one can notice a much larger surface area covered by combustion on the E53/PAC epoxy resin composite sample, which confirms the higher linear combustion rate, with lower temperature values than those obtained for the E53/ET epoxy resin composites.

Based on the conducted tests, it was determined that the values of the linear burning rate obtained for E53/PAC epoxy resin composites were significantly, even more than twice, higher than those obtained for E53/ET epoxy resin composites, which proves that the base material of E53/PAC epoxy resin composites is more susceptible to combustion. At the same time, a beneficial effect of the impact of the applied materials was observed for both tested materials. The use of the modifying agent manifested in a decrease in the linear combustion rate with increases in its content. With a modifier content of 1.5%, an increase in the linear combustion rate was observed, but the reason for this is difficult to determine. In turn, Hergenrother et al. [62] investigated flame retardant epoxy resins containing phosphorus and several formulations, which showed excellent flame retardancy with low phosphorus content as low as 1.5% by weight. Hamciuc et al. [25] showed that the introduction of 2 wt. boron and 2% by weight phosphorus significantly reduced the flammability of the thermosetting epoxy. For this sample, THR and HRC decreased significantly, while the amount of residue increased significantly. Compared to pure EP-0, MCC results showed that the values of PHRR, THR, and HRC decreased by 55.03, 22.93, and 58.365%, respectively. The authors [25] demonstrated that the fire resistance of epoxy resin was significantly improved by the simultaneous introduction of DOPO (9,10-Dihydro-oxa-10-phosphophenanthrene-10-oxide) and H_3BO_3 derivatives. Kumar et al. [34] also reported that flammability decreased with the addition of flame retardant additives, comparing WPC composites containing APP 10%/Ba-Bx 5% and pure composites. In turn, Murat Unlu et al. [55] investigated the effect of three different boron compounds (BA, ZB, and MB) on the fire retardant properties of an APP-based intumescent coating. According to the TGA results, the addition of boron compounds increased the decarburization efficiency, with the most significant increase achieved after the addition of ZB. The authors showed that as the amount of added boron compounds increases, the fire retardant properties of the intumescent coating decrease due to the reduction in the height of the intumescent char.

Taking into account the obtained values of the linear burning rate and the deviations from the standard requirements, all tested samples can be classified into the HB flammability class, which, however, would require the application of the measurement conditions specified in the standard. The maximum temperature in the burning area of E53/ET epoxy resin composite samples generally increases with burning time, which indicates that the fire flares up and is an unfavorable phenomenon. The introduction of modifiers also results in an increase in the maximum temperature in the burning area of samples from this series. The maximum temperature in the burning area of samples of E53/PAC epoxy resin composites also increases with the passage of burning time, but it is lower than that of E53/ET epoxy resin composites, which is a favorable phenomenon for fire conditions. It should be noted, however, that in terms of fire hazards, a greater rate of fire spread, defined in the reported studies as a greater linear combustion rate, poses a greater threat. The recorded thermal images confirm the relationships obtained in the measurements, enabling the assessment of the extent and spread of the burning process.

5. Conclusions

In the present study, the effect of the inclusion of boric acid (used as a flame retardant) in a styrene-modified epoxy resin on its thermal and compressive mechanical properties was investigated. From the analysis of the obtained results, and taking into account the current literature, the following conclusions can be drawn:

- First, analyzing the effect of the type of curing agent, amine-cured epoxy composites exhibit higher compressive strength values and greater glass transition temperatures than polyamide-cured ones, regardless of the type of epoxy resin;
- Higher glass transition temperatures using the amine curing agent are also achieved when boric acid is added to the epoxy resin. The only exception is represented by the epoxy composite containing the highest boric acid content;
- The results of mechanical tests in compression mode do not indicate a clear trend of the influence of boric acid content on the mechanical characteristics of the tested epoxy composites. Similarly, even the Tg values, calculated by DSC analysis, appear to be unaffected by the presence and content of this additive. From an economic point of view, therefore, lower boric acid content could be used to obtain the same characteristics at a lower cost for raw materials;
- Since the results of the thermogravimetric (TGA) tests did not provide conclusive results on the effect of the presence of boric acid on the thermal degradation of the resin, the flame behavior of the epoxy systems under analysis was analyzed;
- The values of the linear burning rate obtained for the epoxy resin composites cured using the polyamide agent were significantly greater (i.e., more than twice as high) than those measured on epoxy resin composites cured using an amine, which proves that the epoxy system cured using a polyamide is more susceptible to combustion;
- A beneficial effect of the addition of boric acid on epoxy composites has been observed, manifested in a decrease in the linear burning rate as the content of this compound increases;
- The maximum temperature in the combustion area of polyamide-cured epoxy resin composites increases with burning time, but it remains lower than that measured for amine-cured epoxy resin composites; this represents an advantage in the event of fire.

As a future perspective, epoxy composites containing boric acid and cured using an amine curing agent could be tested as floor lining. On the other hand, composites based on epoxy resins and boric acid hardened using a polyamide could be tested as adhesives for joining construction materials.

Author Contributions: Conceptualization, A.R., M.F. and V.B.; methodology, M.F.; software, A.R.; validation, A.R., M.F., L.D.L., V.B. and E.O.-K.; formal analysis, A.R., M.F., A.S., L.D.L., V.B. and E.O.-K.; investigation, A.R., A.S. and V.B.; resources, A.R., V.B. and E.O.-K.; data curation, M.F.; writing—original draft preparation, A.R.; writing—review and editing, M.F., V.B. and E.O.-K.; visualization, A.R., M.F., A.S., L.D.L., V.B. and E.O.-K.; supervision, M.F., V.B. and E.O.-K.; project administration, A.R. and M.F.; funding acquisition, M.F. All authors have read and agreed to the published version of the manuscript.

Funding: This research received no external funding.

Institutional Review Board Statement: Not applicable.

Informed Consent Statement: Not applicable.

Data Availability Statement: The data supporting the findings of this study are available from the corresponding author upon reasonable request.

Conflicts of Interest: The authors declare no conflicts of interest.

References

1. Lee, H.L.; Neville, H. *Handbook of Epoxy Resins*; McGraw-Hill: New York, NY, USA, 1988.
2. Pertie, E.M. *Epoxy Adhesive Formulations*; McGraw-Hill: New York, NY, USA, 2006.
3. Licari, J.J.; Swanson, D.W. Chemistry, Formulation, and Properties of Adhesives. In *Adhesives Technology for Electronic Applications*, 2nd ed.; Licari, J.J., Swanson, D.W., Eds.; William Andrew Publishing: Waltham, MA, USA, 2011; Volume 3, pp. 75–141.
4. Fang, M.; Qian, J.; Wang, X.; Chen, Z.; Guo, R.; Shi, Y. Synthesis of a Novel Flame Retardant Containing Phosphorus, Nitrogen, and Silicon and Its Application in Epoxy Resin. *ACS Omega* **2021**, *6*, 7094–7105. [CrossRef]
5. Sukanto, H.; Raharjo, W.W.; Ariawan, D.; Triyono, J.; Kaavesina, M. Epoxy resins thermosetting for mechanical engineering. *Open Eng.* **2021**, *11*, 797–814. [CrossRef]
6. Chen, C.; Li, B.; Kanari, M.; Lu, D. Epoxy adhesives. In *Adhesives and Adhesive Joints in Industry Applications*; Rudawska, A., Ed.; IntechOpen: London, UK, 2019; Volume 3. [CrossRef]
7. Rudawska, A. Epoxy adhesives. In *Handbook of Adhesive Technology*; Pizzi, A., Mittal, K.Z., Eds.; CRS Press Taylor & Frances Group: Boca Raton, FL, USA, 2018; pp. 415–442.
8. Yang, S.; Zhang, Q.; Hu, Y. Synthesis of a novel flame retardant containing phosphorus, nitrogen and boron and its application in flame retardant epoxy resin. *Polym. Degrad. Stab.* **2016**, *133*, 358–366. [CrossRef]
9. Liu, W.C.; Varley, R.J.; Simon, G.P. Phosphorus-Containing Diamine for Flame Retardancy of High Functionality Epoxy Resins. Part II. The Thermal and Mechanical Properties of Mixed Amine Systems. *Polymer* **2006**, *47*, 2091–2098. [CrossRef]
10. Hamciuc, C.; Vlad-Bubulac, T.; Serbezeanu, D.; Carja, I.-D.; Hamciuc, E.; Anghel, I.; Enciu, V.; Şofran, I.-E.; Lisa, G. New fire-resistant epoxy thermosets: Nonisothermal kinetic study and flammability behavior. *J. Polym. Eng.* **2020**, *40*, 21–29. [CrossRef]
11. Rudawska, A. Mechanical Properties of Epoxy Compounds Based on Bisphenol a Aged in Aqueous Environments. *Polymers* **2021**, *13*, 952. [CrossRef]
12. Garcia, F.G.; Soares, B.G. Determination of the epoxide equivalent weight of epoxy resins based on diglycidyl ether of bisphenol A (DGEBA) by proton nuclear magnetic resonance. *Polym. Test.* **2003**, *22*, 51–56. [CrossRef]
13. Saeedi, I.A.; Andritsch, T.; Vaughan, A.S. On the Dielectric Behavior of Amine and Anhydride Cured Epoxy Resins Modified Using Multi-Terminal Epoxy Functional Network Modifier. *Polymers* **2019**, *11*, 1271. [CrossRef]
14. Frigione, M.; Maffezzoli, A. Curing agents. In *Encyclopedia of Composites*, 2nd ed.; Nicolais, L., Borzacchiello, A., Lee, S.M., Eds.; John Wiley & Sons: Hoboken, NJ, USA, 2012; Volume 1, pp. 643–648.
15. Rudawska, A.; Frigione, M. Cold-cured bisphenolic epoxy adhesive filled with low amounts of CaCO$_3$: Effect of the filler on the durability to aqueous environments. *Materials* **2021**, *14*, 1324. [CrossRef]
16. Bereska, B.; Iłowska, J.; Czaja, K.; Bereska, A. Hardeners for epoxy resins. *Chem. Ind.* **2014**, *4*, 443–448.
17. Ellis, B. (Ed.) *Chemistry and Technology of Epoxy Resins*, 1st ed.; Springer Science + Business Media: Dordrecht, The Netherlands, 1993.
18. Hamerton, I. *Recent Developments in Epoxy Resins, Rapra Review Reports*, 2nd ed.; Rapra Technology Limited: Shropshire, UK, 1997.
19. Gotto, J. Epoxy Curing Agents—Part 1: Amines. Available online: https://polymerinnovationblog.com/epoxy-curing-agents-part-1-amines/ (accessed on 10 September 2023).
20. May, C.A. *Epoxy Resins: Chemistry and Technology*; M. Dekker: New York, NY, USA, 1988.
21. Ignatenko, V.Y.; Ilyin, S.O.; Kostyuk, A.V.; Bondarenko, G.N.; Antonov, S.V. Acceleration of epoxy resin curing by using a combination of aliphatic and aromatic amines. *Polym. Bull.* **2020**, *77*, 1519–1540. [CrossRef]
22. Miturska, I.; Rudawska, A.; Müller, M.; Valášek, P. The Influence of Modification with Natural Fillers on the Mechanical Properties of Epoxy Adhesive Compositions after Storage Time. *Materials* **2020**, *13*, 291. [CrossRef]
23. He, H.; Li, K.; Wang, J.; Sun, G.; Li, Y.; Wang, J. Study on thermal and mechanical properties of nano-calcium carbonate/epoxy composites. *Mater. Des.* **2011**, *32*, 4521–4527. [CrossRef]
24. Rudawska, A. Experimental study of mechanical properties of epoxy compounds modified with calcium carbonate and carbon after hygrothermal exposure. *Materials* **2020**, *13*, 5439. [CrossRef]
25. Hamciuc, C.; Vlad-Bubulac, T.; Serbezeanu, D.; Macsim, A.-M.; Lisa, G.; Anghel, I.; Şofran, I.-E. Effects of Phosphorus and Boron Compounds on Thermal Stability and Flame Retardancy Properties of Epoxy Composites. *Polymers* **2022**, *14*, 4005. [CrossRef]
26. Jeencham, R.; Suppakarn, N.; Jarukumjorn, K. Effect of flame retardants on flame retardant, mechanical, and thermal properties of sisal fiber/polypropylene composites. *Compos. Part B Eng.* **2014**, *56*, 249–253. [CrossRef]
27. Rudawska, A.; Sarna-Boś, K.; Rudawska, A.; Olewnik-Kruszkowska, E.; Frigione, M. Biological Effects and Toxicity of Compounds Based on Cured Epoxy Resins. *Polymers* **2022**, *14*, 4915. [CrossRef]
28. Jin, Z.; Liu, H.; Wang, Z.; Zhang, W.; Chen, Y.; Zhao, T.; Meng, G.; Liu, H.; Liu, H. Enhancement of anticorrosion and antibiofouling performance of self-healing epoxy coating using nano-hydrotalcite materials and bifunctional biocide sodium pyrithione. *Prog. Org. Coat.* **2022**, *172*, 107121. [CrossRef]
29. Bourbigot, S.; Duquesne, S. Fire retardant polymers: Recent developments and opportunities. *J. Mater. Chem.* **2007**, *17*, 2283–2300. [CrossRef]
30. Hull, T.R.; Kandola, B.K. *Fire Retardancy of Polymers: New Strategies and Mechanisms*; RSC Publishing: Cambridge, UK, 2009.
31. Shen, K.K. Recent Advances in Boron-Based Flame Retardants. In *Flame Retardant Polymeric Materials*; Hu, Y., Wang, X., Eds.; CRC Press: London, UK, 2019; Volume 6, pp. 97–119.

32. Levchik, S.; Piotrowski, A.; Weil, E.D.; Yao, Q. New Developments in Flame Retardancy of Epoxy Resins. *Polym. Degrad. Stab.* **2005**, *88*, 57–62. [CrossRef]
33. Chen, Y.; Duan, H.; Ji, S.; Ma, H. Novel phosphorus/nitrogen/boron-containing carboxylic acid as co-curing agent for fire safety of epoxy resin with enhanced mechanical properties. *J. Hazard. Mater.* **2021**, *402*, 123769. [CrossRef]
34. Kumar, R.; Gunjal, J.; Chauhan, S. Effect of borax-boric acid and ammonium polyphosphate on flame retardancy of natural fiber polyethylene composites. *Maderas Cienc. Tecnol.* **2022**, *24*, 1–10. [CrossRef]
35. Camino, G.; Delobel, R. Intumescence. In *Fire Retardancy of Polymeric Materials*; Grand, A.F., Wilkie, C.A., Eds.; Marcel Dekker: New York, NY, USA, 2000; Chapter 7, pp. 217–243.
36. Demirhan, Y.; Yurtseven, R.; Usta, N. The effect of boric acid on flame retardancy of intumescent flame retardant polypropylene composites including nanoclay. *J. Thermoplast. Compos. Mater.* **2023**, *36*, 1187–1214. [CrossRef]
37. Kicko-Walczak, E.; Jankowski, P. Effect of halogen free modification of epoxy resins on their fire retardancy levels. *Polimery* **2005**, *50*, 860–867. [CrossRef]
38. Shen, J.; Liang, J.; Lin, X.; Lin, H.; Yu, J.; Wang, S. The Flame-Retardant Mechanisms and Preparation of Polymer Composites and Their Potential Application in Construction Engineering. *Polymers* **2022**, *14*, 82. [CrossRef]
39. Nazarenko, O.B.; Bukhareva, P.B.; Melnikova, T.V.; Visakh, P.M. Effect of Boric Acid on Volatile Products of Thermooxidative Degradation of Epoxy Polymers. *J. Phys. Conf. Ser.* **2016**, *671*, 012041. [CrossRef]
40. Shen, K.K.; Kochesfahani, S.H.; Jouffret, F. Boron Based Flame Retardants and Flame Retardancy. In *Fire Retardancy of Polymeric Materials*, 2nd ed.; Wilkie, C.A., Morgan, A.B., Eds.; CRC Press: Boca Raton, FL, USA, 2009.
41. Schartel, B.; Kebelmann, K. Fire Testing for the Development of Flame Retardant Polymeric Materials. In *Flame Retardant Polymeric Materials: A Handbook*; Hu, Y., Wang, X., Eds.; CRC Press: London, UK, 2019; Volume 3, pp. 35–55. [CrossRef]
42. Visakh, P.M.; Nazarenko, O.B.; Amelkovich, Y.A.; Melnikova, T.V. Effect of zeolite and boric acid on epoxy-based composites. *Polym. Adv. Technol.* **2016**, *27*, 1098–1101. [CrossRef]
43. Schartel, B. Phosphorus-based Flame Retardancy Mechanisms—Old Hat or a Starting Point for Future Development? *Materials* **2010**, *3*, 4710–4745. [CrossRef]
44. Kandola, B.K.; Magnoni, F.; Ebdon, J.R. Flame retardants for epoxy resins: Application-related challenges and solutions. *J. Vinyl Addit. Technol.* **2022**, *28*, 17–49. [CrossRef]
45. Konstantinova, A.; Yudaev, P.; Orlov, A.; Loban, O.; Lukashov, N.; Chistyakov, E. Aryloxyphosphazene-Modified and Graphite-Filled Epoxy Compositions with Reduced Flammability and Electrically Conductive Properties. *J. Compos. Sci.* **2023**, *7*, 417. [CrossRef]
46. Orlov, A.; Konstantinova, A.; Korotkov, R.; Yudaev, P.; Mezhuev, Y.; Terekhov, I.; Gurevich, L.; Chistyakov, E. Epoxy Compositions with Reduced Flammability Based on DER-354 Resin and a Curing Agent Containing Aminophosphazenes Synthesized in Bulk Isophoronediamine. *Polymers* **2022**, *14*, 3592. [CrossRef]
47. Terekhov, I.V.; Chistyakov, E.M.; Filatov, S.N.; Deev, I.S.; Kurshev, E.V.; Lonskii, S.L. Factors Influencing the Fire-Resistance of Epoxy Compositions Modified with Epoxy-Containing Phosphazenes. *Inorg. Mater. Appl. Res.* **2019**, *10*, 1429–1435. [CrossRef]
48. Benin, V.; Cui, X.; Morgan, A.B.; Seiwert, K. Synthesis and Flammability Testing of Epoxy Functionalized Phosphorous-Based Flame Retardants. *J. Appl. Polym. Sci.* **2015**, *132*, 1–10. [CrossRef]
49. Nagieb, Z.A.; Nassar, M.A.; El-Meligy, M.G. Effect of Addition of Boric Acid and Borax on Fire-Retardant and Mechanical Properties of Urea Formaldehyde Saw Dust Composites. *Int. J. Carbohydr. Chem.* **2011**, *2011*, 146763. [CrossRef]
50. Rudawska, A.; Worzakowska, M.; Bociąga, E.; Olewnik-Kruszkowska, E. Investigation of selected properties of adhesive compositions based on epoxy resins. *Int. J. Adhes. Adhes.* **2019**, *92*, 23–36. [CrossRef]
51. Ciech Sarzyna Catalogue. Available online: https://sarzynachemical.pl/ (accessed on 28 September 2023).
52. Awada, H.; Montplaisir, D.; Daneault, C. The Development of a Composite Based on Cellulose Fibres and Polyvinyl Alcohol in the Presence of Boric Acid. *BioResources* **2014**, *9*, 3439–3448. [CrossRef]
53. Bagci, M.; Imrek, H. Solid particle erosion behaviour of glass fibre reinforced boric acid filled epoxy resin composites. *Tribol. Int.* **2011**, *44*, 1704–1710. [CrossRef]
54. Shen, S.; Sun, Y.; Wang, D.; Zhang, Z.; Shi, Y.-E.; Wang, Z. Efficient blue TADF-type organic afterglow material via boric acid-assisted confinement. *Chem. Commun.* **2022**, *58*, 11418–11421. [CrossRef]
55. Murat Unlu, S.; Tayfun, U.; Yildirim, B.; Dogan, M. Effect of boron compounds on fire protection properties of epoxy based intumescent coating. *Fire Mater.* **2017**, *41*, 17–28. [CrossRef]
56. Dagdag, O.; Hamed, O.; Erramli, H.; El Harfi, A. Anticorrosive Performance Approach Combining an Epoxy Polyaminoamide–Zinc Phosphate Coatings Applied on Sulfo-tartaric Anodized Aluminum Alloy 5086. *J. Bio-Tribo-Corros.* **2018**, *4*, 52. [CrossRef]
57. Rabiej, M. *Statistics with the Statistica Program (Statystyka z Programem Statistica)*; Helion: Gliwice, Poland, 2012. (In Polish)
58. Rudawska, A. Mechanical properties of epoxy compounds based on unmodified epoxy resin modified with boric acid as an antiseptic. *Materials* **2024**, *17*, 259. [CrossRef] [PubMed]
59. Mays, C.G.; Hutchinson, A.R. *Adhesives in Civil Engineering*; Cambridge University Press: Cambridge, UK, 1992; ISBN 0-521-32677-X.
60. Kostrzewa, M.; Bakar, M.; Pawelec, Z.; Smoliński, N. Mechanical properties of polymer mixtures with interpenetrating spatial. *Tribologia* **2011**, *239*, 89–100. (In Polish)

61. Kostrzewa, M.; Bakar, M.; Białkowska, A.; Szymańska, J.; Kucharczyk, W. Structure and properties evaluation of epoxy resin modified with polyurethane based on polymeric MDI and different polyols. *Polym. Polym. Comp.* **2019**, *27*, 35–42. [CrossRef]
62. Hergenrother, P.M.; Thompson, C.M.; Smith, J.G.; Connell, J.W.; Hinkley, J.A.; Lyon, R.E.; Moulton, R. Flame retardant aircraft epoxy resins containing phosphorus. *Polymer* **2005**, *46*, 5012–5024. [CrossRef]

Disclaimer/Publisher's Note: The statements, opinions and data contained in all publications are solely those of the individual author(s) and contributor(s) and not of MDPI and/or the editor(s). MDPI and/or the editor(s) disclaim responsibility for any injury to people or property resulting from any ideas, methods, instructions or products referred to in the content.

Article

Self-Crosslinkable Pressure-Sensitive Adhesives from Silicone-(Meth)acrylate Telomer Syrups

Mateusz Weisbrodt * and Agnieszka Kowalczyk

Department of Chemical Organic Technology and Polymeric Materials, Faculty of Chemical Technology and Engineering, West Pomeranian University of Technology in Szczecin, 70-322 Szczecin, Poland
* Correspondence: mateusz.weisbrodt@zut.edu.pl

Abstract: In this study, a novel and environmentally friendly method for the preparation of photoreactive pressure-sensitive adhesives (PSAs) was demonstrated. Adhesive binders based on n-butyl acrylate, methyl methacrylate, acrylic acid, and 4-acryloyloxy benzophenone were prepared with a UV-induced telomerization process in the presence of triethylsilane (TES) as a telogen and acylphosphine oxide (APO) as a radical photoinitiator. The influence of TES (0–10 wt. parts) and APO (0.05–0.1 wt. parts/100 wt. parts of monomer mixtures) concentrations on the UV telomerization process kinetics was investigated using a photodifferential scanning calorimetry method and selected physicochemical features of the obtained silicone-(met)acrylate telomeric syrups (K-value, solid content, glass-transition temperature, and dynamic viscosity), as well as properties of the obtained PSAs (T_g, adhesion, tack, and cohesion), were studied. An increase in TES content caused a significant decrease in the T_g values (approx. 10 °C) and K-value (up to approximately 25 a.u.) of the dry telomers, as well as the dynamic viscosity of the telomeric syrups. PSAs were obtained through UV irradiation of thin polymer films consisting only of silicone-(meth)acrylate telomer solutions (without the use of additional chemical modifiers or of a protective gas atmosphere and protective layers). PSAs were characterized by very good adhesion (12.4 N/25 mm), cohesion at 20 °C (>72 h) and 70 °C (>72 h), and low glass-transition temperature (−25 °C).

Keywords: pressure-sensitive adhesives; telomerization; bulk photopolymerization; polyacrylates; silicone acrylates

Citation: Weisbrodt, M.; Kowalczyk, A. Self-Crosslinkable Pressure-Sensitive Adhesives from Silicone-(Meth)acrylate Telomer Syrups. *Materials* **2022**, *15*, 8924. https://doi.org/10.3390/ma15248924

Academic Editors: Marco Lamberti and Aurelien Maurel-Pantel

Received: 9 November 2022
Accepted: 11 December 2022
Published: 14 December 2022

Publisher's Note: MDPI stays neutral with regard to jurisdictional claims in published maps and institutional affiliations.

Copyright: © 2022 by the authors. Licensee MDPI, Basel, Switzerland. This article is an open access article distributed under the terms and conditions of the Creative Commons Attribution (CC BY) license (https://creativecommons.org/licenses/by/4.0/).

1. Introduction

Pressure-sensitive adhesives (PSAs) are viscoelastic materials that remain permanently adhesive and can adhere even under light pressure [1]. Among the many materials used in the preparation of PSAs, the most common are poly(acrylates); in particular, because of their excellent oxidation resistance, high transparency, high water resistance, and lack of yellowing. However, acrylic PSAs have disadvantages, such as low adhesion to low-energy substrates and low thermal stability [2–5]. Silicone PSAs do not have these disadvantages, but they are used as solvent-based systems (50 wt.%) associated with high emissions of volatile organic compounds (VOCs) during coating [6,7]. To obtain materials with the advantages of both while minimizing their disadvantages, acrylic adhesive binders can be modified with organosilicon compounds. For this purpose, polydimethylsiloxane (PDMS) is often used due to its significantly lower surface energy, non-toxicity, and non-flammable properties. Unfortunately, because of the low compatibility of these materials, creating a homogeneous mixture is difficult. Synthesis of silicone-acrylate resins is generally carried out by emulsion polymerization with the use of appropriate emulsifiers, but this results in the formation of a large amount of hard-to-clean wastewater [2,8,9]. Another approach is to obtain silicone-acrylic pressure-sensitive adhesives through polymerization of (meth)acrylate monomers using silicone macroinitiators. To increase the content of polydimethylsiloxane in the silicone-acrylic structure, the silicone chain is modified with urethane dimethacrylates. These structures can form a semi-interpenetrating polymer network. However, they are obtained by polymerization in an organic solvent, which results in high emission of VOCs in the drying channel, and the reaction products also have broad polydispersity (PDI > 6) [2,3].

To achieve high cohesion in PSAs, they are often crosslinked or formed into structures capable of creating physical interactions. The most popular physical interactions between polymer chains are van der Waals forces, those associated with a microphase separation, interactions between polar monomer units (e.g., acrylic acid, acrylamide), and ionic interactions. The bonds thus formed are non-permanent and can be reversed at the elevated temperature. Another way to increase the cohesion of a PSA is to disperse the filler in the adhesive binder. Unfortunately, this leads to an increase in viscosity and a decrease in adhesion and tack. Commonly used fillers are halloysite and silica [10]. In contrast, chemical crosslinking of adhesive binders can take place in many ways. Branched and then three-dimensional structures (gel formation), which are usually industrially undesirable, can already be formed in the reactor during the polymerization process when a high polymer content and a low monomer concentration are used. If desired, polyfunctional compounds (e.g., triacrylate of trimethylopropane) can be used in the polymerization reaction. The incorporation of epoxy groups, N-alkoxy amides, and organosilanes into the polymer chain enables the crosslinking of the chains with the use of appropriate catalysts. One-component systems (e.g., containing organometallic crosslinkers, metal salts, or hydrazines) and two-component systems (e.g., containing polyfunctional isocyanates, amino resins, polyfunctional azirdines, or peroxides), which enable crosslinking of the binder under the influence of various factors (i.e., temperature, humidity or time), have been studied [11,12]. Radiation-induced crosslinking is a dynamically developing field for the crosslinking of adhesives. In this way, the standard peroxides can be replaced by type I or type II photoinitiators. One special kind of photoinitiators are unsaturated (polymerizable) type II photoinitiators, which require a proton donor to form radicals. In the photocrosslinking processes involving their participation, no by-products are produced, which is of particular importance in the preparation of self-adhesive materials for medical purposes. Crosslinking of this type of photoinitiators is most often initiated with mercury lamps characterized by the emission of UV-C radiation with a length of 220–280 nm [11,13].

Telomerization may be a new, waste-free method of obtaining acrylic PSAs with embedded organosilicon particles. It is a chain reaction of a monomer (taxogen, M) with a reactive compound (telogen, YZ)—also called chain transfer agents (CTAs)—in which $Y(M)_nZ$ molecules (telomers/oligomers) are formed [14,15]. Telogens can be divided into three groups. The first group consists of halo compounds, the second of organic compounds containing an active center bound to the carbon atom (i.e., alcohols, carboxylic acids, amines, etc.), and the third group consists of compounds containing S-H, Si-H, or P-H bonds (sulfur, silicon, and phosphorus compounds) [16]. The telomerization process can be initiated by many factors; e.g., thermal initiators (organic peroxides, hydroperoxides, azo compounds, etc.), UV radiation, γ radiation accompanying beta decay of ^{60}Co to nonradioactive ^{60}Ni, and redox processes involving metal ions with variable valences [17]. However, thermally initiated reactions have the greatest importance (the process takes place at an elevated temperature of $60 \div 90\ ^\circ$C over several hours), with 2,2′-azobisisobutyronitrile being most often used for initiators [18–24]. There are also reports of UV-induced telomerization, which is the main research interest of the authors of the present article [25–28]. The telomerization process should be distinguished from controlled polymerization processes (e.g., ATRP, CRP), although the common feature is the presence of CTAs. What distinguishes controlled polymerization from telomerization is the CTA content (below 3 wt.% in the case of controlled polymerization and from 3 to 50 wt.% in telomer systems [29]) and the nature of the resulting macromolecules (in the case of polymerization, they are polymers and, less frequently, oligomers; in the case of telomerization, they are oligomers with characteristic end groups derived from telogen). As emphasized by the leading researchers studying the telomerization process, telomerization is not a living polymerization (such as ATRP or RAFT) but a partially controlled reaction to obtain macromolecules. The advantage of telomerization over controlled polymerization processes is its high efficiency and the ease of carrying out the process at an industrial scale (lower requirements for the purity of the reagents) [17].

This work describes a new, environmentally friendly method of obtaining pressure-sensitive adhesives with very good adhesive properties that are capable of self-crosslinking under the influence of UV radiation. The UV telomerization of (met)acrylate monomers with triethylsilane as the telogen in the presence of acylphosphine oxide as the radical photoinitiator was performed and is described in detail. Moreover, the influence of telogen concentration on the properties of the obtained silicone-(met)acrylate telomer syrups and PSAs in comparison with the materials obtained through photopolymerization was investigated.

2. Materials and Methods

2.1. Materials

For the preparation of silicone-(met)acrylate telomers (Si telomers), the following components were employed: n-buthyl acrylate (BA), acrylic acid (AA), and methyl methacrylate (MMA; BASF; Ludwigshafen, Germany) were used as monomers; 4-acryloylooxybenzophenone (ABP, Chemitec, Scandiccy, Italy) was used as the copolymerizable photoinitiator; ethyl (2,4,6-trimethylbenzoyl)-phenyl-phosphinate (APO; Omnirad TPOL, IGM Resins, Waalwijk, the Netherlands) was used as a radical photoinitiator; and triethylsilane (TES; Merck, Warsaw, Poland) was used as the telogen. The components were used without additional purification.

2.2. Synthesis and Characterization of Syrups

The UV telomerization of BA, AA, MMA, and ABP was initiated using the radical photoinitiator APO (0.05, 0.075, and 0.1 wt. parts/100 wt. parts of monomer mixture) and TES (3.5, 7, and 10 wt. parts/100 wt. parts of monomer mixture) as a telogen. As reference samples (R), (met)acrylic syrups were obtained. The reaction mechanism for the cotelomerization is shown in Figure 1.

The UV telomerization processes were realized at 20 °C for 30 min in a glass reactor (250 mL) equipped with a mechanical stirrer and thermocouple under argon as an inert gas. A mixture of monomers (50 g) was introduced into the reactor and purged with argon. A high-intensity UV lamp emitting UV-A radiation (UVAHAND 250, Dr. Hönle AG UV Technology, Gräfelting, Germany) was used as a UV light source. The UV irradiation inside the reactor (15 mW/cm^2) was controlled with an SL2W UV radiometer (UV-Design, Brachttal, Germany). The compositions of the mixtures and their symbols are presented in Table 1.

The kinetics studies of the UV telomerization process were conducted at 25 °C using the photo-DSC method with a differential scanning calorimeter with UV equipment (Q100, TA Instruments, New Castle, DE, USA). During the measurements, samples of 5 mg were UV-irradiated (320–390 nm) with an intensity of 15 mW/cm^2 in an argon atmosphere. The polymerization rate (R_p, %/s) and photoinitiation index (I_p)—i.e., the ability of the initiation reaction in the tested systems (TES/APO)—were calculated according to Equations (1) and (2), respectively [30]:

$$R_p = \frac{\left(\frac{dH}{dt}\right)}{H_0} \ [\%/s] \tag{1}$$

$$I_p = \frac{R_p^{max}}{t_{max}} \tag{2}$$

where dH/dt is the heat flow recorded during UV irradiation; H_0 is the theoretical heat value for the complete degree of conversion (ΔH = 78.0 kJ/mol for acrylates and ΔH = 54.8 kJ/mol for methacrylates); ΔH_t is the reaction heat that has evolved at time t; R_p^{max} is the maximum polymerization rate; and t_{max} is the time when the maximum polymerization rate occurs.

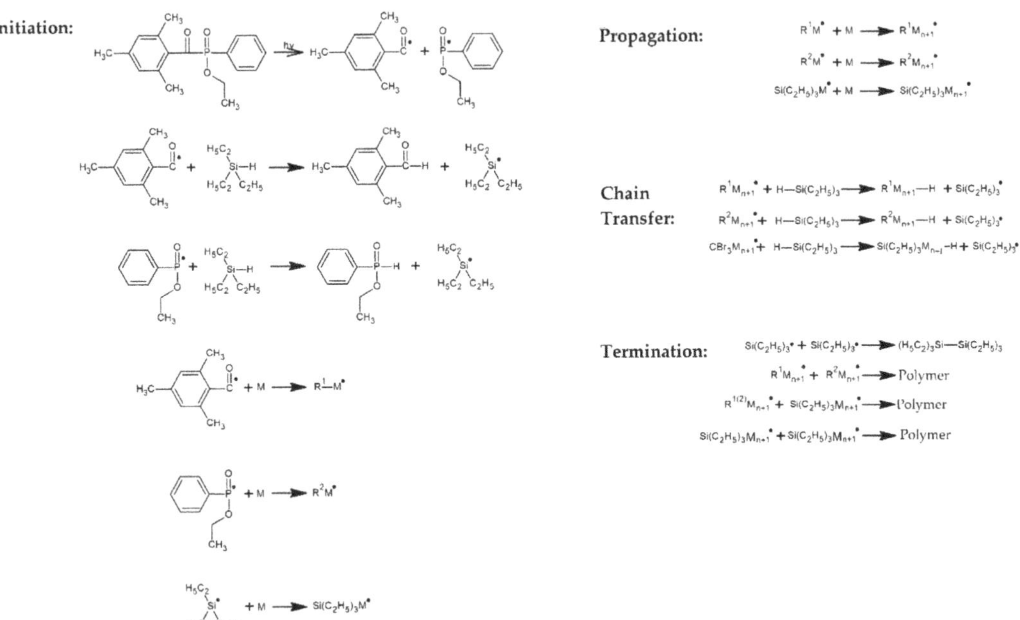

Figure 1. Schematic mechanism of the UV telomerization process (adapted from [17], where M is a taxogen).

Table 1. Compositions of the monomers, photoinitiator, and telogen for the preparation of silicone-(meth)acrylic telomer syrups (Si) and (meth)acrylic syrups (R).

Syrups Acronym	Monomers (wt.%)				APO (wt. part) *	TES (wt. parts) *
	BA	AA	MMA	ABP		
R/APO-5					0.05	0
R/APO-7.5					0.075	0
R/APO-10					0.1	0
Si3.5/APO-5					0.05	3.5
Si3.5/APO-7.5					0.075	3.5
Si3.5/APO-10					0.1	3.5
Si7/APO-5	86.5	7.5	5	1	0.05	7
Si7/APO-7.5					0.075	7
Si7/APO-10					0.1	7
Si10/APO-5					0.05	10
Si10/APO-7.5					0.075	10
Si10/APO-10					0.1	10

* per 100 wt. parts of monomer mixtures.

The solid content (SC) in the syrups was determined using a thermobalance (MA 50/1.X2.IC.A; Radwag, Radom, Poland). Syrups samples (ca. 2 g) were heated in an aluminum pan at 105 °C for 4 h. The SC parameter was calculated according to Equation (3):

$$SC = \frac{m_2}{m_1} \cdot 100 \ (wt\%) \tag{3}$$

where m_1 is the initial weight of a sample and m_2 is the residual weight after the evaporation process.

The dynamic viscosity of the syrups was measured at 25 °C using a DV-II Pro Extra viscometer (spindle #6 or #7, 50 rpm; Brookfield, New York, NY, USA). The K-values for the dry telomers or copolymers were determined using an Ubbelohde viscometer according to the EN ISO 1628-1:1998 standard and the Fikentscher equation (Equation (4)) [31]:

$$K = 1000 \cdot k = 1000 \cdot \frac{1.5 \log \eta_r - 1 + \sqrt{1 + \left(\frac{2}{c} + 2 + 1.5 \log \eta_r\right) 1.5 \log \eta_r}}{150 + 300c} \tag{4}$$

where $\eta_r = \eta/\eta_0$; η is the viscosity of the telomer/copolymer solution; η_0 is the viscosity of the pure auxiliary diluent (i.e., tetrahydrofurane); and c is the telomer/polymer concentration (g/cm^3).

To determine the presence of TES in the dry telomers, post-reaction mixtures were heated in a vacuum dryer at 60 °C and 10 mm Hg for 1 h. In the spectra of the obtained materials, absorbance was searched for at a wavelength of about 720 cm^{-1} (characteristic for vibrations of Si-C bonds) using FTIR spectroscopy (Nexus FT-IR, Thermo Nicolet, New Castle, DE, USA) [32].

2.3. Preparation and Characterization of Self-Crosslinkable Pressure-Sensitive Adhesives

The self-crosslinkable PSAs were composed of only the silicone-(meth)acrylate telomer syrups (i.e., the solutions of the Si telomers in unreacted monomers obtained in the UV telomerization process). As reference samples, PSA films based on (meth)acrylate syrups were used (i.e., the solutions of the (meth)acrylate copolymers in unreacted monomers obtained in the UV photopolymerization process). The syrups were applied onto polyester foils and UV-irradiated (UV irradiation doses were 1, 2, 3, and 4 J/cm^2) using a medium pressure mercury lamp (UV-ABC; Hönle UV-Technology, Gräfelfing, Germany). The UV exposure was controlled with a radiometer (Dynachem 500; Dynachem Corp., Westville, IL, USA). The base weight of the PSA layers was 60 g/m^2.

The photocrosslinking process in the tested systems took place with the participation of an ABP photoinitiator (hydrogen transfer photoinitiator). This process consists of the production of free radicals through the detachment of the hydrogen atom (from the hydrogen donor molecule) by the triplet ketone (benzophenone group in ABP). When ABP is used, the hydrogen atom is often abstracted from the tertiary carbon atom present in the structure of the comonomers [33]. In this article, we disclose for the first time that the MMA molecule can be the hydrogen donor (such as the case where the hydrogen donor is $(CH_3)_2C=C\,(CH_3)_2$) [34,35]. A proposed course for the photocrosslinking process in the prepared PSAs is shown in Figure 2.

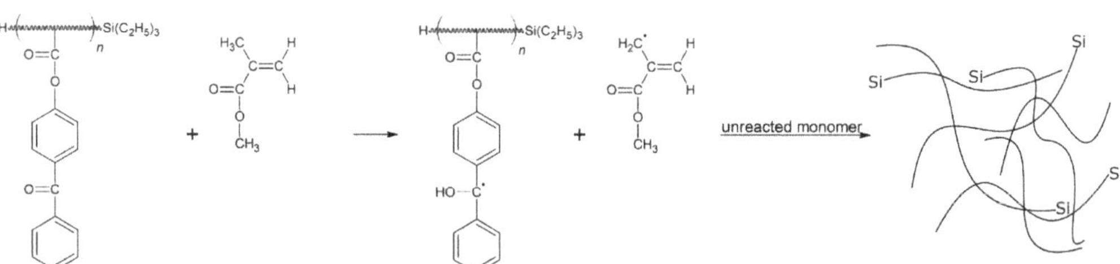

Figure 2. Probable course of photocrosslinking process for Si telomer-based PSAs with pendant benzophenone groups.

Self-adhesive tests (adhesion to a steel, tack, and cohesion at 20 °C and 70 °C) of the UV-crosslinked PSAs were performed at 23 ± 2 °C and 50% ± 5% relative humidity. The values of adhesion to a steel (also called the peel adhesion) at an angle of 180° were determined according to the AFERA 5001 standard developed by the European Association des Fabricants Europeens de Rubans Auto-Adhesifs (AFERA) using a Zwick/Roell Z010 testing machine (Zwick/Roell, Ulm, Germany). A one-sided PSA film with dimensions of 175 × 25 mm was applied to the degreased steel plate and pressed with a rubber roller weighing 2 kg. The test was performed 20 min after the application of the film to the plate with a peeling speed of 300 mm/min. The tack values were determined with the loop method in accordance with the AFERA 5015 standard using a Zwick/Roell Z010 testing machine (Zwick/Roell, Ulm, Germany). A PSA film with dimensions of 175 × 25 mm was mounted in the upper jaws to obtain loops with the adhesive layer on the outside. The sample was lowered perpendicularly to the degreased steel plate placed in the lower jaws at a speed of 100 mm/min. The contact area was about 6.25 cm². The machine recorded the force needed to detach the adhesive film after brief contact with the steel surface without external forces. The values of cohesion (i.e., the static shear adhesion) were determined in accordance with the AFERA 5012 standard using a device designed by the International Laboratory of Adhesives and Self-Adhesive Materials of the West Pomeranian University of Technology in Szczecin, which enables automatic measurement of the time of joint-crack occurrence. A one-sided adhesive film was applied to the degreased steel plate to form a 25 × 25 mm (6.25 cm²) joint and pressed with a 2 kg rubber roller to improve wettability. A 1 kg weight was attached to the free end of the film. The setup was then placed in a tripod so that the force of gravity was exerted on the weld at an angle of 180°. The cohesion value was defined as the time needed for the weld to crack. The test was carried out at temperatures of 20 °C and 70 °C. These parameters were evaluated using three samples for each adhesive film. During self-adhesive properties tests, three types of damage failures may occur: adhesive failure (af), when the adhesive layer remains on the carrier (the cohesion forces are higher than the adhesion forces); cohesive failure (cf), when the adhesive layer remains on both the carrier and the substrate; and mixed failure (mf), when both adhesive and cohesive failures occur. Moreover, the conversion of C=C bonds (DB) in the PSAs after the UV-crosslinking process was analyzed using the FTIR technique (Nexus FT-IR, Thermo Nicolet, New Castle, DE, USA); variations in the absorbance value at 1635 cm⁻¹ (C=C bond) and reference value at 1730 cm⁻¹ (C=O bond) were monitored according to Equation (5):

$$\mathrm{DB} = \frac{A_c^{1635} / A_c^{1730}}{A_u^{1635} / A_u^{1730}} \cdot 100 \ (\%) \tag{5}$$

where A_c^{1635} is the absorbance of the crosslinked sample at 1635 cm⁻¹; A_u^{1635} is the absorbance of the uncrosslinked sample at 1635 cm⁻¹; A_c^{1730} is the absorbance of the crosslinked sample at 1730 cm⁻¹; and A_u^{1730} is the absorbance of the uncrosslinked sample at 1730 cm⁻¹.

The total conversion of C=C bonds (TC) was determined as the total solids content (assuming that the volatiles constituted 100% of the unreacted monomers in the syrup, as confirmed by comparing the NMR conversion to SC [36]) and the double-bond conversion for the crosslinked films was evaluated according to Equation (6):

$$\mathrm{TC} = ((\mathrm{SC}/100) + ((1 - (\mathrm{SC}/100)) * (\mathrm{DB}/100)) \cdot 100 \ (\%) \tag{6}$$

Moreover, the glass-transition temperature (T_g) values of the UV-crosslinked PSAs were determined with the DSC method (DSC250 differential scanning calorimeter, TA Instruments, New Castle, DE, USA). Samples (ca. 10 mg) were analyzed using hermetic aluminum pans at temperatures from −80 to 200 °C (heating rate of 10 °C/min).

3. Results and Discussion

3.1. Kinetics of UV-Telomerization Process

First, the influence of the telogen and photoinitiator concentration on the UV-telomerization process in the selected monomer systems was investigated with the photo-DSC method. The results of the kinetic studies for the systems containing 0.05, 0.075, or 0.1 wt. parts APO and 0, 3.5, 7, or 10 wt. parts TES are presented in Figure 3.

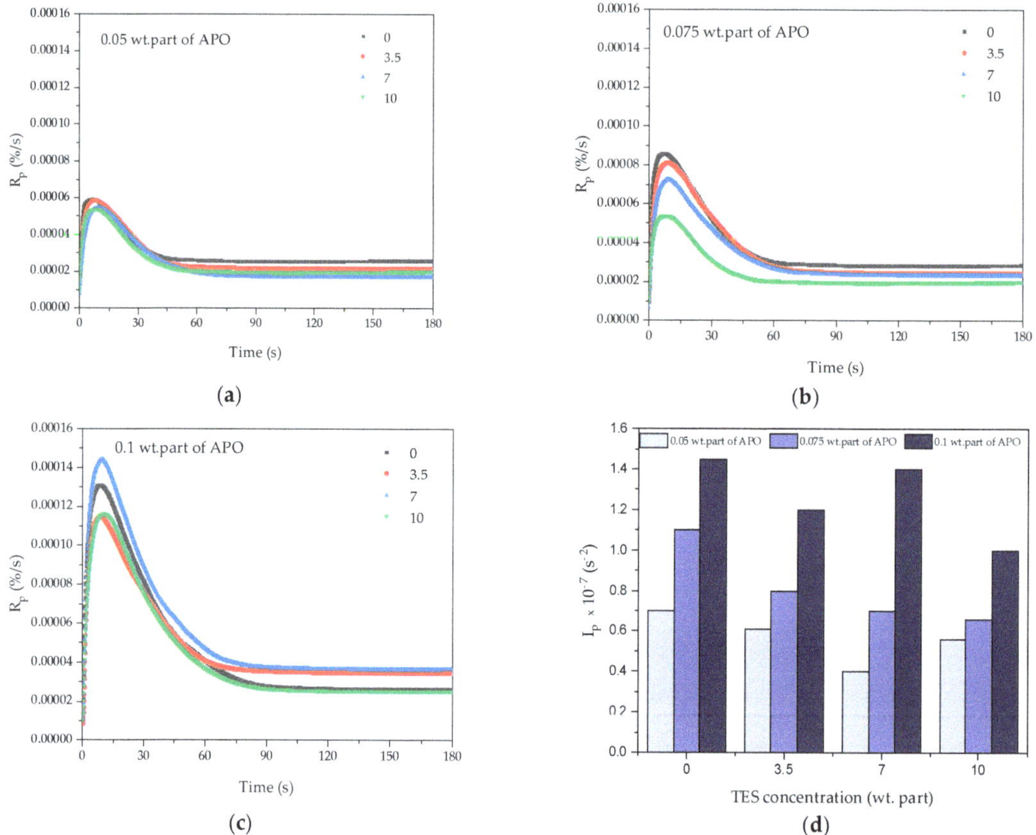

Figure 3. The kinetic curves recorded during UV telomerization of BA, AA, MMA, and ABP in the presence of TES as a telogen (0; 3.5, 7, or 10 wt. parts) initiated by (**a**) 0.05 wt. parts, (**b**) 0.07 wt. parts, and (**c**) 0.1 wt. parts APO and the photoinitiation index of the APO/TES systems (**d**).

Kinetic studies revealed that the reaction rate was strongly dependent on the APO and TES concentrations. As can be seen (Figure 3a–c), the higher the APO concentration in the reaction mixture was, the faster the reaction rate. At a low concentration of APO, there were no significant differences in the effects of TES on the rate of reaction (Figure 3a). With an average amount of APO, more TES in the system led to a slower reaction, and the reaction was fastest without telogen (Figure 3b). Interestingly, there was no increasing or decreasing tendency in the influence of the TES concentration on the reaction rate in the case of mixtures with 0.1 wt. parts APO. In these systems, the highest R_p was recorded with 7 wt. parts TES, and the smallest R_p values for 3.5 wt. parts TES. However, the highest concentration of TES tended to slow down the reaction rate (Figure 3a,d). Moreover, the kinetic studies highlighted a characteristic feature of the telomerization process with TES; namely, that after 60 s of irradiation, there was a rapid decrease in the reaction rate. However, in systems with medium TES content (3.5 or 7 wt. parts) and 0.1 wt. parts APO, the process ran further but only at a low rate (Figure 3c). Regarding the abilities of APO and TES to initiate the process, it can be seen (Figure 3d) that the more photoinitiator there was, the higher the Ip value (regardless of the TES concentration). Moreover, the highest Ip values were found in the arrangement without TES. With the increase in telogen concentration, the initiating abilities of the APO/TES system decreased. One exception was the mixture with 0.1 wt. parts APO/7 wt. parts TES (similar Ip values as for 0.1 wt. parts APO without TES).

3.2. The Physicochemical Properties of Syrups

The courses of the UV-telomerization process (with TES) and UV-photopolymerization (reference samples without TES) in the glass reactor (with the desired mixing speed for the reactants) were investigated by recording the time dependences of the mixture temperature; thermographs for the systems with different contents of APO and TES are presented in Figure 4.

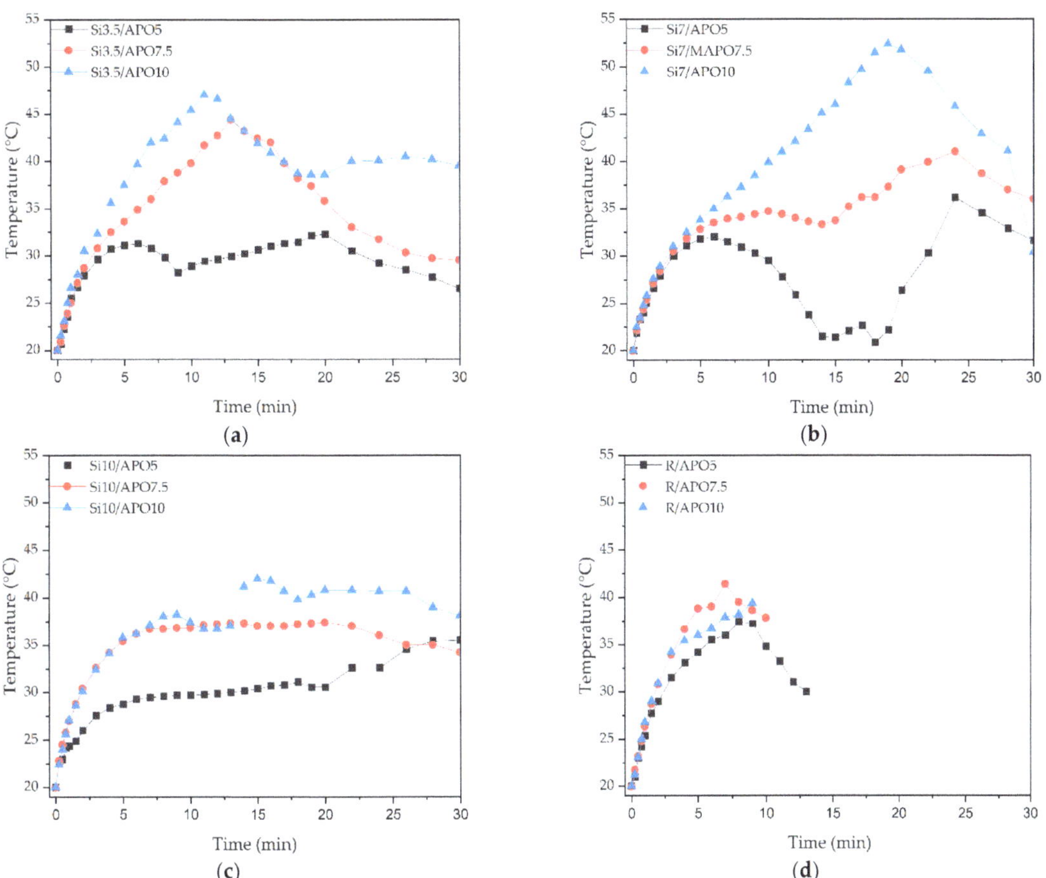

Figure 4. The temperature change in the reactor during the UV-telomerization process for 3.5 (**a**), 7 (**b**), or 10 (**c**) wt. parts TES and during the photopolymerization process (**d**).

The presented thermograms revealed, first of all, that the UV-telomerization process could be carried out for a longer duration (up to 30 min) than the photopolymerization process with the same APO content (up to 9–14 min). The shorter reaction time for the reference syrups (R, Figure 4d) was caused by the winding up of the reaction products on the stirrer (gel formation), which did not occur for the telomer syrups. The maximum temperature values for photopolymerization were 37–42 °C. However, the values of the temperature peaks in the telomeric systems were generally higher (reaching 53 °C for Si7/APO-10), which was in agreement with the photo-DSC results (for this system, the reaction rate and Ip values were the highest; see Figure 3c,d). The presented thermographs also confirmed that, with the increase in the APO content in the telomeric mixtures, the recorded peak temperature became higher and occurred faster. An interesting observation was the occurrence of a double (Si3.5/APO-10; Si7/APO-7.5) or even triple temperature peak (Si10/APO-10), which in some arrangements was unobservable (Si7/APO-10; probably due to the high first peak) and may indicate a process with two or more stages. This was confirmed by the kinetic studies results, which did not show a simple correlation between TES concentration and the course of the telomerization reaction. Additionally, the presented thermograms revealed a multistage course for the reaction. It should be emphasized that this was also influenced by the mechanical mixing of the reactants during irradiation. Nevertheless, the UV-telomerization process allowed the production of liquid syrups.

The results for the solid content and dynamic viscosity of the obtained syrups, as well as the K-values and glass-transition temperatures (T_g) of the dry silicone-(meth)acylate telomers and (meth)acrylate copolymers, prepared with different APO and TES concentrations are presented in Figure 5.

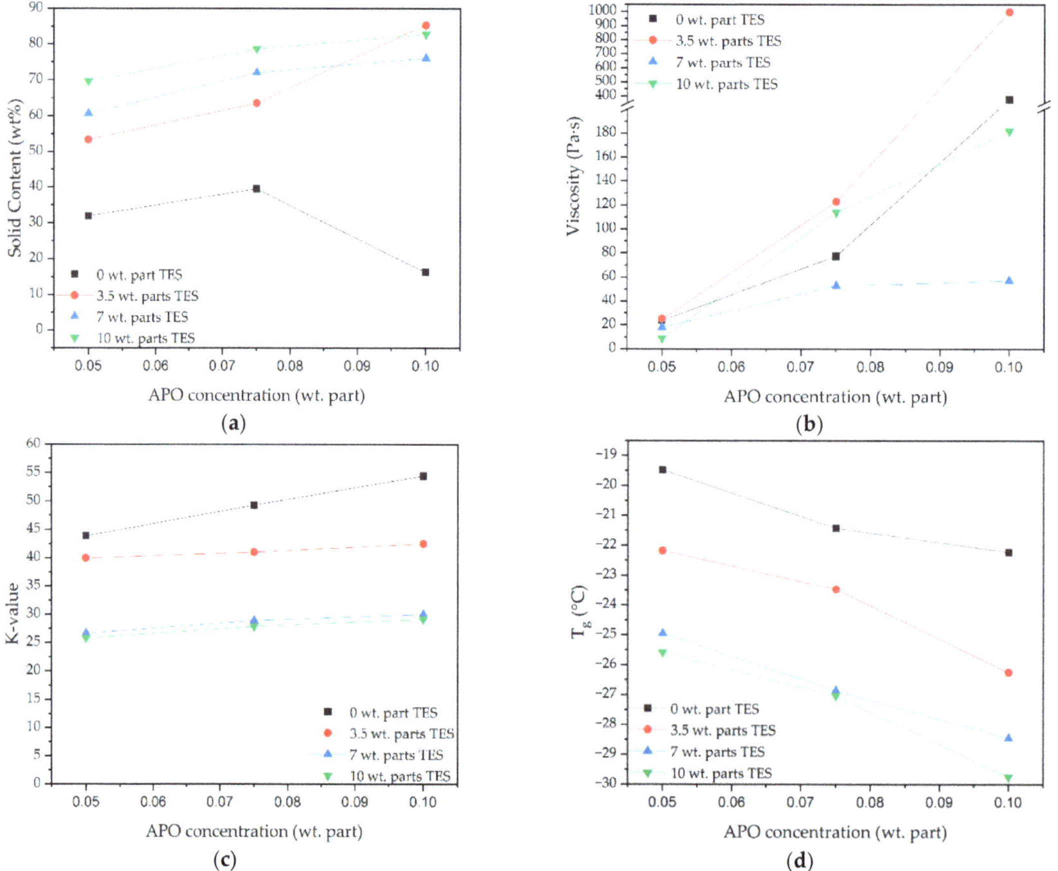

Figure 5. Solid content (**a**) and dynamic viscosity (**b**) of syrups and K-values (**c**) and T_g (**d**) of silicone-(meth)acrylate telomers and (meth)acrylate copolymers.

As can be seen, the UV-telomerization products (i.e., the silicone-(meth)acrylate telomer solutions) were characterized by higher SC values than the photopolymerization process products (i.e., the (meth)acrylate copolymer solutions); specifically, 53–87%

and <40%, respectively. Additionally, as the contents of APO and TES increased, the SC values increased. A decrease in the SC value was recorded only for the R/APO-10 sample (18%). It was mentioned earlier that, during the photopolymerization, the viscosity of the reaction mixture increased very quickly (the gel effect), which indirectly indicated the high molecular weights of the products. However, the SC values were low for (meth)acrylate syrups, so the monomer conversion during photopolymerization was also lower than during telomerization. This result for photopolymerization was expected, as the overall reaction rate of this process was higher than that of the telomerization process, resulting in a gel effect and low monomer conversion. The high molecular weights of the photopolymerization products were confirmed by their greater K-values, ranging from 45 to 55 a.u. (Figure 5c).

The influence of the concentrations of APO and TES on the dynamic viscosity values of the syrups was determined. Only when the APO concentration was increased did the dynamic viscosity values increase. However, there was no correlation with the TES concentration in the system. The lowest APO concentration (0.05 wt. part) led to low-viscosity liquid-process products (10–30 Pa·s) suitable for coating carriers in the production of PSAs. In addition, for the same systems, it was observed that more telogen resulted in lower dynamic viscosity in the telomer syrups. The highest values for the dynamic viscosity (>900 Pa·s) were revealed for Si3.5/APO-10 syrups. Unexpectedly, the Si7 syrups showed much lower dynamic viscosity than the others (20 Pa·s for Si7/APO-5 and ca. 50 Pa·s for Si7/APO-7.5 and Si7/APO-10). In these systems, the telomerization process also took place differently from the others, which was confirmed by the thermograms (Figure 4b). Nevertheless, it should be mentioned that the obtained dynamic viscosity values were influenced by both the telomer/oligomer content (or copolymers, as in the case of the reference samples) and their molecular weights, as well as the amounts of unreacted monomers in the syrups. The determined K-values were indirectly indicative of the molecular weight values of the obtained oligomers and copolymers (Figure 5c). Generally, higher K-values were observed for copolymers than for telomeres, as expected. Additionally, with increasing APO concentrations, K-values increased. However, with more telogen in the system, lower K-values were obtained, and the results for Si7 and Si10 syrups were similar (26–28 a.u.).

The new silicone-(meth)acrylate telomers obtained were also characterized in terms of their glass-transition temperature (Figure 5d). The Si telomers exhibited significantly lower T_g values (-22 °C for Si3.5/APO-5 and -29.5 °C for Si10/APO-10) than those for the reference copolymers (-19.5 °C for R/APO-5 and -21.5 °C for R/APO-10). The T_g values decreased with increases in both TES and APO concentrations. Thus, telomeres with lower molecular weights (lower K-values) and higher amounts of silicon atoms in the structure were characterized by lower T_g. This type of telomer could have a positive effect on the applicability of the prepared PSAs.

To confirm the incorporation of triethylsilane into the oligomeric chains, FTIR analysis of the dry telomers and reference copolymer was performed. For this purpose, unreacted monomers were evaporated from the syrups by heating them in a vacuum dryer (at 60 °C and 10 mm Hg) for 1 h. The IR absorption band corresponding to the stretching vibration of the Si-C bond and located around 720 cm^{-1} [32] was detected both for pure TES and the synthesized telomers. However, it was not detected for the reference sample. This confirmed the incorporation of the TES moiety into the oligomeric chains of the Si telomers. The FT-IR spectra are shown in Figure 6.

Due to the high solids content (>60%) and low glass-transition temperature, as well as their appropriate viscosity, the syrups based on 0.075 wt. parts APO (i.e., Si3.5/APO-7.5, Si7/APO-7.5 and Si10/APO-7.5) could be used to produce self-crosslinkable pressure-sensitive adhesives.

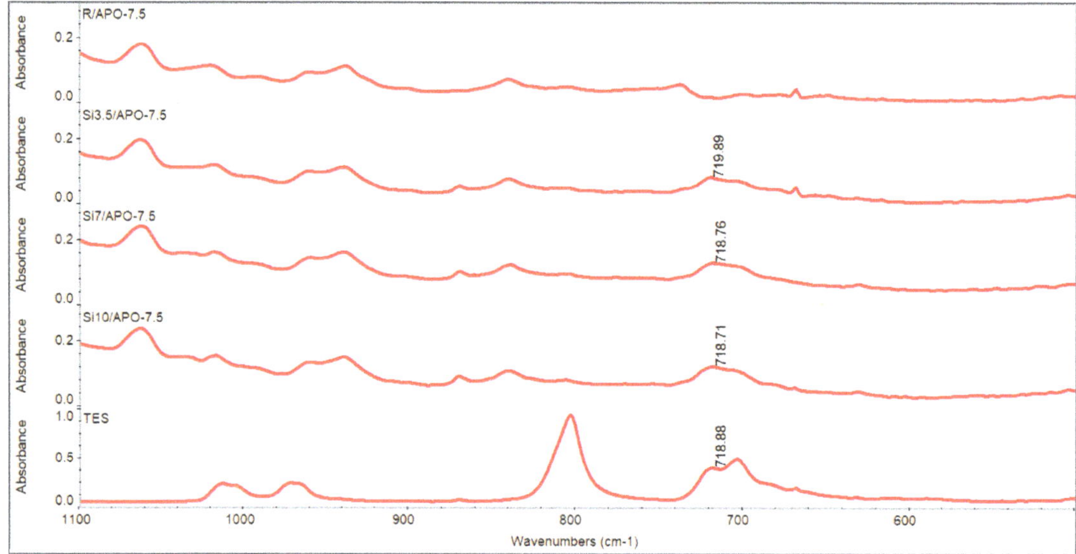

Figure 6. FT-IR spectra of TES, dry systems consisting of Si telomers (with the same content of 0.075 wt. parts APO), and dry reference with the same APO concentration.

3.3. Properties of UV-Crosslinked PSA

The self-adhesives properties of PSA films based on silicone-(meth)acrylate syrups or (meth)acrylate copolymer syrups after the UV-crosslinking process (with UV energy doses of 1, 2, 3, or 4 J/cm^2) are presented in Figure 7. As is known, the crosslinking process is critical in the preparation of self-adhesive materials and can provide them with the desired peculiarities; in particular, cohesion. Moreover, as can be seen in Figure 2, the proposed mechanism for crosslinking both in telomeric and copolymeric syrups involves abstracting a hydrogen atom from an unreacted MMA molecule using a carbonyl moiety of an ABP unit to generate reactive radicals able to further chemically combine with the unreacted monomer in the system, with the final result of telomer/copolymer branching and then crosslinking. The results for the self-adhesive properties, which indirectly prove the correctness of the process, are presented in Figure 7.

Thus, the values for the adhesion to a steel surface for the PSAs without Si atoms (i.e., PSAs based on the R/APO-7.5 syrup) were relatively low in comparison with the PSAs based on Si telomers. The highest value (9 N/25 mm) was recorded for the sample subjected to 2 J/cm^2 UV-dose irradiation. In the case of the PSA systems based on the Si telomer syrups, the maximum value for adhesion was found to be 12.4 N/25 mm for Si7/APO-7.5 at a UV dose of 4 J/cm^2. However, it should be noted that the adhesion values depended on the UV dose (the higher the UV dose, the higher the adhesion). It is known that a higher UV dose promotes cleavage of the π bonds of the C=C bonds belonging to the unreacted monomer molecules and the formation of a denser polymer network. In the case of the PSAs based on R/APO-7.5, the lowest adhesion values were recorded after irradiation with 3 and 4 J/cm^2 (1.5 and 0.8 N/25 mm, respectively). As shown above, this system contained (meth)acrylate copolymer (40% SC) with a relatively high molecular weight (high K-value) and a large amount of unreacted monomers (60%). Thus, at the irradiation stage, a dense polymer network was formed (more monomers were involved in its formation than in the telomer systems). Hence, the stiffness of the formed polymer network limited its adhesion to the steel substrate. In the case of PSAs based on telomeric syrups, generally higher adhesion values were found for those with the highest proportions of Si atoms and the smallest molecular weights (lowest K-values); i.e., Si10/APO-7.5 (9–11 N/25 mm). On the other hand, a significant decrease in adhesion (down to 3 N/25 mm) was noticed for PSAs based on Si3.5/APO-7.5 at the maximum UV-dose irradiation, and a more considerable proportion of telomers and higher molecular weights (SC ca. 60%, K-value ca. 40 a.u.) were observed than for the other telomeric systems. The tack results for the obtained PSAs were interesting. Generally, the tack values decreased with increasing UV dose due to the increase in the crosslinking density of the systems, which has been described previously in the literature [11,37]. However, only the PSAs from the Si7/APO-7.5 syrup showed high tack values (11.5 N after 1 J/cm^2 UV dose). All the other systems displayed lower tack values below 2.5 N, with the lowest value recorded for the reference sample (<0.5 N).

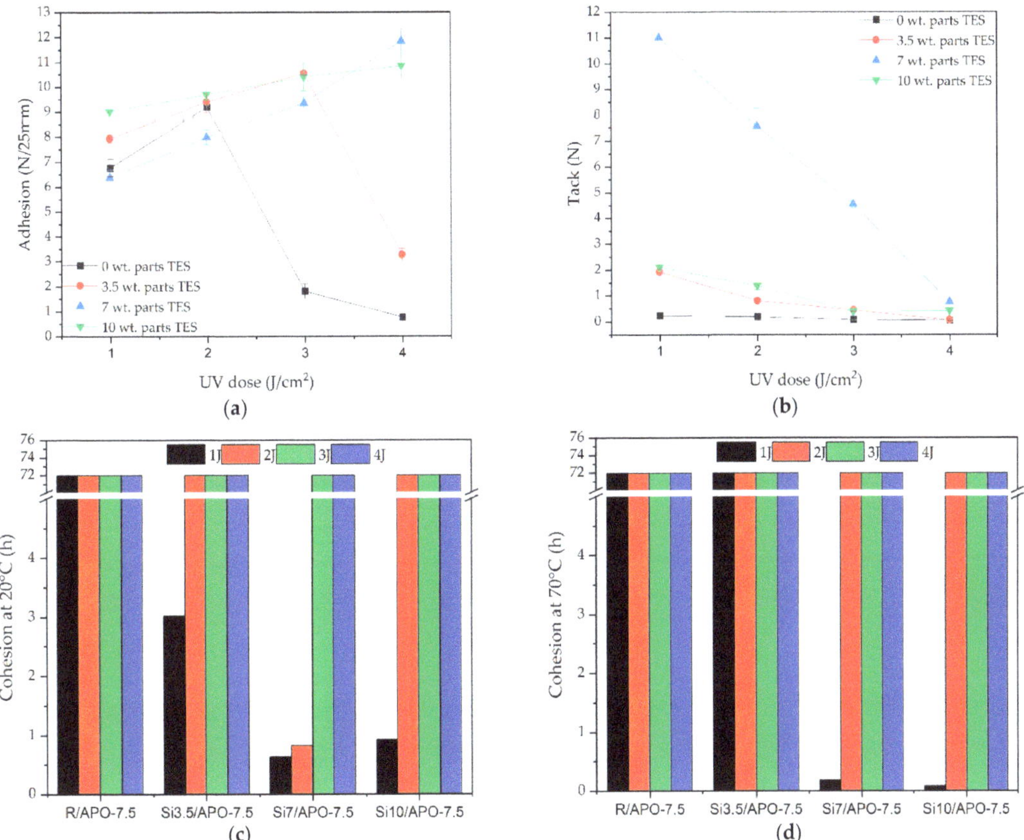

Figure 7. Self-adhesive properties of PSAs based on Si telomere syrups with the same amounts of APO (0.075 wt. parts): (**a**) adhesion to steel, (**b**) tack, and (**c**) cohesion at 20 °C and (**d**) 70 °C.

The cohesion at both 20 °C and 70 °C for most of the PSAs reached very high values (above 72 h). However, this value decreased when the TES and APO concentrations increased. This was due to the decreasing K-value (lower molecular weights), as well as the production of increasingly stiffer films, which can exhibit wetting problems, as indicated by the presence of adhesion failure when tested at 20 °C. Excellent cohesion at 70 °C also supports this finding, as the wettability of the adhesive films at elevated temperature was higher. During the tests of the self-adhesive properties, only adhesive cracks were observed. Due to the very good self-adhesive properties found for most of the compositions, only samples crosslinked with the dose of 3 J/cm^2 were further investigated.

Double-bond conversions in the crosslinked adhesive films determined with FT-IR, SC, and total conversion are shown in Table 2.

Table 2. Double-bond conversion determined with FT-IR spectroscopy and total monomer conversion.

PSA	Content of Unreacted Monomers in the Adhesive Film before UV Crosslinking [1]	Conversion of C=C Bonds in PSAs during UV-Crosslinking Process (DB)	Total Conversion of C=C Bonds (TC)
	(%)	(%)	(%)
R/APO-7.5	60	84	90
Si3.5/APO-7.5	36	75	91
Si7/APO0-7.5	28	72	92
Si10/APO7.5	21	81	96

[1]—Based on SC calculation.

The conversion of unreacted monomers during the UV crosslinking of PSAs (DB) was closely related to the amounts of unreacted monomers in the telomeric syrups from which the adhesive films were obtained. As can be seen, the highest DB value (84%) was recorded for R/APO-7.5 (reference sample), but this was due to the large amount of unreacted monomers in the starting syrup (60%) and the low linear (meth)acrylate copolymer content, which facilitated the migration of the radicals formed. In the case of the telomeric systems, the DB values were lower (76–79%). Considering the total conversion of the C=C bonds in the PSAs after the UV-crosslinking process, the highest value was recorded for Si10/APO-7.5 (96%). It should be noted that the conversion of unreacted monomers—and, thus, the UV-photocrosslinking process—was most effective in the system with the highest silicon moiety content (10 wt. parts). This was due to the system having the lowest molecular weights for the Si10 telomers containing benzophenone groups (from ABP) in their side chains (the lowest K-values), which made the mobility of such macroradicals higher and the photocrosslinking process more effective.

Eventually, the values of the glass-transition temperatures measured for the UV-crosslinked PSAs based on Si telomers and (meth)acrylate copolymers (Figure 8) reached accordance with the content of the TES moiety in the final telomer products.

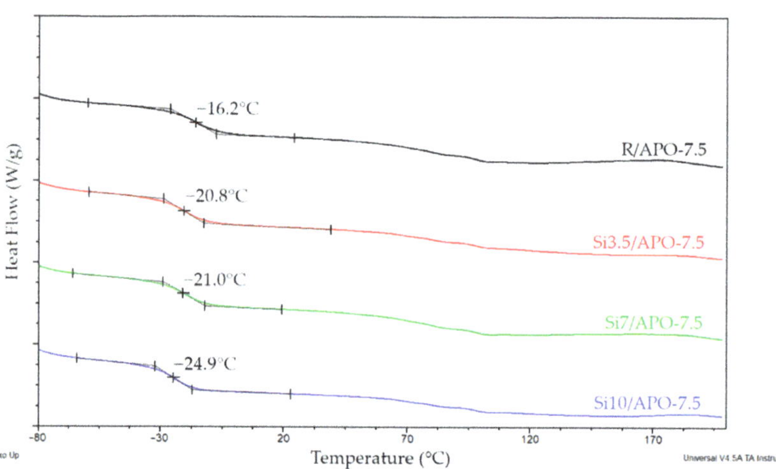

Figure 8. DSC thermograms of the selected PSAs crosslinked under UV radiation with a dose of 3 J/cm^2.

As can be seen, PSAs based on Si telomers were characterized by significantly lower T_g values compared to the reference sample (a difference of almost 9 °C). This was because of the incorporation of TES molecules into the chain; the more TES residues in the system, the higher the chain flexibility and, consequently, the lower the T_g. This result is very satisfactory for self-adhesive materials.

4. Conclusions

Silicone-(meth)acrylic telomers with terminal Si atoms were prepared via a UV-telomerization process using triethylsilane (TES) as a telogen. The prepared Si telomer syrups were used as adhesive binders to obtain self-crosslinkable pressure-sensitive adhesives without any additional crosslinking agent. The main conclusions are as follows:

- The UV telomerization of the tested (meth)acrylate monomers was strongly dependent on the radical photoinitiator concentration, and there was no trend observed in the TES concentration; only at the average APO content was a decrease in the reaction rate found with an increase in the concentration of telogen;
- Silicone-(meth)acrylate telomer solutions were characterized by higher solid content values (>60%) and lower dynamic viscosity than the reference (meth)acrylate syrups;
- Si telomers exhibited significantly lower K-values and T_g values ($-29.5\,^\circ\text{C}$) than the (meth)acrylate copolymers ($-21.5\,^\circ\text{C}$);
- Si telomer-based self-crosslinking PSAs were distinguished by excellent adhesion to steel (up to 12.4 N/25 mm) and tack (11 N) and cohesion both at $20\,^\circ\text{C}$ and $70\,^\circ\text{C}$ (>72 h); in addition, they demonstrated low T_g compared to the reference sample (difference of up to almost $9\,^\circ\text{C}$).

The results of the studies carried out and described here encourage the authors to work further on the preparation of new Si telomeres (with other silicon telogens) and analyze their structures and the thermo-mechanical properties of the PSAs obtained from them.

Author Contributions: Conceptualization, M.W. and A.K.; methodology, M.W. and A.K.; investigation, M.W. and A.K.; writing—review and editing, M.W. and A.K.; visualization, M.W. and A.K.; supervision, A.K. All authors have read and agreed to the published version of the manuscript.

Funding: This work was supported by the Rector of the West Pomeranian University of Technology in Szczecin for PhD students of the Doctoral School, grant number: ZUT/16/2022.

Institutional Review Board Statement: Not applicable.

Informed Consent Statement: Not applicable.

Data Availability Statement: The data presented in this study are available on request from the corresponding author.

Conflicts of Interest: The authors declare no conflict of interest.

References

1. Benedek, I.; Feldstein, M.M. (Eds.) *Fundamentals of Pressure Sensitivity*, 1st ed.; CRC Press: Boca Raton, FL, USA, 2008; ISBN 9781420059380.
2. Park, H.W.; Park, J.W.; Lee, J.H.; Kim, H.J.; Shin, S. Property Modification of a Silicone Acrylic Pressure-Sensitive Adhesive with Oligomeric Silicone Urethane Methacrylate. *Eur. Polym. J.* **2019**, *112*, 320–327. [CrossRef]
3. Seok, W.C.; Park, J.H.; Song, H.J. Effect of Silane Acrylate on the Surface Properties, Adhesive Performance, and Rheological Behavior of Acrylic Pressure Sensitive Adhesives for Flexible Displays. *J. Ind. Eng. Chem.* **2022**, *111*, 98–110. [CrossRef]
4. Dastjerdi, Z.; Cranston, E.D.; Dubé, M.A. Pressure Sensitive Adhesive Property Modification Using Cellulose Nanocrystals. *Int. J. Adhes. Adhes.* **2018**, *81*, 36–42. [CrossRef]
5. Lee, S.W.; Park, J.W.; Lee, Y.H.; Kim, H.J.; Rafailovich, M.; Sokolov, J. Adhesion Performance and UV-Curing Behaviors of Interpenetrated Structured Pressure Sensitive Adhesives with 3-MPTS for Si-Wafer Dicing Process. *J. Adhes. Sci. Technol.* **2012**, *26*, 1629–1643. [CrossRef]
6. White, C.C.; Tan, K.; Wolf, A.T.; Carbary, L.D. *Advances in Structural Silicone Adhesives*; Woodhead Publishing Limited: Sawston, UK, 2010; ISBN 9781845694357.
7. Lin, S.B.; Durfee, L.D.; Ekeland, R.A.; McVie, J.; Schalau, G.K. Recent Advances in Silicone Pressure-Sensitive Adhesives. *J. Adhes. Sci. Technol.* **2007**, *21*, 605–623. [CrossRef]
8. Cheng, J.; Li, M.; Cao, Y.; Gao, Y.; Liu, J.; Sun, F. Synthesis and Properties of Photopolymerizable Bifunctional Polyether-Modified Polysiloxane Polyurethane Acrylate Prepolymer. *J. Adhes. Sci. Technol.* **2015**, *30*, 2–12. [CrossRef]

9. Balaban, M.; Antić, V.; Pergal, M.; Godjevac, D.; Francolini, I.; Martinelli, A.; Rogan, J.; Djonlagić, J. Influence of the Chemical Structure of Poly(Urea-Urethane-Siloxane)s on Their Morphological, Surface and Thermal Properties. *Polym. Bull.* **2013**, *70*, 2493–2518. [CrossRef]

10. Kostyuk, A.; Ignatenko, V.; Smirnova, N.; Brantseva, T.; Ilyin, S.; Antonov, S. Rheology and Adhesive Properties of Filled PIB-Based Pressure-Sensitive Adhesives. I. Rheology and Shear Resistance. *J. Adhes. Sci. Technol.* **2014**, *29*, 1831–1848. [CrossRef]

11. Benedek, I.; Feldstein, M.M. *Technology Pressure-Sensitive Adhesives and Products*; CRC Press: Boca Raton, FL, USA, 2019; ISBN 9781420059397.

12. Satas, D. *Handbook of Pressure Sensitive Adhesive Technology*; Springer Science & Business Media: Berlin, Germany, 1989; ISBN 9781475708684.

13. Czech, Z.; Kowalczyk, A.; Górka, K.; Głuch, U.; Shao, L.; Świderska, J. Influence of the Unsaturated Photoinitiators Kind on the Properties of Uv-Crosslinkable Acrylic Pressure-Sensitive Adhesives. *Pol. J. Chem. Technol.* **2012**, *14*, 83–87. [CrossRef]

14. Bolshakov, A.I.; Kuzina, S.I.; Kiryukhin, D.P. Tetrafluoroethylene Telomerization Initiated by Benzoyl Peroxide. *Russ. J. Phys. Chem. A* **2017**, *91*, 482–489. [CrossRef]

15. Chen, J.; Chalamet, Y.; Taha, M. Telomerization of Butyl Methacrylate and 1-Octadecanethiol by Reactive Extrusion. *Macromol. Mater. Eng.* **2003**, *288*, 357–364. [CrossRef]

16. Starks, C.M. *Free Radical Telomerization*; Academic Press: Cambridge, MA, USA, 1974; ISBN 9780126636505.

17. Boutevin, B. From Telomerization to Living Radical Polymerization. *J. Polym. Sci. A Polym. Chem.* **2000**, *38*, 3235–3243. [CrossRef]

18. Loubat, C.; Boutevin, B. Telomerization of Acrylic Acid with Mercaptans: Part 2. Kinetics of the Synthesis of Star-Shaped Macromolecules of Acrylic Acid. *Polym. Int.* **2001**, *50*, 375–380. [CrossRef]

19. Loubat, C.; Boutevin, B. Telomerization of Acrylic Acid with Thioglycolic AcidEffect of the Solvent on the CT Value. *Polym. Bull.* **2000**, *44*, 569–576. [CrossRef]

20. Boyer, C.; Boutevin, G.; Robin, J.; Boutevin, B. Synthesis of Macromonomers of Acrylic Acid by Telomerization: Application to the Synthesis of Polystyrene-g-poly (Acrylic Acid) Copolymers. *J. Polym. Sci. A Polym. Chem.* **2007**, *45*, 395–415. [CrossRef]

21. Li, G.H.; Yang, P.P.; Zhao, Z.J.; Gao, Z.S.; He, L.Q.; Tong, Z.F. Synthesis of Poly (Tert-Butyl Acrylate) Telomer by Free-Radical Telomerization. *CIESC J.* **2011**, *62*, 2061–2066.

22. Jeanmaire, T.; Brondino, C.; Hervaud, Y.; Boutevin, B. Synthesis of new phosphonic derivatives bearing a perfluorinated chain and their adhesive properties on steel. *Phosphorus Sulfur Silicon Relat. Elem.* **2002**, *177*, 2331–2343. [CrossRef]

23. Tan, B.; Pan, H.; Irshad, H.; Chen, X. Synthesis of Ultraviolet-Curable Modified Polysiloxane and Its Surface Properties. *J. Appl. Polym. Sci.* **2002**, *86*, 2135–2139. [CrossRef]

24. Weisbrodt, M.; Kowalczyk, A.; Kowalczyk, K. Structural Adhesives Tapes Based on a Solid Epoxy Resin and Multifunctional Acrylic Telomers. *Polymers* **2021**, *13*, 3561. [CrossRef]

25. Ghosh, P.; Mitra, P.S. Novel Solvent Effects in the Photopolymerization of Methyl Methacrylate by Use of Quinoline–Bromine Charge-transfer Complex as Photoinitiator. *J. Polym. Sci. Polym. Chem. Ed.* **1977**, *15*, 1743–1758. [CrossRef]

26. Ghosh, P.; Banerjee, H. Photopolymerization of Methyl Methacrylate With the Use of Bromine As Photoinitiator. *J. Polym. Sci. Polym. Chem. Ed.* **1973**, *11*, 2021–2030. [CrossRef]

27. Kowalczyk, A.; Weisbrodt, M.; Schmidt, B.; Kraśkiewicz, A. The Effect of Type-I Photoinitiators on the Kinetics of the UV-Induced Cotelomerization Process of Acrylate Monomers and Properties of Obtained Pressure-Sensitive Adhesives. *Materials* **2021**, *14*, 4563. [CrossRef] [PubMed]

28. Kowalczyk, A.; Weisbrodt, M.; Schmidt, B.; Gziut, K. Influence of Acrylic Acid on Kinetics of UV-Induced Cotelomerization Process and Properties of Obtained Pressure-Sensitive Adhesives. *Materials* **2020**, *13*, 5661. [CrossRef] [PubMed]

29. Spitzer, J.J.; Midgley, A.; Takamura, K.; Lok, K.P.; Xanthopoulo, V.G.; Ditrich, U. Novel Polytelomerization Process and Polytelomeric Products. WO Patent 94/04575, 3 March 1994.

30. Kabatc, J. The Influence of a Radical Structure on the Kinetics of Photopolymerization. *J. Polym. Sci. A Polym. Chem.* **2017**, *55*, 1575–1589. [CrossRef]

31. Coelho, J.F.J.; Gonçalves, P.M.F.O.; Miranda, D.; Gil, M.H. Characterization of Suspension Poly(Vinyl Chloride) Resins and Narrow Polystyrene Standards by Size Exclusion Chromatography with Multiple Detectors: Online Right Angle Laser-Light Scattering and Differential Viscometric Detectors. *Eur. Polym. J.* **2006**, *42*, 751–763. [CrossRef]

32. Omar, M.F.; Ismail, A.K.; Sumpono, I.; Alim, E.A.; Nawi, M.N.; Rahim Mukri, M.'A.; Othaman, Z.; Sakrani, S. FTIR Spectroscopy Characterization of Si-C Bonding in SiC Thin Film Prepared at Room Temperature by Conventional 13.56 MHz RF PECVD. *Malays. J. Fundam. Appl. Sci.* **2012**, *8*, 242–244. [CrossRef]

33. Czech, Z.; Loclair, H. Investigations of UV-Crosslinkable Water-Soluble Acrylic Pressure-Sensitive Adhesives. *Polimery* **2005**, *50*, 65. [CrossRef]

34. Turro, N.J. *Modern Molecular Photochemistry*; Benjamin/Cummings Publishing Company, Inc.: London, UK, 1978; ISBN 0805393544.

35. Pączkowski, J. (Ed.) *Fotochemia Polimerów: Teoria i Zastosowanie*; Wydawnictwo Naukowe Uniwersytetu Mikołaja Kopernika: Toruń, Poland, 2003; ISBN 83-231-1615-6.

36. Gziut, K.; Kowalczyk, A.; Schmidt, B. Free-Radical Bulk-Photopolymerization Process as a Method of Obtaining Thermally Curable Structural Self-Adhesive Tapes and Effect of Used Type I Photoinitiators. *Polymers* **2020**, *12*, 2191. [CrossRef]

37. Czech, Z.; Kowalczyk, A.; Kabatc, J.; Świderska, J. UV-Crosslinkable Acrylic Pressure-Sensitive Adhesives for Industrial Application. *Polym. Bull.* **2012**, *69*, 71–80. [CrossRef]

Article

Influence of Physical Modification of the Adhesive Composition on the Strength Properties of Aerospace Aluminum Alloy Sheet Adhesive Joints

Izabela Miturska-Barańska *, Anna Rudawska and Elżbieta Doluk

Department of Production Computerisation and Robotisation, Faculty of Mechanical Engineering, Lublin University of Technology, Nadbystrzycka 36, 20-618 Lublin, Poland
* Correspondence: i.miturska@pollub.pl

Abstract: One of the most important design factors in the constitution of adhesive joints is the correct choice of adhesive. Currently, there is a full range of options on the commercial market in this regard, but there is increasing research into modifying adhesives for specific engineering applications. The aim of this study was to analyze the effect of physical modification with fillers on the properties of the adhesive composition and the adhesive joints. The adhesives used in the study were a composition of Epidian 5 epoxy resin and PAC curing agent modified with 1% montmorillonite, 5% calcium carbonate and 20% activated carbon. The adhesive compositions in the cured state were subjected to strength tests and SEM and DSC analyses. Using these compositions, adhesive joints of EN AW 2024 T3 aluminum alloy sheets were also made. The tests carried out showed that, due to the use of different fillers, their effects on certain properties of the adhesive compositions are different types. It was shown that physical modification of the adhesive composition does not always result in positive effects. The study also attempted to determine the correlation between the properties of the adhesive compositions in the cured state and the strength of the adhesive joints.

Keywords: mechanical properties of adhesives; adhesive testing; adhesive joints

Citation: Miturska-Barańska, I.; Rudawska, A.; Doluk, E. Influence of Physical Modification of the Adhesive Composition on the Strength Properties of Aerospace Aluminum Alloy Sheet Adhesive Joints. *Materials* **2022**, *15*, 7799. https://doi.org/10.3390/ma15217799

Academic Editors: Marco Lamberti and Aurelien Maurel-Pantel

Received: 12 October 2022
Accepted: 31 October 2022
Published: 4 November 2022

Publisher's Note: MDPI stays neutral with regard to jurisdictional claims in published maps and institutional affiliations.

Copyright: © 2022 by the authors. Licensee MDPI, Basel, Switzerland. This article is an open access article distributed under the terms and conditions of the Creative Commons Attribution (CC BY) license (https://creativecommons.org/licenses/by/4.0/).

1. Introduction

Assembly joints, i.e., made by welding, sealing, soldering, riveting or bonding, are used in almost every industry. One of the most recently developed methods of joining materials is bonding, which makes it possible to obtain structures that are much larger than could be made as a single component [1]. According to many authors described in their publications, such as Messler or Saboori [2–4], adhesive joints are also an alternative to other assembly joints in engineering applications because they have several advantages over conventional joining methods, while not altering the microstructure of the parts being bonded. The use of bonding technology allows for structures that are lighter in weight, yet have high strength, high fatigue life and reliability [5–11]. Structural bonding is, in many cases, one of the methods of relatively fast integration of machine parts, installations, vehicles or aircraft [9,10,12–14]. In addition to these mentioned fields, bonding technology is used in many industries such as construction, electrical engineering, medicine or light industries [15,16]. As Kinloch mentioned in his paper [17], with bonding technology being so widely used, adhesives are being increasingly challenged to develop new and beneficial performance properties. Key manufacturers of adhesives, as well as various research institutions, carry out continuous research work with the aim of obtaining adhesives with the most advantaged properties possible for a specific field [6,18–22]. Bonding technology is inherently an interdisciplinary field, requiring a fundamental understanding of mechanics, surface engineering and materials engineering, and the topics are still relevant and being intensively developed.

The strength properties of adhesive joints significantly depend on the technology used to make them, to a greater extent than in other joints used in engineering [8,23]. The large selection of adhesives produced by various manufacturers creates problems in selecting the right adhesive for the designed structure, especially as the properties of adhesives presented by manufacturers do not always characterize their most important features and are not always clear to the potential user. Researchers dealing with adhesion issues in the context of obtaining the highest possible properties for adhesive joints have at their disposal the main key variable characteristics, the modification of which leads to an increase in or control of the adhesion force between the adhesive and the component to be joined. Among these dependent variables are the chemical composition and properties of the adhesive composition and the surface stereometrics structure of the material to be bonded. The adhesive can be selected from commercially available compositions that exhibit the ability for Lifshitz–van der Waals intermolecular interactions, but also involving as little contribution as possible to permanent dipole interactions [24,25]. The adhesive can also be formulated as an acidic, basic or bifunctional bonding agent. Then, its properties, such as surface tension or viscosity, can be altered through the use of different types of additives, such as thixotropy agents [26]. Most adhesives used as structural adhesives are polymer compositions. For example, adhesives made from epoxy resins are designed so that internal crosslinking occurs leading to increased cohesive strength, but can also produce covalent bonds [27–29]. The durability of an adhesive joint depends mainly on the way it is loaded and the environment in which it is exposed [11,23,30]. Due to these aspects, numerous experimental, often destructive, experiments are conducted that address the strength of the adhesive compositions themselves, as well as adhesive joints made under varying structural and technological factors. The strength of an adhesive joint is one measure of the properties of adhesives [31,32]. An important direction of modern technology research is to subject adhesive compositions to modifications, particularly through the use of nanofillers, even a small addition of which can improve certain characteristics of adhesive materials [29,33–36]. Three types of modification in the literature are distinguished [11,37,38]: chemical, physical and physicochemical.

This paper presents the results of physical modification of epoxy adhesive compositions. Physical modification occurs through physical phenomena. Modified adhesives differ from those before modification in structure, physical properties, functional properties, visual properties, etc. The most common methods of physical modification are the addition of fillers. The performance properties of modified materials significantly depend on the type of filler used (particle shape and size, specific surface area, dispersed phase concentration) [39–41]. The best properties are obtained when the smallest possible fillers are introduced, preferably with particle sizes measured on the nanometer scale [42]. The authors of papers [43–47] have researched modifications related to bonding technology, but these mainly relate to surface modification issues and surface preparation of bonded components. Zheng, in one of his papers [48], states that the strength of the joint depends on the properties of the adhesive, but also on the adhesion between the adhesive and the binder. The effect of introduced fillers on the properties of adhesive compositions was presented in the work of Miturska et al. [34] where researchers described the results of a study on the effect of modification with natural fillers on the mechanical properties of epoxy adhesive compositions after storage time. In this study, the authors use two epoxy resins, Epidian 5 and Epidian 53, which were modified with a 2% addition of Montmorillonite, calcium carbonate and activated carbon, which were cured with the Mannich base. The compositions were cured for 7 days and seasoned for 4 months. Rudawska and Frigione in their work [49] present the effect of the aqueous environment on the mechanical properties of the epoxy resin Epidian 5 modified with calcium carbonate in amounts of 1%, 2% and 3% by weight of the resin. Since epoxy adhesives are one of the most common types of adhesives used in mechanical engineering, it seems reasonable to undertake research to determine the effect of physical modification of an epoxy adhesive composition on its mechanical and physical properties, as well as its effect on the strength properties of adhesive joints. In the various ways of modifying epoxy compositions, it should be taken into consideration that, while influencing the change in certain properties, others can be improved or degraded at the same time. As epoxy adhesives are one of the most commonly used structural adhesives, an attempt was made to modify them. The focus was on the use of modified adhesives in the context of bonding material that is used for aerospace structures. The aerospace industry is constantly developing and a lot of research is being carried out in this area, which makes this topic current and interesting from the point of view of both researchers and manufacturers. Aerospace manufacturers are looking for solutions to achieve lighter but stronger structures compared to conventional joints, such as riveted or welded joints. Structural bonding is currently widely used in the construction of aircraft airframes in the manufacture of components consisting of thin sheets and profiles, in the manufacture of sandwich structures, aircraft control components (rudders, ailerons), wing mechanization components, as well as in aircraft structures made of composites, where together with mechanical joints they form so-called hybrid joints. The large-scale introduction of structural adhesive joints has enabled aircraft weight reductions in the range of 10–15%. Examples of the use of adhesive joints in aviation include the Airbus A380 and Boeing 787 Dreamliner aircraft, as well as the bonding of the rotor blades of the Mi-2 helicopter's carrier. This paper deals with the problem of epoxy adhesive compositions modification with fillers of various origin in the aspect of changing the mechanical properties of adhesive compositions, as well as some mechanical properties of aerospace aluminum alloy sheet adhesive joints. The adhesive compositions used in the study were made on the basis of the Epidian 5 epoxy resin. The choice of this resin is based on its properties, as it has excellent adhesion to most construction materials and, moreover, it is a basic resin, not pre-modified, which makes it an ideal base for physically modified adhesive compositions. The main purpose of using fillers in adhesive compositions is to improve certain performance properties. In addition to this, an important aspect, given the dynamic technological development, is also the environmental aspect, which is why three types of fillers were used in the study, both from the group of organic fillers: activated carbon (CWZ-22) and inorganic fillers—calcium carbonate ($CaCO_3$) and montmorillonite (ZR2).

2. Materials and Methods

In this study, a number of tests were carried out on both adhesive compositions and adhesive joints made using these compositions. A flowchart of the sample making and testing procedure is shown in the Figure 1.

The detailed description and parameters of the various steps shown in the flowchart are described in the following subsections.

2.1. Adhesives

The adhesive used in the study was an epoxy adhesive based on bisphenol A epoxy resin with epichlorohydrin, cured with a polyamide curing agent and modified with three fillers: aluminosilicate modified with quaternary ammonium salt, calcium carbonate and particulate activated carbon.

Epoxy resin used in this study with the trade name Epidian 5 epoxy resin (CIECH S.A., Sarzyna, Poland) is a pure form of epoxy resin, which is a product of the reaction of bisphenol A with epichlorohydrin [38,50]. Adhesives prepared on the basis of this resin are used in metal bonding, in building structures as anti-corrosion and electro-insulating coatings. It is characterized by good dielectric and mechanical properties, minimal contraction during curing and high chemical resistance. The epoxy number of this resin is 0.48–0.52 mol/100 g, the viscosity measured at 25 °C is in the range 20,000–30,000 mPa·s and the density measured at 20 °C is 1.16 g/cm^3 [51,52].

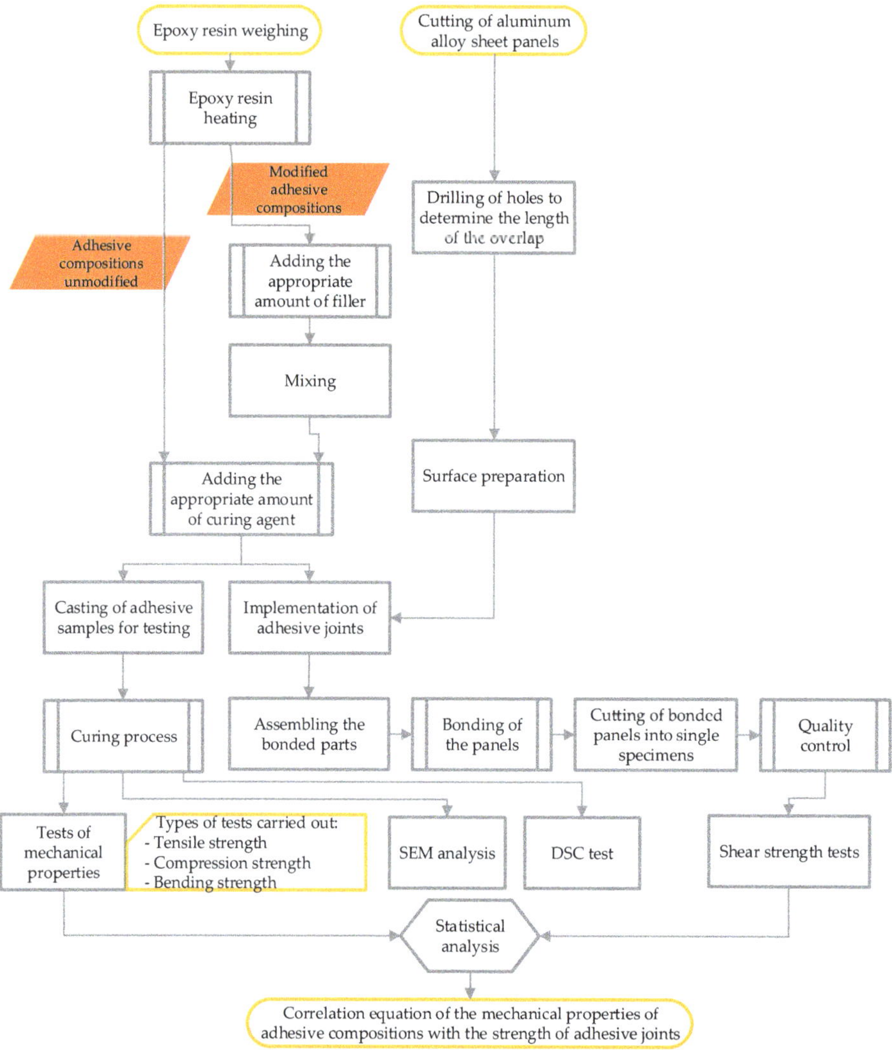

Figure 1. Flowchart of the sample making and testing procedure.

The curing agent used in the study was a polyamide curing agent with the trade name PAC curing agent (CIECH S.A., Sarzyna, Poland) consisting of fatty acids, C18-unsaturated, dimers, polymeric reaction products with triethylenetetramine. This curing agent increases the elasticity and impact strength of the composition. It belongs to the group of slow-reacting curing agents and is, therefore, an excellent component for curing modified adhesive compositions. The amine number of this curing agent is 290–360 mg KOH/g, the viscosity measured at 25 °C is in the range 10,000–25,000 mPa·s and the density measured at 20 °C is 1.10–1.20 g/cm^3 [52,53]. The curing agent in the compositions used in this study was added at a stoichiometric ratio of 80 parts by weight per 100 parts by weight of resin.

Three types of fillers were used in the study. The choice of fillers was determined by the wide range of application possibilities. The first filler used was a filler with a high degree of fineness (i.e., a filler with a particle size on the micro and nano scale) with the trade name ZR2 NanoBent (Zakłady Górniczo-Metalowe "Zębiec" S.A., Zębiec, Poland).

NanoBent ZR2 is an aluminosilicate modified with quaternary ammonium salt. It can be used as a dual-action additive: thixotropic and biocidal. The bulk density of ZR2 filler is less than 5×10^6 g/m^3. The ZR2 filler was 1 part by weight per 100 parts by weight of epoxy resin.

The second filler used in the study was calcium carbonate $CaCO_3$ in powder form (Zakłady Przemysłu Wapienniczego Trzuskawica S.A., Siatkówka, Poland). The $CaCO_3$ filler used in the study is free of any chemical structures associated with explosive properties, does not contain excess oxygen or any structural group tending to react exothermically with combustible material and is, therefore, classified as a non-explosive material without oxidizing properties. The bulk weight of $CaCO_3$ is $(0.7–1.4) \times 10^6$ g/m^3. The $CaCO_3$ filler was 5 parts by weight per 100 parts by weight of epoxy resin.

The third filler used in the study was CWZ-22 activated carbon (PPH STANLAB SP. Z O.O., Lublin, Poland) in particulate form with a molar mass of 12.01 g/mol. Due to its properties, CWZ-22 is used as a catalyst and solid support for other catalysts as a component of gas scavengers and as a material to achieve large capacities in supercapacitors. The presence of carbon as a powder filler in a cured polymer matrix can significantly alter not only its thermal properties, but also its strength properties. Therefore, CWZ-22 is very widely used in many industries. The bulk weight of CWZ-22 is approximately 4×10^5 g/m^3. The CWZ-22 filler was 20 parts by weight per 100 parts by weight of epoxy resin.

For easier identification of the adhesives, in this paper the designations used are presented in Table 1.

Table 1. Designation of adhesive compositions used in the tests.

Epoxy Resin	Curing Agent	Filler	Amount of Filler (by Weight of Resin)	Designation of the Epoxy Composition
		-	-	E5/PAC/100:80
Epidian 5	PAC	ZR2	1%	E5/PAC/ZR2/100:80:1
(100 g)	(80 g)	CaCO$_3$	5%	E5/PAC/CaCO$_3$/100:80:5
		CWZ-22	20%	E57/PAC/CWZ-22/100:80:20

The amounts of fillers used were selected on the basis of own experimental studies and a review of the literature [29,34,54–61]. The adhesive compositions were prepared immediately before use. In order to achieve a proper mixing of the components of the adhesive compositions used, the mixing stage was carried out in several steps:

1. The components of the mixtures were carefully weighed using a KERN CKE 3600-2 laboratory scale (Kern, Albstadt, Germany).
2. Heating of the epoxy resin using an electric heater—DEPILUX 400 (Activ, Wroclaw, Poland) to 50 °C in order to reduce its viscosity. The temperature of the heated resin was controlled using an electric thermometer (Amarell Electronic, Kreuzwertheim, Germany).
3. Addition of accurately weighed quantity of filler (for modified compositions).
4. Mechanical mixing using a Güde GTB 16/5 A mixer (Güde, Wolpertshausen, Germany) with a turbine dispersing disc mixer at 1170 rpm for 2 min with simultaneous venting to remove gas bubbles formed during mixing.
5. Addition of an accurately weighed quantity of curing agent.
6. Mechanical mixing carried out according to the technology described in step 4.

The mixing process was carried out under laboratory conditions at a temperature of 23 °C $\pm$ 2 °C and a relative humidity of 23% $\pm$ 3%. The adhesive compositions were used to prepare samples for testing the properties of the adhesive in the cured state and to prepare adhesive joints.

2.2. Adherend

The material used for the adhesive joints under study was the EN AW 2024 T3 aluminum alloy. This alloy has lower corrosion resistance and lower weldability compared to other aluminum alloys, but contains a high amount of copper and has very high strength—compared to, for example, AW 2014 alloy and high fatigue strength—which is why it is often used in aviation. The chemical composition of the alloy used is shown in Table 2. The mechanical properties of the used adherend are presented in Table 3.

Table 2. Chemical composition of EN AW 2024 T3 aluminum alloy [62].

The Element	Si	Fe	Cu	Mn	Mg	Cr	Zn	Ti	Al
Contents, %	0.1671	0.2153	4.0975	0.4281	1.4405	0.0053	0.0154	0.0191	93.5699

Table 3. Mechanical properties of EN AW 2024 T3 aluminum alloy [62].

Mechanical Properties	Values
Tensile strength	447.2 MPa
Yield strength	302.5 MPa
Elongation	16.5%
Hardness	123 HB
Thermal conductivity	170 W/mK
Density	2.78 g/cm^3

- Yield strength—302.5 MPa;
- elongation—16.5%;
- hardness—123 HB;
- thermal conductivity—170 W/mK;
- thermal conductivity—2.78 g/cm^3.

2.3. Adhesive Test Samples

All the samples used in the testing of the adhesive compositions' properties were obtained in a casting process using specially prepared molds with shapes and dimensions that corresponded to the required geometry of the samples. The mechanical properties of the adhesive compositions were tested in tensile, compression and bending tests.

For the tensile tests, dump-bell type 1B samples were used, in accordance with PN EN ISO 527-2 standard [63]. The dimensions of the samples used are shown in Figure 2.

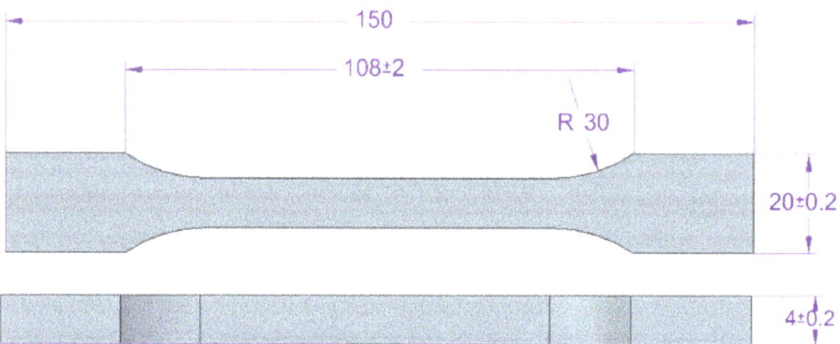

Figure 2. Shape and dimensions of the dump-bell type 1B sample of adhesive compositions for the tensile strength tests (all units are in millimeters).

Tensile testing of the adhesive compositions was carried out on a Zwick Roell Z150 testing machine (Zwick/Roell, Wroclaw, Poland), in accordance with PN EN ISO 527-1 standard [63]. The crosshead speed during the test was 5 mm/min. The initial tensile force was 30 N.

Cylindrical samples were used for compressive strength testing, with a height-to-base ratio (3:1) [64]. Special care was taken during sample preparation to ensure that the bases of the samples were perpendicular to the direction of force application and parallel to each other. The dimensions of the cylindrical samples used are shown in Figure 3.

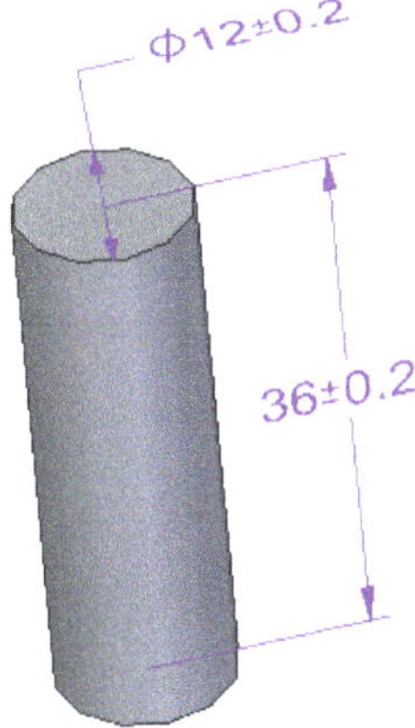

Figure 3. Shape and dimensions of the cylindrical sample of adhesive compositions for the compressive strength tests (all units are in millimeters).

Compressive strength tests of the adhesive compositions were also carried out on a Zwick Roell Z150 testing machine (Zwick/Roell, Wroclaw, Poland). These tests were carried out in accordance with ISO 604 standard [65]. The assumed crosshead traverse speed during the test was 10 mm/min. The pre-test force was 20 N.

For bending strength testing, beam-shaped samples were used. The dimensions of the bending strength test specimens are present in Figure 4.

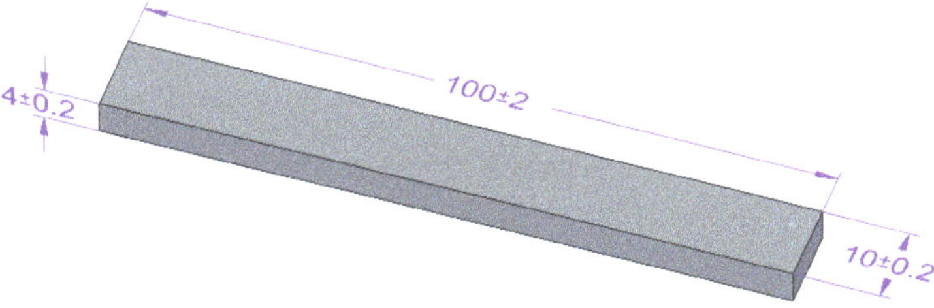

Figure 4. Shape and dimensions of the beam sample of adhesive compositions for the bending strength tests, in accordance with ISO 178:2003 standard [66] (all units are in millimeters).

According to the specifications [66], the dimensions used of $100 \times 10 \times 4$ mm are suitable for the bending strength test in the three-point bending test, since the support spacing was 80 mm and, in addition, the recommended proportions for powder-filled enriched plastics: 10 mm $\leq$ b $\leq$ 25 mm and l $\geq$ 20 h were also followed.

Bending strength tests were carried out on a Zwick Roell Z2.5 testing machine (Zwick/Roell, Wroclaw, Poland), according to DIN-EN ISO 178 standard [66]. The test speed was 10 mm/min, and the initial test force was 5 N.

A series of 10 specimens was prepared for each adhesive composition.

In addition to the strength tests, the scanning electron microscopy (SEM) and differential scanning calorimetry (DSC) analyses were also carried out.

For SEM microscopic tests of the adhesive compositions, beam samples measuring $100 \times 10 \times 4$ mm were used. The breakthroughs of the samples obtained by the percussion method were studied, which were then sputtered with gold using a Quorum Q150R ES—Spreading Deposition Rate (Quorum, Laughton, UK). The results of scanning electron microscopy (SEM) analysis were performed on the Tescan MIRA3 microscope (Tescan Orsay Holding, Brno-Kohoutovice, Czech Republic).

Samples of the adhesive compositions in the cured state for DSC differential scanning calorimetry analysis, weighing 6–12 mg, were taken from $100 \times 10 \times 4$ mm beams. The analysis of modified epoxy composition using differential scanning calorimetry was carried out in accordance with EN ISO 11357-1 standard [67]. Special calibration files and samples made of light metals, such as indium, zinc, tin and bismud $D > T$, were used to calibrate the temperature scale. The tests were carried out in the air atmosphere in the temperature range from 20 °C to 220 °C, with a heating rate of 10 K/min and a cooling rate of 5 K/min. The tests were carried out using a DSC Phox 200 P instrument (NETZSCH, Selb, Germany). SEM tests were carried out at an accelerating voltage of 5.0 kV, an SE secondary electron detector was used to image the adhesive compositions and the samples were sputtered with gold for 10 min.

All samples were cured and seasoned under laboratory conditions identical to those of mixing for 7 days. The technological conditions for curing process have been selected on the basis of manufacturers' guidelines and research literature [68,69].

2.4. Adhesive Joint Test Specimens

The adhesive joints used in the study were made in accordance with the requirements of ASTM D1002 standard [70]. The preparation plan for the adhesive joints according to the guidelines included several steps:

1. Cutting of aluminum sheets.
   The components to be bonded were cut from EN AW 2024 T3 sheet metal with a thickness of 2.00 ± 0.12 mm to a dimension of 101.60 ± 0.25 mm $\times$ 177.80 ± 3.17 mm. Cutting was carried out in a hydroabrasive cutting process using a Waterjet Eckert Combo portal cutting machine (Eckert AS Sp. z o.o., Legnica, Poland). Cutting process speed—200 mm/min, water pressure—3500–105 Pa, nozzle distance from the material being cut—3 mm, abrasive flow rate during the cutting process—approx. 0.4 kg/min and abrasive material—Garnet sand of mesh 80 granulation.

2. Drilling of holes to determine the length of the overlap.
   Two ϕ2.5-mm holes were drilled in each cut sheet for ground fixing pins, which enabled the panels to be assembled in a defined geometry in the next stage, while maintaining a constant overlap length of 12.7 ± 0.25 mm, as specified by the guidelines.

3. Surface preparation of the plates to be bonded.

   The surface of the plates was prepared for bonding by sandblasting and degreasing with acetone. Sandblasting was carried out on a cabin sandblaster (Cormak, Siedlce, Poland) using Garnet abrasive granulation mesh 80 (Garnet Poland, Elbląg, Poland) with the following parameters:
   - Distance of the nozzle from the sample—$h = 97$ mm;
   - sandblasting speed—$V = 53$ mm/min;
   - pressure—$P = 5{-}10^5$ Pa;
   - angle between specimen and direction of jet—$90°$;
   - number of specimen displacements—2.

   Sandblasted parts were subjected to degreasing with technical acetone in a bath for 20 min and wiped with clean cleaning cloth, followed by a second washing and drying—self-drying in 10 min.

4. Assembling the panels.

    After measuring the length of the overlap on the sheets to be bonded, the surface was secured with self-adhesive Teflon tape, which made it easier to remove excessive adhesive bleed. The adhesive was applied in a thin, homogeneous layer to the surfaces of the parts to be bonded using a spatula over the target length of the overlap.

5. Bonding of the panels.

    The assembled components were bonded using the vacuum bag method with a constant pressure of $0.6\text{---}10^5$ Pa, which was realized using an SVAGG vacuum pump (Schunk, Lauffen/Neckar, Germany). The adhesive joints were cured for 7 days under laboratory conditions similar to the adhesive mixing.

6. Cutting of bonded panels into single specimens.

    Cutting of the panels into specimens for testing the properties of the adhesive joints was carried out on a Waterjet Eckert Combo machine. The cut single-lap specimens were 25.4 ± 0.25 mm $\times$ 190.50 ± 0.25 mm.

7. Quality control.

    The cut-out single specimens of the adhesive joints were visually controlled to ensure that the bonding was correct and that excessive adhesive bleed was removed. The required dimensions were also checked. The required measurement accuracy and tolerances of the dimensions were obtained. In Figure 5, the dimensions of the tested adhesive joints are shown. The average thickness of the adhesive joint was 0.100 mm ± 0.025 mm.

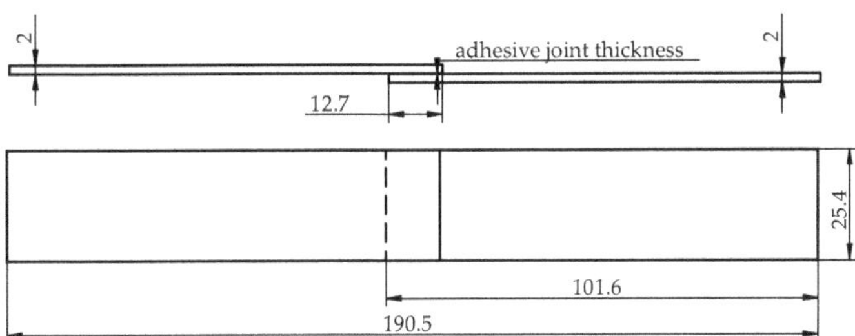

Figure 5. Geometry and dimensions of the single-lap adhesive joint used in the tests performed in accordance with ASTM D1002 standard (all units are in millimeters).

The single-lap adhesive joints were subjected to strength tests on a Zwick Roell Z150 testing machine (Zwick/Roell, Wroclaw, Poland), in accordance with ASTM D1002 standard [70]. The crosshead speed during the test was 1.5 mm/min with an initial force of 5 N. The shear tensile strength was determined. Ten adhesive joint specimens were made for each adhesive composition.

3. Results

3.1. Strength Properties Test Results of Adhesive Compositions

To determine the effect of the filler on selected properties of the adhesive compositions, their properties were analyzed and compared with those of unmodified adhesives. The obtained test results are presented in Tables 4, 6 and 8. Statistical analysis of the obtained results was additionally carried out for a more accurate analysis and interpretation; the results of which are presented in Tables 5, 7 and 9. The statistical analysis presented in this paper was performed using Statistica 13.1 software.

Table 4 shows the results of the tensile strength test of the adhesive compositions.

Table 4. Tensile strength of adhesive compositions.

Epoxy Adhesive Composition	Tensile Strength σ_m [MPa]				
	Average	Median	Quartile Range	Variation	Standard Deviation
E5/PAC/100:80	53.86	54.44	2.38	9.61	3.10
E5/PAC/ZR2/100:80:1	55.70	55.79	2.61	3.61	1.90
E5/PAC/CaCO$_3$/100:80:5	54.91	54.88	1.92	4.47	2.11
E5/PAC/CWZ-22/100:80:20	46.79	46.35	0.85	0.97	0.99

Analyzing the tensile strength results of the tested adhesive compositions, it can be seen that the physical modification had a positive effect on the results obtained for two modified compositions. The highest average tensile strengths were obtained for E5/PAC/ZR2/100:80:1 adhesive composition—55.70 MPa. A lower strength of only 1.36% was obtained for E5/PAC/CaCO$_3$/100:80:5 compositions—54.94 MPa. The reference (unmodified) E5/PAC/100:80 composition had an average strength of 53.86 MPa. The lowest strength was obtained for the E5/PAC/CWZ-22/100:80:20 composition—46.79 MPa—but in this case the repeatability of the results was at the highest level. The lowest repeatability of results was obtained with the reference E5/PAC/100:80 composition. However, in order to clearly assess the results obtained from the study, a more thorough statistical analysis was carried out. The normality of distribution and homogeneity of variance assumption was met, so a post hoc parametric test of significant differences was performed. The results of this test are shown in Table 5.

Table 5. Significant difference in average tensile strength test results.

Epoxy Adhesive Composition		Tukey's HSD Test for Average Values of Tensile Strength σ_m [MPa] at $\alpha = 0.05$			
		{1} 53.86	{2} 55.70	{3} 54.91	{4} 46.79
E5/PAC/100:80	{1}		0.546	0.867	0.000
E5/PAC/ZR2/100:80:1	{2}	0.546		0.935	0.000
E5/PAC/CaCO$_3$/100:80:5	{3}	0.867	0.935		0.000
E5/PAC/CWZ-22/100:80:20	{4}	0.000	0.000	0.000	

Based on the statistical analysis of the adhesive compositions' tensile strength results, it can be seen that no statistically significant differences are found for compositions E5/PAC/100:80, E5/PAC/ZR2/100:80:1 and E5/PAC/CaCO$_3$. At the assumed significance level $\alpha = 0.05$, a statistically different average tensile strength was obtained for the E5/PAC/CWZ-22/100:80:20 composition. In order to clearly assess the results obtained from the study, a more thorough statistical analysis was carried out. The normality of distribution and homogeneity of variance assumption was met, so a post hoc parametric test of significant differences was performed. The results of this test are shown in Table 5.

Table 6 summarizes the compression strength test results.

Analyzing the compression test results, the positive effect of physical modification of the tested adhesive compositions can be seen. The highest average compression strength was obtained for the E5/PAC/CWZ-22/100:80:20—81.43 MPa. A lower average compression strength was obtained for the E5/PAC/CaCO$_3$/100:80:5—77.04 MPa. A 1.13% lower average compression strength was obtained for the E5/PAC/ZR2/100:80:1—76.17 MPa. However, the results for samples of this composition had the highest repeatability, with a standard deviation of 0.98%. The lowest compression strength was obtained for the reference E5/PAC/100:80 composition—71.72 MPa—as well as the lowest repeatability of results—a standard deviation of 5.44%. The assumption of normality of distribution and homogeneity of variance was met, so the parametric Tukey statistical test was used to determine significant differences at the assumed significance level of $\alpha = 0.05$. The results of this test are presented in Table 7.

Table 6. Compression strength of adhesive compositions.

Epoxy Adhesive Composition	Compression Strength σ_c [MPa]				
	Average	Median	Quartile Range	Variation	Standard Deviation
E5/PAC/100:80	71.72	73.79	3.09	15.24	3.90
E5/PAC/ZR2/100:80:1	76.17	76.41	1.18	0.56	0.75
E5/PAC/CaCO$_3$/100:80:5	77.04	77.41	1.28	0.90	0.95
E5/PAC/CWZ-22/100:80:20	81.43	80.88	2.28	1.67	1.29

Table 7. Significant difference in average compression strength test results.

Epoxy Adhesive Composition	Tukey's HSD Test for Average Values of Compression Strength σc [MPa] at $\alpha = 0.05$			
	{1} 71.72	{2} 76.17	{3} 77.04	{4} 81.43
E5/PAC/100:80 {1}		0.021	0.006	0.000
E5/PAC/ZR2/100:80:1 {2}	0.021		0.916	0.023
E5/PAC/CaCO$_3$/100:80:5 {3}	0.006	0.916		0.006
E5/PAC/CWZ-22/100:80:20 {4}	0.000	0.006	0.023	

Statistical analysis of the compression strength results of the adhesive compositions showed significant differences between the reference E5/PAC/100:80 composition and the others, and the E5/PAC/CWZ-22/100:80:20 composition and the others. No significant differences were observed for the E5/PAC/ZR2/100:80:1 composition and E5/PAC/CaCO$_3$/100:80:5 composition. Thus, it can be concluded that modification with these two fillers causes a similar effect in terms of the compression strength of the compositions, while the greatest effect of the modification on the compression strength of the adhesive compositions was observed for the E5/PAC/CWZ-22/100:80:20 composition.

Table 8 shows the bending strength results obtained for the analyzed adhesive compositions.

Table 8. Bending strength of adhesive compositions.

Epoxy Adhesive Composition	Bending Strength σ_f [MPa]				
	Average	Median	Quartile Range	Variation	Standard Deviation
E5/PAC/100:80	81.92	81.73	3.97	4.01	2.00
E5/PAC/ZR2/100:80:1	74.47	74.29	2.45	1.60	1.26
E5/PAC/CaCO$_3$/100:80:5	79.29	79.64	1.10	0.70	0.84
E5/PAC/CWZ-22/100:80:20	74.35	75.34	4.14	12.52	3.54

Based on tests results, it can be seen that the highest bending strength was obtained for the E5/PAC/100:80 reference composition. E5/PAC/CaCO$_3$/100:80:5 epoxy composition had a slightly lower bending strength by 3.2% of the average bending strength—79.29 MPa. The lowest failure strength in the three-point bending test was obtained for E5/PAC/ZR2/100:80:1 composition and E5/PAC/CWZ-22/100:80:20 composition. These compositions achieved strengths of 74.47 MPa and 74.35 MPa, respectively. In addition, the E5/PAC/CWZ-22/100:80:20 composition had the largest statistical dispersion of results, with a standard deviation of 4.76%. In order to further verify the differences between the results for each group, statistical analysis was carried out. The assumption that the distribution of test results conformed to a normal distribution was met. Likewise, the assumption of homogeneity of variance was verified with the appropriate statistical test—Levene's test. Therefore, the Tukey post hoc test was used for further analysis.

The results of this test are presented in Table 9.

Table 9. Significant difference in average bending strength test results.

Epoxy Adhesive Composition		Tukey's HSD Test for Average Values of Bending Strength σ_f [MPa] at $\alpha = 0.05$			
		{1} **81.92**	{2} **74.47**	{3} **79.29**	{4} **74.35**
E5/PAC/100:80	{1}		0.000	0.259	0.000
E5/PAC/ZR2/100:80:1	{2}	0.000		0.013	0.999
E5/PAC/CaCO$_3$/100:80:5	{3}	0.259	0.013		0.011
E5/PAC/CWZ-22/100:80:20	{4}	0.000	0.999	0.011	

Statistical analysis of the bending strength test results showed that, at an assumed significance level of $\alpha = 0.05$, there were no significant differences between the reference E5/PAC/100:80 composition and the E5/PAC/CaCO$_3$/100:80:5 composition, and between the E5/PAC/ZR2/100:80:1 composition and E5/PAC/CWZ-22/100:80:20 composition.

3.2. SEM Analysis

From the research results of Fu et al. [42], it was concluded that the mechanical properties of particulate composites depend on the appropriate type of filler, on the interfacial interaction between matrix and filler, on the size of the particles used, on their distribution in the composite system and on their concentration of course.

The structure of the analyzed adhesive compositions in the cured state was studied by scanning electron microscopy (SEM). Figures 6–9 show SEM images of the compositions studied.

SEM images in Figure 6 show that the unmodified E5/PAC/100:80 adhesive composition is characterized by a homogeneous, solid structure. Few gas bubbles are visible on the surface. In the case of the reference composition, the breakthrough is mild and malleable.

Analyzing the SEM images shown in Figure 7, a strong interaction of the filler with the matrix, i.e., the epoxy resin, can be seen. This shows that there is strong wettability at the interfacial surface of the filler and matrix in the structure of the composition. A large variation in the particle size of the filler can be observed, which consists of particles distinguished by an irregular, lamellar shape.

Analyzing the SEM images in Figure 8, it is possible to observe good wettability at the filler–matrix interface, which is due to the filler–matrix interaction seen in Figure 8d. An uneven distribution of the filler in the matrix can also be observed, which is a typical phenomenon for compositions subjected to physical modification with molecular fillers.

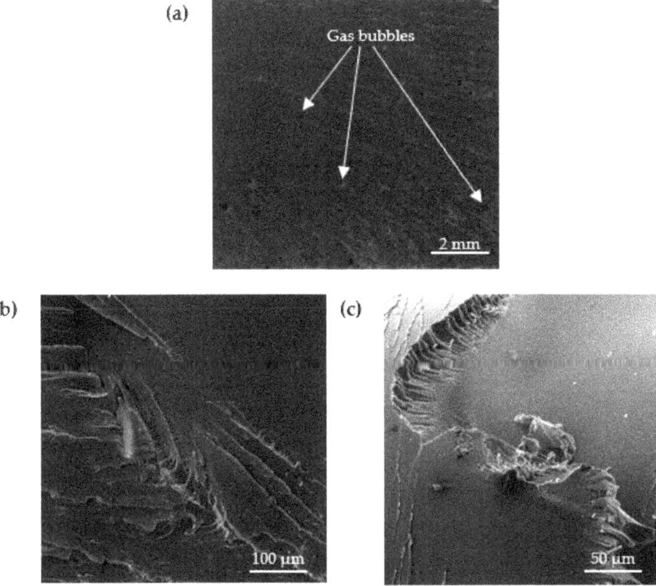

Figure 6. SEM images of fracture surface of E5/PAC/100:80 adhesive composition: (**a**) 9.78-mm view field, (**b**) 549-μm view field and (**c**) 241-μm view field.

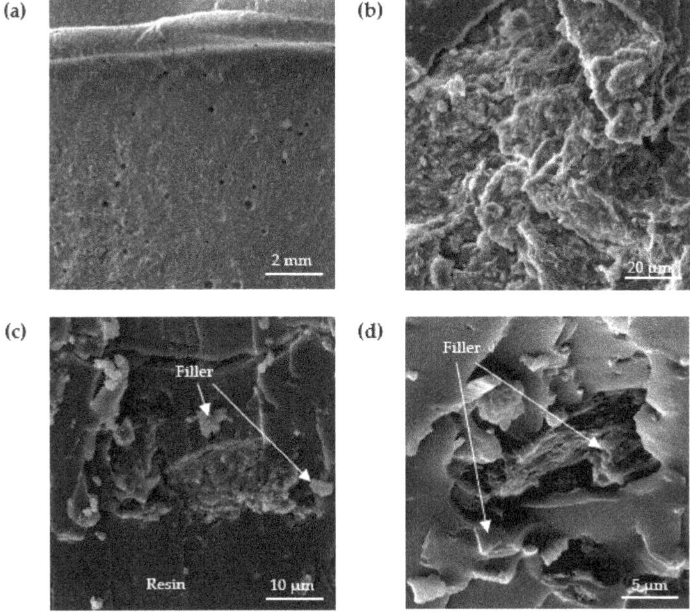

Figure 7. SEM images of fracture surface of E5/PAC/ZR2/100:80:1 adhesive composition: (**a**) 10.6-mm view field, (**b**) 69.1-μm view field, (**c**) 55.4-μm view field and (**d**) 23.1-μm view field.

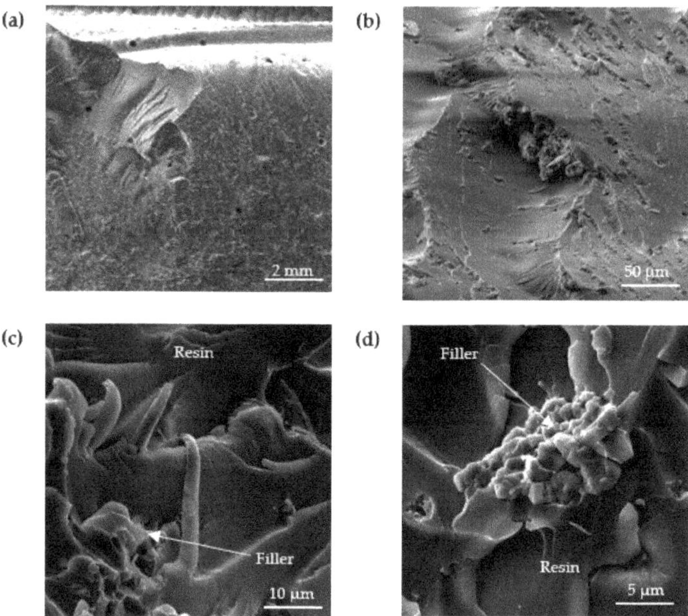

Figure 8. SEM images of fracture surface of E5/PAC/CaCO$_3$/100:80:5 adhesive composition: (a) 10.5-mm view field, (b) 216-μm view field, (c) 43.9-μm view field and (d) 27.7-μm view field.

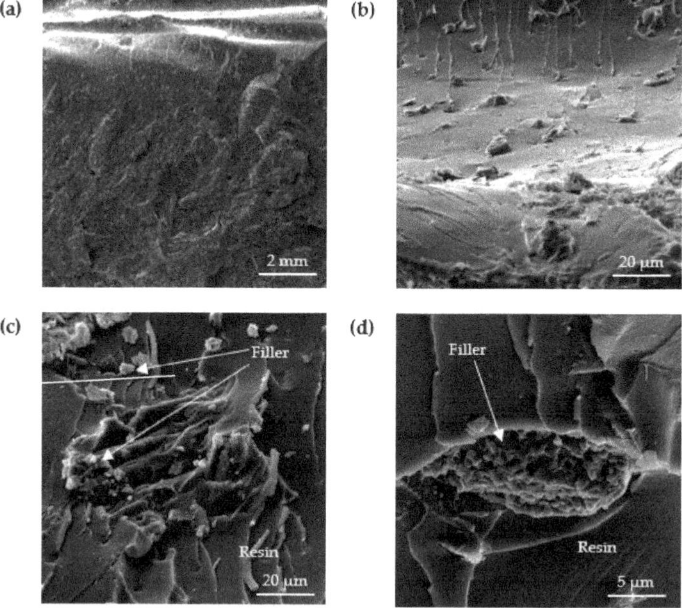

Figure 9. SEM images of fracture surface of E5/PAC/CWZ-22/100:80:20 adhesive composition: (a) 10.7-mm view field, (b) 153-μm view field, (c) 69.3-μm view field and (d) 27.7-μm view field.

Figure 9 shows SEM images of E5/PAC/CWZ-22/100:80:20 composition. It can be seen that the impact test, after which the obtained breakthroughs were subjected to SEM analyses, resulted in the delamination of the dusty part of the carbon filler. In Figure 9c, a detailed image of the interaction between filler and matrix at the interfacial interface can be observed, from which it can be inferred that the filler is well wettable in the resin. It can also be seen from Figure 9d that there is delamination of the filler in the matrix.

3.3. Compositions Physical Properties Test Results

Differential scanning calorimetry (DSC) was carried out to determine the effect of the modification on the physical properties of the adhesive composition, determined by varying the temperature (when the sample is heated and cooled at a specific rate). Differential scanning calorimetry diagrams are presented in Figures 10–13.

The presented diagrams show the process of physicochemical changes in the test compositions under the influence of imposed temperature changes. The green curve in the DSC diagrams indicates the first heating range, the blue curve the cooling range and the purple curve the second heating range. Two heats of the system with the sample are carried out during the test. The first heating (indicated by the green line on the graphs) characterizes the melting effect, while the second heating (indicated by the purple line on the graphs) characterizes the glass transition effect. The test allows the glass transition temperature of the material to be determined, the method of which is shown graphically in Figure 13 (ΔCp).

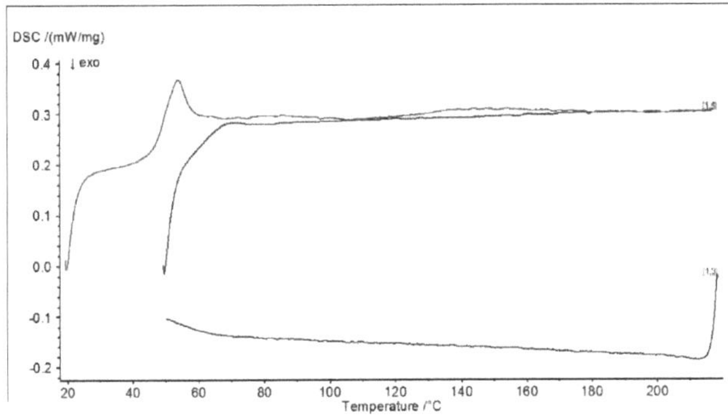

Figure 10. DSC diagram for E5/PAC/100:80 composition.

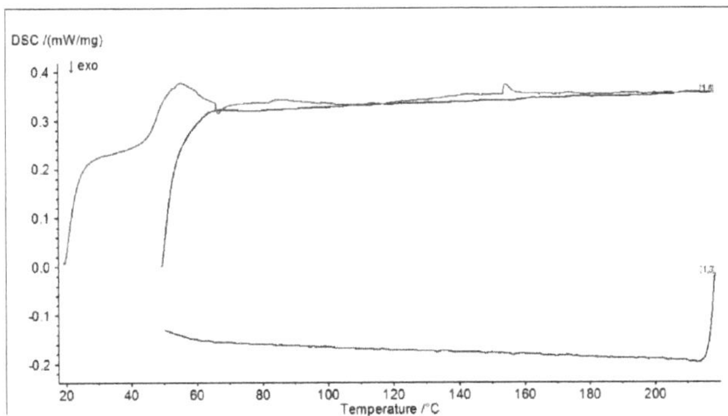

Figure 11. DSC diagram for E5/PAC/ZR2/100:80:1 composition.

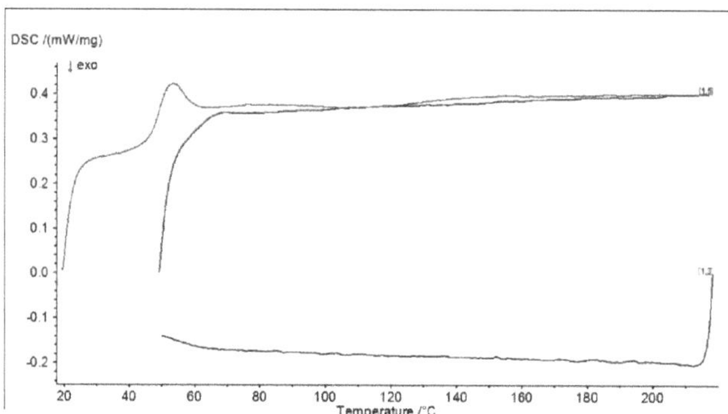

Figure 12. DSC diagram for E5/PAC/CaCO$_3$/100:80:5 composition.

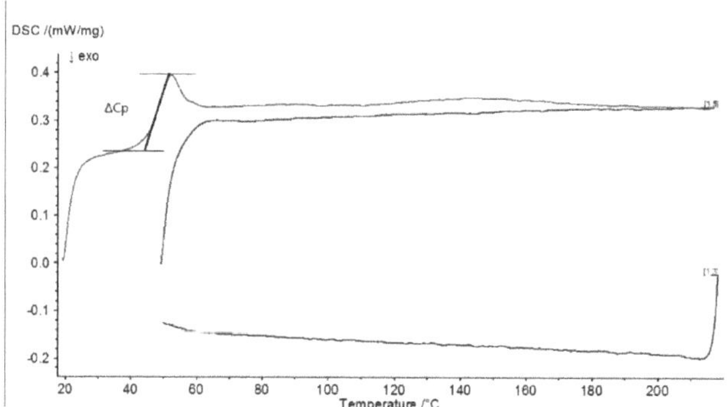

Figure 13. DSC diagram for E5/PAC/CWZ-22/100:80:20 composition.

In the DSC studies carried out, the characteristic temperatures of the modified adhesive compositions were determined. The glass transition temperature is one of the most important quantities characterizing the plastic properties of polymers. In the DSC curves presented, it is possible to distinguish sections of the so-called baseline, which are shifted parallel to the temperature axis. These mark the temperature intervals in which no heat release or absorption processes take place in the sample. When a reaction or phase transition occurs, the baseline changes to a peak—part of the curve deviates from the baseline and then returns to it. A distinction is made between an exothermic peak, when the temperature of the test sample is below the reference sample, and an endothermic peak, when the temperature of the test sample rises above the reference sample. In the case of an exothermic peak, heat must be supplied to the test sample (downwards peak), while in the case of an endothermic peak, the situation is reversed—heat is removed by the circuit, with the peak pointing upwards.

In the case of the reference E5/PAC/100:80 composition, a characteristic exothermic peak at the beginning of the DSC curves can be observed in both I heating and II heating. For the purple scale, II heating is shifted due to the fact that the temperature of the start of heating was higher, so the endothermic peak from I heating, which occurred at about 46 °C, is barely noticeable in II heating at 65 °C, indicating that the heat flux was quickly balanced. In addition, a second endothermic peak at around 120 °C can be observed in the first heating. For all adhesive compositions, characteristic endothermic peaks, i.e., related to heat release, can be observed. For the E5/PAC/ZR2/100:80:1 composition, an endothermic peak can be observed at 43–58 °C (which may indicate combustion or crystallization of filler molecules), followed by a small exothermic peak at 63 °C, and then a second peak can be observed, also endothermic at about 153 °C, which may indicate evaporation of the curing agent or other substance. Its formation may be related to the effect of the montmorillonite introduced into the resin. In the case of the E5/PAC/CaCO$_3$/100:80:5 composition and E5/PAC/CWZ-22/100:80:20 composition, the first endothermic peak can also be observed at a temperature of about 43 °C and the second at about 120 °C.

The shape of the peak indicates the transformation taking place is also important. A sharp peak indicates that the transformation is taking place at a constant temperature; a fuzzy peak characterizes a transformation taking place over a certain temperature range.

3.4. Shear Strength of Single-Lap Adhesive Joints Test Results

The aim of this study was also to determine the effect of modifying the epoxy adhesive composition on selected mechanical properties of EN AW 2024 T3 aluminum alloy sheet adhesive joints. The results of the adhesive joints' shear strength test are shown in Figure 14.

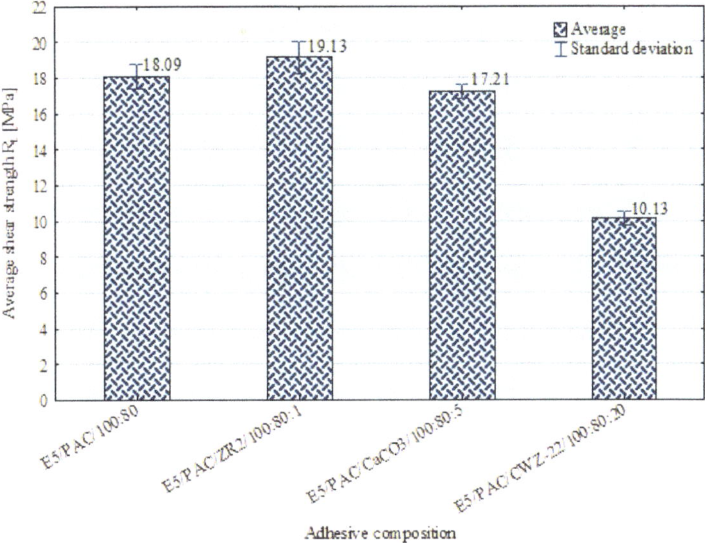

Figure 14. Shear strength (average values) of EN AW 2024 T3 aluminum alloy sheets single-lap adhesive joints.

Analyzing the obtained results of the shear strengths of single-lap adhesive joints, it can be seen that the highest average strength values, above 19 MPa, were obtained for joints made with a E5/PAC/ZR2/100:80:1 composition. The lowest, almost twice lower, strengths were characterized by joints made with E5/PAC/CWZ-22/100:80:20 composition—10.13 MPa.

One of the problems considered in the study was whether the filler introduced into the adhesive composition increased the strength of the adhesive joints, so to be able to make an unambiguous assessment, the results obtained were subjected to more thorough statistical analysis. In the statistical analysis, a significance test was used to compare the average values of the test characteristics. As a first step, the Shapiro–Wilk W-test was used to check whether the distribution of the results in the separate groups follows a normal distribution. The results of this test are presented in Table 10.

Table 10. Shapiro–Wilk W-test results for single-lap adhesive joints of EN AW 2024 T3 aluminum alloy sheets in groups.

Adhesive Composition	The Value of the W Statistic Shapiro–Wilk	The *p*-Value for the Shapiro–Wilk W-Test
E5/PAC/100:80	0.943	0.587
E5/PAC/ZR2/100:80:1	0.892	0.180
E5/PAC/CaCO$_3$/100:80:5	0.813	0.021
E5/PAC/CWZ-22/100:80:20	0.859	0.074

Based on the results obtained, it can be seen that the condition of normality of distribution in all groups was not met ($p < 0.05$). Therefore, a non-parametric test for comparing multiple independent samples was used in further analysis. The Kruskal–Wallis test and the median test were applied. Assuming a significance level of $\alpha = 0.05$, it was checked whether the average shear strength values of the adhesive joints for the different compositions did not differ significantly. The results of these tests are shown in Tables 11 and 12.

Table 11. Results of non-parametric ANOVA rank Kruskal–Wallis test of the shear strength of single-lap adhesive joints of EN AW 2024 T3 aluminum alloy sheets divided according to the adhesive composition used.

Dependent: Average Shear Strength R_t [MPa]	Kruskal–Wallis Rank ANOVA; Average Shear Strength R_t [MPa] The Kruskal–Wallis Test: H (3, N = 40) = 32.34732 p = 0.0000		
	N Significant	Sum of Rank	Average Rank
E5/PAC/100:80	10	263.00	26.30
E5/PAC/ZR2/100:80:1	10	167.00	16.70
E5/PAC/CaCO$_3$/100:80:5	10	55.00	5.50
E5/PAC/CWZ-22/100:80:20	10	335.00	33.50

Table 12. Test results of the median shear strength of single-lap adhesive joints of EN AW 2024 T3 aluminum alloy sheets, divided according to the adhesive composition used.

Dependent: Average Shear Strength R_t [MPa]	Median Test, Overall Median = 17.5759; Average Tensile Shear Strength R_t [MPa] Chi-Square = 23.20000 df = 3 p = 0.0000				
	E5/PAC/100:80	E5/PAC/ZR2/100:80:1	E5/PAC/CaCO$_3$/100:80:5	E5/PAC/CWZ-22/100:80:20	Total
≤medians: observ.	3.00	0.00	7.00	10.00	20.00
expected	5.00	5.00	5.00	5.00	
obs.-exp.	−2.00	−5.00	2.00	5.00	
>medians: observ.	7.00	10.00	3.00	0.00	20.00
expected	5.00	5.00	5.00	5.00	
obs.-exp.	2.00	5.00	−2.00	−5.00	
Total: observed	10.00	10.00	10.00	10.00	40.00

For the Kruskal–Wallis test, the calculated significance level is less than the assumed $\alpha = 0.05$, so it can be concluded that the obtained shear strength results differ significantly between groups. The median test can be interpreted similarly. Therefore, a test of multiple comparisons of average ranks for all trials was used to indicate where significant differences exist. The results of this test are shown in Table 13.

Based on the statistical analysis of the shear strength test results of adhesive joints, it can be concluded that significant differences exist between:

- E5/PAC/100:80 composition and E5/PAC/CWZ-22/100:80:20 composition;
- E5/PAC/ZR2/100:80:1 composition and E5/PAC/CaCO$_3$/100:80:5 composition;
- E5/PAC/ZR2/100:80:1 composition and E5/PAC/CWZ-22/100:80:20 composition.
- No significant differences, however, were observed between:
- E5/PAC/100:80 composition and E5/PAC/ZR2/100:80:1 composition;
- E5/PAC/100:80 composition and E5/PAC/CaCO$_3$/100:80:5 composition;
- E5/PAC/CaCO$_3$/100.80.5 composition and E5/PAC/CWZ-22/100:80:20 composition.

Analysis of the results, therefore, shows that the addition of filler in the form of montmorillonite ZR2 and calcium carbonate CaCO$_3$ in specific amounts in the adhesive composition used to make adhesive joints of EN AW 2024 T3 aluminum alloy sheets does not significantly alter the strength of the adhesive joints compared to the unmodified composition. The addition of activated carbon, on the other hand, significantly worsened the shear strength of the adhesive joints compared to the values determined for joints made with the reference composition. Considering the statistical analysis of the results of adhesive joints made with the modified compositions, it can be assumed that the most favorable results were obtained for adhesive joints made with E5/PAC/ZR2/100:80:1 composition.

Table 13. p-value for multiple comparisons test of average ranks for all shear strength tests of EN AW 2024 T3 aluminum alloy sheet adhesive joints.

Adhesive Composition	p-Value for Multiple Comparisons Kruskal–Wallis Test: H (3, N = 40) = 32.34732 p = 0.0000			
	E5/PAC/100:80	E5/PAC/ZR2/100:80:1	E5/PAC/CaCO$_3$/100:80:5	E5/PAC/CWZ-22/100:80:20
E5/PAC/100:80		1.000	0.397	0.000
E5/PAC/ZR2/100:80:1	1.000		0.007	0.000
E5/PAC/CaCO$_3$/100:80:5	0.397	0.007		0.193
E5/PAC/CWZ-22/100:80:20	0.000	0.000	0.193	

4. Discussion

Based on the test results presented above and analyzing the results of the statistical analyses, it can be seen that the addition of CWZ-22 activated carbon in the composition in the amount of 20% adversely affects the tensile strength of the adhesive composition made based on Epidian 5 resin and PAC curing agent. For compositions with 1% montmorillonite ZR2 and 5% CaCO$_3$ calcium carbonate filler, the modification resulted in an increase in tensile strength compared to the reference composition. The increase in the tensile strength values of the compositions and its change are very important, particularly for later structural joints, as the increase in the value of the longitudinal elastic coefficient is associated with an increase in the shear stresses in adhesive joints.

In compression strength tests, the addition of CWZ-22 activated carbon in the composition in the amount of 20% positively influences the compression strength of the adhesive composition made based on Epidian 5 resin and PAC curing agent. Similarly, in the case of compositions with 1% ZR2 montmorillonite and 5% $CaCO_3$ calcium carbonate filler, the physical modification of the composition resulted in an increase in compression strength compared to the reference composition. It can also be seen that the introduction of the modifying additives improved the repeatability of the results, as the lowest repeatability of the results was obtained for the reference composition E5/PAC/100:80, as in the case of tensile strength.

The results of the bending strength determined by the three-point bending test and their statistical analysis show that the physical modification does not have a beneficial effect on the properties of the adhesive compositions. Only for the composition with 5% $CaCO_3$ calcium carbonate filler the strength remained at a similar level. In addition, an additive of 20% activated carbon CWZ-22 in dust form reduced the repeatability of the test results. It can be concluded that the high filler content results in particle agglomeration, which, due to the lower contact surface area of the binder with the filler, contributes to a reduction in bonding capacity between filler and matrix. Therefore, this can affect significant changes in strength parameters in tests where the force is applied halfway along the sample length.

As the authors notice in their works [34,71], there is no known universal filler, the addition of which has only a positive effect on all the parameters of the adhesives evaluated according to the criterion adopted. In the opinion of Samal [72], the shape and size of the filler has a significant effect on the obtained properties of adhesive compositions. Kundie et al. [73] presents similar conclusions in their work, that the properties of modified compositions strongly depend on many factors, such as the type and mechanical properties of the fillers themselves, as well as the uniform dispersion of nanofillers in the polymer matrix. This can also be seen in the SEM microphotographs included in this paper.

For all modified adhesive compositions, there was a good interaction between the filler and the epoxy resin matrix, which is a basic assumption for the correct physical modification of the adhesive composition. The reason for this phenomenon can be seen in the relatively low viscosity, which was characterized by the matrix of the plastics, i.e., the epoxy resin Epidian 5. Lowering the viscosity of the resin was achieved by heating it to 50 °C during the preparation of the composition. According to Michels et al. [68], pre-exposure of epoxy to high temperatures accelerates curing and allows for a much faster development of strength and stiffness. This resulted in sufficient wetting of the filler in the epoxy resin, which facilitated the penetration of the resin particles between the filler particles. As also noted by authors such as Bittmann [74] in their paper, the wettability of filler surfaces plays an important role during physical modification. Matykiewicz [75] in her paper described the important role not only of the type of fillers, but also of the modification of the epoxy matrix to ensure good adhesion between all components in the composition to provide enhanced mechanical and thermomechanical properties of the hybrid composite.

The addition of modifying fillers reduces the amount of air bubbles in the structure of adhesive compositions in the cured state, which can affect the strength properties determined in static tests in which external forces are applied parallel to the specimen axis.

It can also be observed that the introduced fillers differ in particle shape: ZR2 montmorillonite has a lamellar structure, while $CaCO_3$ calcium carbonate and CWZ-22 activated carbon particles are spherical. The fillers introduced into the composition tend to form agglomerates. The reasons for this phenomenon may lie in the interaction between the individual filler particles and in the filler–resin interaction and may affect the properties of the adhesive compositions.

In this paper, differential calorimetry studies were also carried out. As described by Moussa et al. [76] in their paper, DSC provides a method to effectively determine the glass transition temperature dependence of the degree of cure for structural adhesives. From the DSC plots presented, a shift in the glass transition temperature (purple line) towards a lower temperature (by about 5 °C) can also be observed for compositions modified with ZR-2 montmorillonite and CWZ-22 activated carbon compared to the reference composition. In the case of the composition modified with $CaCO_3$ calcium carbonate, no changes were observed in the glass transition temperature region. In conclusion, it can be stated that the tested adhesive compositions are thermally stable, and in the case of the composition modified with 1% montmorillonite—E5/PAC/ZR2/100:80:1—there is a certain tendency for the curve to shift towards a higher temperature.

The results of the shear strength tests and their statistical analysis presented In this paper allow us to conclude that, for modified adhesive compositions intended for adhesive joints of aluminum alloy sheets, the type of filler used is important. In the case of the ZR2 NanoBent lamellar filler, a positive effect of the adhesive composition modification on the adhesive joint strength was observed. Considering the relatively low viscosity of the heated Epidian 5 resin, which is the matrix in the modified compositions, it can be concluded that adequate dispersion of the filler in the epoxy resin occurred, which consequently allows the penetration of resin particles between the filler particles. In the case of the composition modified with activated carbon, the strength of the joints deteriorated significantly compared to the other compositions. The reason for this may be too much modifier.

Due to the lack of a description in the literature to date of issues related to the correlation of the strength properties of the adhesive itself in relation to the strength of constituted adhesive joints, in the presented paper it was checked whether the strength of adhesive joints made with these compositions could be predicted from the properties of epoxy compositions in the cured state. An attempt was made to determine the correlation between the properties of the compositions in the cured state and the strength of adhesive joints. For this purpose, a linear multiple regression model was used. The purpose of multiple regression is to test the relationship between multiple independent variables and the dependent variable, so that it is possible to determine which independent variables have a significant effect on the dependent variable. Model verification in this case involves checking that the following model assumptions are met:

- the significance of the linear regression,
- significance of partial regression coefficients,
- no collinearity between independent variables,
- the assumption of constancy of variance, which means that the variance of the random component (residuals εt) is the same for all observations,
- no autocorrelation of the residuals,
- normality of the distribution of the residuals,
- the random component has an expected value equal to 0.

When there are strong correlations between the independent variables, the multiple regression function is statistically significant. This significance is verified by the F test, and for this test the probability level p should be less than the assumed significance level α.

The dependent variable in the case analyzed in this paper was the average shear strength of the adhesive joints (R_t). The average strength test results of the adhesive compositions obtained in static tests, i.e., tensile strength (σ_m), compression strength (σ_c) and bending strength (σ_f), were treated as the independent variables. It was assumed that there is a linear relationship between the variables, and the relationship has the formula:

$$R_t = b_0 + b_1 \cdot \sigma_m + b_2 \cdot \sigma_c + b_3 \cdot \sigma_f \pm S_e \tag{1}$$

The task was to build a linear regression model, determining the coefficients of this equation—b_0, b_1, b_2, b_3—the standard error of the estimate S_e. The number of estimated parameters was four, and the number of data was four. The results of the analyses are shown in Table 14.

A linear correlation coefficient R close to one was obtained, which indicates a linear relationship between the variables. The *p*-value for the F test was 0.018, which is less than the accepted significance test of $\alpha = 0.05$, indicating the significance of the regression equation. The independent variables in Table 12 highlighted in red (i.e., compression strength and bending strength) are characterized by non-significant parameters $p > 0.05$, i.e., they are not correlated with the dependent variable. This may mean that these variables are collinear with another variable or are weakly correlated with the dependent variable. Therefore, these variables should be removed from the regression equation.

Analyzing the results obtained, it can be seen that the F value = 51.794, $p < 0.1877$, i.e., the regression equation is significant. The multivariate correlation coefficient is 0.99 and means that there is a strong linear relationship between the tensile strength of the adhesive compositions and the shear strength of the adhesive joints.

Based on the test performed, a multiple regression equation can be derived to determine the shear strength of the analyzed EN AW 2024 T3 aluminum alloy sheet adhesive joints from the tensile strength of the adhesive composition used to make the joints. However, it must be considered that this relationship applies only to compositions and joints prepared according to the technology described in this paper. The equation will be of the form:

$$R_t = 0.9807 \cdot \sigma_m - 35.65 \pm 0.96 \tag{2}$$

Table 14. Results of multiple regression.

N = 4	Summary of Dependent Variable Regression: Tensile Shear Strength R = 0.98123447 R^2 = 0.96282108 Correct. R^2 = 0.94423163 F(1,4) = 51.794 p < 0.01877 Estimation std. Error: 0.96454					
	b*	Err. std. of b*	b	Err. std. of b	t (2)	*p*
Free expression			−35.6	7.212	−4.942	0.038
Tensile strength of adhesive composition σ_m [MPa]	0.981	0.136	0.98	0.136	7.196	0.018
Compression strength of adhesive composition σ_c [MPa]	−0.807	0.417	−0.82	0.428	−1.935	0.192
Bending strength of adhesive composition σ_f [MPa]	0.446	0.632	0.488	0.691	0.705	0.553
	Legend:					
b*	- standardized regression coefficients					
Err. std. of b*	- standard error of the coefficients b*					
b	- coefficients a_0 and a_1 of the regression equation $\hat{y}_i = a_1 x + a_0 + u_i$					
Err. std. of b	- standard error of calculated coefficients;					
t (2)	- quotient b/(Error std. of b);					
p	- computer significance level of the coefficients.					

The data and the fitted surface are presented in Figure 15.

The graph presents the results of the tests, which were used to determine a multiple regression equation showing the dependence of the shear strength of the adhesive joints on the tensile strength of the adhesive compositions for the specific adhesive compositions. This makes it possible to predict the shear strength of the analyzed adhesive joints of aluminum alloy sheets, considering the tensile strength of the adhesive composition used to make the adhesive joint analyzed in this paper.

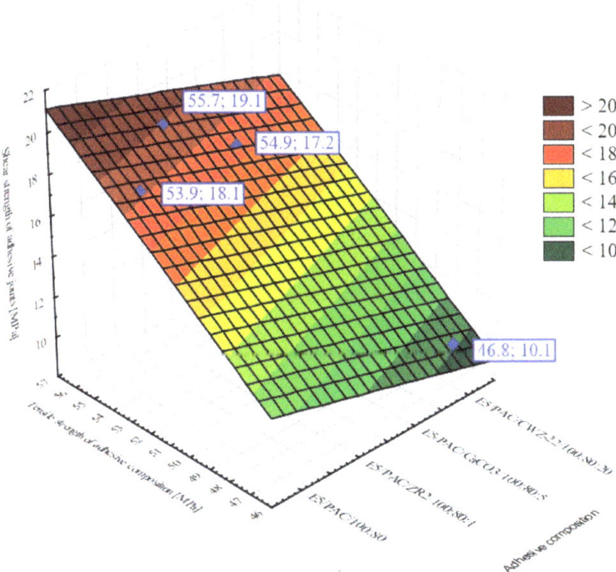

Figure 15. Area graph 3W of the dependence of the average shear strength of adhesive joints on the average tensile strength of adhesive compositions.

5. Conclusions

The experimental studies presented in this paper concerned the analysis of the mechanical and physical properties of selected modified adhesive compositions and the strength properties of EN AW 2024 T3 aluminum alloy sheet adhesive joints. Due to the use of different fillers, it was observed that their effects on specific properties of adhesive compositions are of different nature. It was shown that physical modification of the compositions does not always result in positive effects. Positive results of physical modification of adhesive compositions expressing an increase in strength parameters can be mentioned when the tensile strength and compression strength in the cured state are considered. In this case, the addition of fillers increased the strength parameters. When the impact and bending strength of the adhesive compositions were tested, the addition of the CWZ-22 activated carbon filler caused a deterioration in the results compared to the reference composition, and in addition, the results were also characterized by the lowest repeatability. It can be concluded that this is due to the high filler content, which can result in particle agglomeration. For the tensile and compression strength tests, the lowest repeatability was obtained for the unmodified composition. This may be due to the higher amount of air bubbles present in the plastic structure compared to the modified compositions, which affects the results of tests where external forces act axially to the test samples.

During the study, microscopic tests were also carried out on the structure of the compositions in the cured state. For all modified compositions, a good interaction between the filler and the epoxy resin matrix was evident, which is a basic assumption for the correct physical modification of an adhesive composition. It was observed that the best dispersion was obtained with the filler in the form of CWZ-22 activated carbon, as the filler in the matrix was delaminated in the structure of the plastic.

Thermal property tests conducted to determine the effect of temperature on the properties of the modified epoxy plastics showed that all adhesive compositions were thermally stable.

Tests on the strength properties of the adhesive joints showed that the addition of activated carbon had a significant effect on the deterioration of the shear strength of the adhesive joints compared to the results obtained for joints made with the reference composition. The analysis also showed that the addition of a filler in the form of ZR2 montmorillonite and $CaCO_3$ calcium carbonate did not significantly alter the strength of adhesive joints of EN AW 2024 T3 aluminum alloy sheets compared to the unmodified composition. Composition with ZR2 montmorillonite caused the most favorable results.

The final stage of the experimental study was to verify whether the strength of adhesive joints made with these compositions could be inferred from the properties of the epoxy compositions in the cured state. The realized multiple regression model showed that there was a strong correlation between the tensile strength of the adhesive compositions and the shear strength of aluminum alloy sheet adhesive joints made with these compositions. The other strength properties of the adhesive compositions are not correlated with the adhesive joint strength. This correlation is described by relation (2).

By determining this correlation, it is possible to predict the strength of adhesive joints using knowledge of the strength properties of adhesive compositions, using the technology for making the joints as described in this paper, without the need to carry out destructive strength tests. However, it should be borne in mind that this relationship is specific to the materials and preparation technology of the adhesive compositions and adhesive joints used in the presented studies.

Epoxy adhesives are excellent structural adhesives with the advantage that they can be applied to dissimilar materials. Therefore, in future studies, the modified adhesive will be used for bonding other aluminum alloys as well as for bonding other materials.

Conclusions and findings from the studies show the need to intensify research work, especially in modifying epoxy adhesive compositions with fillers of organic and inorganic origin, in the aspect of using them in adhesive joints, and in the direction of determining the effect of different filler contents on the properties of modified epoxy adhesives. Future research will analyze the impact of modifying epoxy adhesives made with other epoxy resins and curing agents to produce more flexible adhesive joints. These adhesives will be used to bond structural materials other than aluminum alloy.

Author Contributions: Conceptualization, I.M.-B.; data curation, I.M.-B.; formal analysis, I.M.-B.; funding acquisition, I.M.-B.; investigation, I.M.-B., A.R. and E.D.; methodology, I.M.-B., A.R. and E.D.; project administration, I.M.-B.; resources, I.M.-B., A.R. and E.D.; software, I.M.-B.; supervision, I.M.-B.; validation, I.M.-B.; visualization, I.M.-B.; writing—original draft, I.M.-B.; writing—review and editing, I.M.-B., A.R. and E.D. All authors have read and agreed to the published version of the manuscript.

Funding: The project/research was financed in the framework of the project Lublin University of Technology-Regional Excellence Initiative, funded by the Polish Ministry of Science and Higher Education (contract no. 030/RID/2018/19).

Institutional Review Board Statement: Not applicable.

Informed Consent Statement: Not applicable.

Data Availability Statement: The raw/processed data required to reproduce these findings cannot be shared at this time due to technical or time limitations. Data can be made available on individual request.

Acknowledgments: We acknowledge Faculty of Mechanical Engineering, Lublin University of Technology, Lublin, Poland.

Conflicts of Interest: The authors declare no conflict of interest.

References

1. Bai, Y.; Yang, X. Novel Joint for Assembly of All-Composite Space Truss Structures: Conceptual Design and Preliminary Study. *J. Compos. Constr.* **2013**, *17*, 130–138. [CrossRef]
2. Messler, R.W. *Joining of Materials and Structures: From Pragmatic Process to Enabling Technology*; Elsevier: Amsterdam, The Netherlands; Boston, MA, USA, 2004; ISBN 978-0-7506-7757-8.

3. Messler, R.W. Joining Composite Materials and Structures: Some Thought-Provoking Possibilities. *J. Thermoplast. Compos. Mater.* **2004**, *17*, 51–75. [CrossRef]
4. Saboori, A.; Aversa, A.; Marchese, G.; Biamino, S.; Lombardi, M.; Fino, P. Application of Directed Energy Deposition-Based Additive Manufacturing in Repair. *Appl. Sci.* **2019**, *9*, 3316. [CrossRef]
5. Uddin, M.A.; Chan, H.P. Adhesive Technology for Photonics. In *Advanced Adhesives in Electronics*; Elsevier: Amsterdam, The Netherlands, 2011; pp. 214–258, ISBN 978-1-84569-576-7.
6. Janda, R.; Roulet, J.-F.; Wulf, M.; Tiller, H.-J. A New Adhesive Technology for All-Ceramics. *Dent. Mater.* **2003**, *19*, 567–573. [CrossRef]
7. Vaillancourt, A.; Abele, T. *Adhesive Technology: Surface Preparation Techniques on Aluminum*; Worcester Polytechnic Institute: Worcester, MA, USA, 2009.
8. Banea, M.D.; da Silva, L.F.M. Adhesively Bonded Joints in Composite Materials: An Overview. *Proc. Inst. Mech. Eng. Part J. Mater. Des. Appl.* **2009**, *223*, 1–18. [CrossRef]
9. Fekete, J.R.; Hall, J.N. Design of Auto Body. In *Automotive Steels*; Elsevier: Amsterdam, The Netherlands, 2017; pp. 1–18, ISBN 978-0-08-100638-2.
10. Słania, J.; Kuk, Ł. Process of Joining Materials to Build Vehicles and Motor-Car Bodies in the Automotive Industry. *Weld. Technol. Rev.* **2014**, *86*, 40–46.
11. Silva, L.F.M.; da Öchsner, A.; Adams, R.D. (Eds.) *Handbook of Adhesion Technology*; Springer reference; Springer: Heidelberg, Germany, 2011; ISBN 978-3-642-01168-9.
12. Mucha, J.; Kaščák, L.; Spišák, E. Joining the Car-Body Sheets Using Clinching Process with Various Thickness and Mechanical Property Arrangements. *Arch. Civ. Mech. Eng.* **2011**, *11*, 135–148. [CrossRef]
13. Barnes, T.A.; Pashby, I.R. Joining Techniques for Aluminium Spaceframes Used in Automobiles. *J. Mater. Process. Technol.* **2000**, *99*, 72–79. [CrossRef]
14. Purr, S.; Meinhardt, J.; Lipp, A.; Werner, A.; Ostermair, M.; Glück, B. Stamping Plant 4.0—Basics for the Application of Data Mining Methods in Manufacturing Car Body Parts. *Key Eng. Mater.* **2015**, *639*, 21–30. [CrossRef]
15. Yoshida, Y.; Inoue, S. Chemical Analyses in Dental Adhesive Technology. *Jpn. Dent. Sci. Rev.* **2012**, *48*, 141–152. [CrossRef]
16. El-Tantawy, F.; Kamada, K.; Ohnabe, H. Electrical Properties and Stability of Epoxy Reinforced Carbon Black Composites. *Mater. Lett.* **2002**, *57*, 242–251. [CrossRef]
17. Kinloch, A.J. Toughening Epoxy Adhesives to Meet Today's Challenges. *MRS Bull.* **2003**, *28*, 445–448. [CrossRef]
18. Szabelski, J.; Karpiński, R.; Machrowska, A. Application of an Artificial Neural Network in the Modelling of Heat Curing Effects on the Strength of Adhesive Joints at Elevated Temperature with Imprecise Adhesive Mix Ratios. *Materials* **2022**, *15*, 721. [CrossRef] [PubMed]
19. Sun, J.; Su, J.; Ma, C.; Göstl, R.; Herrmann, A.; Liu, K.; Zhang, H. Fabrication and Mechanical Properties of Engineered Protein-Based Adhesives and Fibers. *Adv. Mater.* **2020**, *32*, 1906360. [CrossRef] [PubMed]
20. Engels, H.-W.; Pirkl, H.-G.; Albers, R.; Albach, R.W.; Krause, J.; Hoffmann, A.; Casselmann, H.; Dormish, J. Polyurethanes: Versatile Materials and Sustainable Problem Solvers for Today's Challenges. *Angew. Chem. Int. Ed.* **2013**, *52*, 9422–9441. [CrossRef]
21. Rudawska, A. Comparison of the Adhesive Joints' Strength of the Similar and Dissimilar Systems of Metal Alloy/Polymer Composite. *Appl. Adhes. Sci.* **2019**, *7*, 7. [CrossRef]
22. Miturska-Barańska, I.; Rudawska, A.; Doluk, E. The Influence of Sandblasting Process Parameters of Aerospace Aluminium Alloy Sheets on Adhesive Joints Strength. *Materials* **2021**, *14*, 6626. [CrossRef]
23. Ebnesajjad, S. *Adhesives Technology Handbook*, 2nd ed.; William Andrew Pub: Norwich, NY, USA, 2008; ISBN 978-0-8155-1533-3.
24. Speranza, G.; Gottardi, G.; Pederzolli, C.; Lunelli, L.; Canteri, R.; Pasquardini, L.; Carli, E.; Lui, A.; Maniglio, D.; Brugnara, M.; et al. Role of Chemical Interactions in Bacterial Adhesion to Polymer Surfaces. *Biomaterials* **2004**, *25*, 2029–2037. [CrossRef]
25. Page, S.A.; Mezzenga, R.; Boogh, L.; Berg, J.C.; Månson, J.-A.E. Surface Energetics Evolution during Processing of Epoxy Resins. *J. Colloid Interface Sci.* **2000**, *222*, 55–62. [CrossRef]
26. Xie, Y.; Hill, C.A.S.; Xiao, Z.; Militz, H.; Mai, C. Silane Coupling Agents Used for Natural Fiber/Polymer Composites: A Review. *Compos. Part Appl. Sci. Manuf.* **2010**, *41*, 806–819. [CrossRef]
27. von Fraunhofer, J.A. Adhesion and Cohesion. *Int. J. Dent.* **2012**, *2012*, 951324. [CrossRef] [PubMed]
28. Wei, H.; Xia, J.; Zhou, W.; Zhou, L.; Hussain, G.; Li, Q.; Ostrikov, K.K. Adhesion and Cohesion of Epoxy-Based Industrial Composite Coatings. *Compos. Part B Eng.* **2020**, *193*, 108035. [CrossRef]
29. Miturska, I.; Rudawska, A.; Müller, M.; Hromasová, M. The Influence of Mixing Methods of Epoxy Composition Ingredients on Selected Mechanical Properties of Modified Epoxy Construction Materials. *Materials* **2021**, *14*, 411. [CrossRef] [PubMed]
30. Rudawska, A.; Maziarz, M.; Miturska, I. Impact of Selected Structural, Material and Exploitation Factors on Adhesive Joints Strength. *MATEC Web Conf.* **2019**, *252*, 01006. [CrossRef]
31. Sun, S.; Li, M.; Liu, A. A Review on Mechanical Properties of Pressure Sensitive Adhesives. *Int. J. Adhes. Adhes.* **2013**, *41*, 98–106. [CrossRef]
32. Pethrick, R.A. Design and Ageing of Adhesives for Structural Adhesive Bonding—A Review. *Proc. Inst. Mech. Eng. Part J. Mater. Des. Appl.* **2015**, *229*, 349–379. [CrossRef]
33. Khalili, S.M.R.; Jafarkarimi, M.H.; Abdollahi, M.A. Creep Analysis of Fibre Reinforced Adhesives in Single Lap Joints—Experimental Study. *Int. J. Adhes. Adhes.* **2009**, *29*, 656–661. [CrossRef]

34. Miturska, I.; Rudawska, A.; Müller, M.; Valášek, P. The Influence of Modification with Natural Fillers on the Mechanical Properties of Epoxy Adhesive Compositions after Storage Time. *Materials* **2020**, *13*, 291. [CrossRef]

35. Mohan, P. A Critical Review: The Modification, Properties, and Applications of Epoxy Resins. *Polym.-Plast. Technol. Eng.* **2013**, *52*, 107–125. [CrossRef]

36. Keibal, N.A.; Bondarenko, S.N.; Kablov, V.F. Modification of Adhesive Compositions Based on Polychloroprene with Element-Containing Adhesion Promoters. *Polym. Sci. Ser. D* **2011**, *4*, 267–280. [CrossRef]

37. Dunn, D.J. *Engineering and Structural Adhesives*; Rapra review reports; Rapra Technology: Shrewsbury, MA, USA, 2004; ISBN 978-1-85957-436-2.

38. Yoon, I.-N.; Lee, Y.; Kang, D.; Min, J.; Won, J.; Kim, M.; Soo Kang, Y.; Kim, S.; Kim, J.-J. Modification of Hydrogenated Bisphenol A Epoxy Adhesives Using Nanomaterials. *Int. J. Adhes. Adhes.* **2011**, *31*, 119–125. [CrossRef]

39. Avolio, R.; Gentile, G.; Avella, M.; Carfagna, C.; Errico, M.E. Polymer–Filler Interactions in PET/CaCO3 Nanocomposites: Chain Ordering at the Interface and Physical Properties. *Eur. Polym. J.* **2013**, *49*, 419–427. [CrossRef]

40. Kibria, M.A.; Anisur, M.R.; Mahfuz, M.H.; Saidur, R.; Metselaar, I.H.S.C. A Review on Thermophysical Properties of Nanoparticle Dispersed Phase Change Materials. *Energy Convers. Manag.* **2015**, *95*, 69–89. [CrossRef]

41. Park, S.-J.; Jin, F.-L.; Lee, C. Preparation and Physical Properties of Hollow Glass Microspheres-Reinforced Epoxy Matrix Resins. *Mater. Sci. Eng. A* **2005**, *402*, 335–340. [CrossRef]

42. Fu, S.-Y.; Feng, X.-Q.; Lauke, B.; Mai, Y.-W. Effects of Particle Size, Particle/Matrix Interface Adhesion and Particle Loading on Mechanical Properties of Particulate–Polymer Composites. *Compos. Part B Eng.* **2008**, *39*, 933–961. [CrossRef]

43. Acda, M.N.; Devera, E.E.; Cabangon, R.J.; Ramos, H.J. Effects of Plasma Modification on Adhesion Properties of Wood. *Int. J. Adhes. Adhes.* **2012**, *32*, 70–75. [CrossRef]

44. Broyles, A.C.; Pavan, S.; Bedran-Russo, A.K. Effect of Dentin Surface Modification on the Microtensile Bond Strength of Self-Adhesive Resin Cements: Dentin Modification and Self-Adhesive Resin Cements. *J. Prosthodont.* **2013**, *22*, 59–62. [CrossRef]

45. Sundriyal, P.; Pandey, M.; Bhattacharya, S. Plasma-Assisted Surface Alteration of Industrial Polymers for Improved Adhesive Bonding. *Int. J. Adhes. Adhes.* **2020**, *101*, 102626. [CrossRef]

46. Encinas, N.; Oakley, B.R.; Belcher, M.A.; Blohowiak, K.Y.; Dillingham, R.G.; Abenojar, J.; Martínez, M.A. Surface Modification of Aircraft Used Composites for Adhesive Bonding. *Int. J. Adhes. Adhes.* **2014**, *50*, 157–163. [CrossRef]

47. Ostapiuk, M.; Bieniaś, J. Fracture Analysis and Shear Strength of Aluminum/CFRP and GFRP Adhesive Joint in Fiber Metal Laminates. *Materials* **2019**, *13*, 7. [CrossRef]

48. Zheng, R.; Lin, J.; Wang, P.-C.; Zhu, C.; Wu, Y. Effect of Adhesive Characteristics on Static Strength of Adhesive-Bonded Aluminum Alloys. *Int. J. Adhes. Adhes.* **2015**, *57*, 85–94. [CrossRef]

49. Rudawska, A.; Frigione, M. Aging Effects of Aqueous Environment on Mechanical Properties of Calcium Carbonate-Modified Epoxy Resin. *Polymers* **2020**, *12*, 2541. [CrossRef] [PubMed]

50. Czub, P.; Penczek, P. *Epoxy Resin Chemistry and Technology*; WNT: Warsaw, Poland, 2002; ISBN 978-83-204-2611-3.

51. *BN-89 6376-02*; Industry Standard. Epoxy Resins Epidian 1, 2, 3, 4, 5, 6. Wydawnictwa Normalizacyjne "ALFA": Warsaw, Poland, 1989. (In Polish)

52. Information Catalogue of Ciech S.A. Available online: https://Ciechgroup.Com/Produkty/Chemia-Organiczna/Zywice/Zywice-Epoksydowe/ (accessed on 20 February 2022).

53. Bereska, B.; Iłowska, J.; Czaja, K.; Bereska, A. Curing agents for epoxy resins. *Przem. Chem.* **2014**, *93*, 443–448.

54. Zhou, Y.; Pervin, F.; Rangari, V.K.; Jeelani, S. Influence of Montmorillonite Clay on the Thermal and Mechanical Properties of Conventional Carbon Fiber Reinforced Composites. *J. Mater. Process. Technol.* **2007**, *191*, 347–351. [CrossRef]

55. Szymańska, J.; Bakar, M. Mechanical properties of epoxy resin modified with nanobent and polyurethane. *Przetw. Tworzyw* **2015**, *4*, 355–359.

56. Jin, F.-L.; Park, S.-J. Interfacial Toughness Properties of Trifunctional Epoxy Resins/Calcium Carbonate Nanocomposites. *Mater. Sci. Eng. A* **2008**, *475*, 190–193. [CrossRef]

57. Zebarjad, S.M.; Sajjadi, S.A. On the Strain Rate Sensitivity of HDPE/CaCO3 Nanocomposites. *Mater. Sci. Eng. A* **2008**, *475*, 365–367. [CrossRef]

58. He, H.; Li, K.; Wang, J.; Sun, G.; Li, Y.; Wang, J. Study on Thermal and Mechanical Properties of Nano-Calcium Carbonate/Epoxy Composites. *Mater. Des.* **2011**, *32*, 4521–4527. [CrossRef]

59. Chen, S.; Feng, J. Epoxy Laminated Composites Reinforced with Polyethyleneimine Functionalized Carbon Fiber Fabric: Mechanical and Thermal Properties. *Compos. Sci. Technol.* **2014**, *101*, 145–151. [CrossRef]

60. Chen, Q.; Zhao, Y.; Zhou, Z.; Rahman, A.; Wu, X.-F.; Wu, W.; Xu, T.; Fong, H. Fabrication and Mechanical Properties of Hybrid Multi-Scale Epoxy Composites Reinforced with Conventional Carbon Fiber Fabrics Surface-Attached with Electrospun Carbon Nanofiber Mats. *Compos. Part B Eng.* **2013**, *44*, 1–7. [CrossRef]

61. Feller, J.F.; Linossier, I.; Grohens, Y. Conductive Polymer Composites: Comparative Study of Poly(Ester)-Short Carbon Fibres and Poly(Epoxy)-Short Carbon Fibres Mechanical and Electrical Properties. *Mater. Lett.* **2002**, *57*, 64–71. [CrossRef]

62. Inspection Certificate No 1828896CZ of the Material Supplier Adamet-Niemet. Available online: https://www.adamet.com.pl/wp-content/uploads/2019/01/EN-9120-EN.pdf (accessed on 22 September 2022).

63. *PN-EN ISO 527-1*; Plastics—Determination of Mechanical Properties in Static Tension. ISO: Geneva, Switzerland, 2012. (In Polish)

64. Broniewski, T.; Kapko, J.; Płaczek, W.; Thomalla, J. *Test Methods and Evaluation of the Properties of Plastics*, 2nd ed.; Plastics; WNT: Warsaw, Poland, 2000; ISBN 978-83-204-2468-3.
65. *ISO 604:2002*; ISO 604 Standard—Plastics—Determination of Compressive Properties. ISO: Geneva, Switzerland, 2002.
66. *DIN-EN ISO 178*; Plastics—Determination of Bending Properties. ISO: Geneva, Switzerland, 2019.
67. *PN-EN ISO 11357-6:2018*; ISO 11357 Standard—Plastics—Differential Scanning Calorimetry (DSC). ISO: Geneva, Switzerland, 2018.
68. Michels, J.; Sena Cruz, J.; Christen, R.; Czaderski, C.; Motavalli, M. Mechanical Performance of Cold-Curing Epoxy Adhesives after Different Mixing and Curing Procedures. *Compos. Part B Eng.* **2016**, *98*, 434–443. [CrossRef]
69. Rudawska, A.; Czarnota, M. Selected Aspects of Epoxy Adhesive Compositions Curing Process. *J. Adhes. Sci. Technol.* **2013**, *27*, 1933–1950. [CrossRef]
70. *ASTM D1002*; Standard Test Method for Apparent Shear Strength of Single-Lap-Joint Adhesively Bonded Metal Specimens by Tension Loading (Metal-to-Metal). ASTM International: West Conshohocken, PA, USA, 2019.
71. Müller, M.; Valášek, P.; Rudawska, A. Mechanical Properties of Adhesive Bonds Reinforced with Biological Fabric. *J. Adhes. Sci. Technol.* **2017**, *31*, 1859–1871. [CrossRef]
72. Samal, S. Effect of Shape and Size of Filler Particle on the Aggregation and Sedimentation Behavior of the Polymer Composite. *Powder Technol.* **2020**, *366*, 43–51. [CrossRef]
73. Kundie, F.; Azhari, C.H.; Muchtar, A.; Ahmad, Z.A. Effects of Filler Size on the Mechanical Properties of Polymer-Filled Dental Composites: A Review of Recent Developments. *J. Phys. Sci.* **2018**, *29*, 141–165. [CrossRef]
74. Bittmann, B.; Haupert, F.; Schlarb, A.K. Ultrasonic Dispersion of Inorganic Nanoparticles in Epoxy Resin. *Ultrason. Sonochem.* **2009**, *16*, 622–628. [CrossRef]
75. Matykiewicz, D. Hybrid Epoxy Composites with Both Powder and Fiber Filler: A Review of Mechanical and Thermomechanical Properties. *Materials* **2020**, *13*, 1802. [CrossRef]
76. Moussa, O.; Vassilopoulos, A.P.; Keller, T. Experimental DSC-Based Method to Determine Glass Transition Temperature during Curing of Structural Adhesives. *Constr. Build. Mater.* **2012**, *28*, 263–268. [CrossRef]

Article

On the Durability Performance of Two Adhesives to Be Used in Bonded Secondary Structures for Offshore Wind Installations

Khaoula Idrissa [1,2], Aurélien Maurel-Pantel [1], Frédéric Lebon [1,*] and Noamen Guermazi [2]

[1] Aix-Marseille Université, CNRS, Centrale Marseille, LMA, CEDEX 13, 13453 Marseille, France; khawla.drissa@gmail.com (K.I.); maurel@lma.cnrs-mrs.fr (A.M.-P.)

[2] LGME Laboratory, National Engineering School of Sfax, University of Sfax, Sfax B.P. 1173-3038, Tunisia; noamen.guermazi@enis.tn

* Correspondence: lebon@lma.cnrs-mrs.fr

Abstract: The development of offshore wind farms requires robust bonding solutions that can withstand harsh marine conditions for the easy integration of secondary structures. This paper investigates the durability performance of two adhesives: Sikadur 30 epoxy resin and Loctite UK 1351 B25 urethane-based adhesive for use in offshore wind environments. Tensile tests on adhesive samples and accelerated aging tests were carried out under a variety of temperatures and environmental conditions, including both dry and wet conditions. The long-term effects of aging on adhesive integrity are investigated by simulating the operational life of offshore installations. The evolution of mechanical properties, studied under accelerated aging conditions, provides an important indication of the longevity of structures under normal conditions. The results show significant differences in performance between the two adhesives, highlighting their suitability for specific operating parameters. It should also be noted that for both adhesives, their exposure to different environments (seawater, distilled water, humid climate) over a prolonged period showed that (i) Loctite adhesive has a slightly faster initial uptake than Sikadur adhesive, but the latter reaches an asymptotic plateau with a lower maximum absorption rate than Loctite adhesive; and (ii) a progressive deterioration in the tensile properties occurred following an exponential function. Therefore, aging behavior results showed a clear correlation with the Arrhenius law, providing a predictive tool for the aging process and the aging process of the two adhesives followed Arrhenius kinetics. Ultimately, the knowledge gained from this study is intended to inform best practice in the use of adhesives, thereby improving the reliability and sustainability of the offshore renewable energy infrastructure.

Keywords: adhesives; thermal aging; hygrothermal aging; durability; mechanical properties

Citation: Idrissa, K.; Maurel-Pantel, A.; Lebon, F.; Guermazi, N. On the Durability Performance of Two Adhesives to Be Used in Bonded Secondary Structures for Offshore Wind Installations. *Materials* **2024**, *17*, 2392. https://doi.org/10.3390/ma17102392

Academic Editor: Luca Sorrentino

Received: 10 April 2024
Revised: 7 May 2024
Accepted: 12 May 2024
Published: 16 May 2024

Copyright: © 2024 by the authors. Licensee MDPI, Basel, Switzerland. This article is an open access article distributed under the terms and conditions of the Creative Commons Attribution (CC BY) license (https://creativecommons.org/licenses/by/4.0/).

1. Introduction

The growing global demand for sustainable energy solutions has led to the rapid expansion of offshore wind farms, which have become key elements in the renewable energy landscape [1]. These installations, characterized by their massive scale and exposure to harsh marine environments, require technologies to ensure their longevity and structural integrity. At the heart of the wind turbine structure is the critical role played by bonded secondary structures in withstanding the harsh conditions of salt water, fluctuating temperatures, and long service life [2,3]. As offshore installations expand into deeper waters and more challenging environments, the importance of reliable bonding technology for secondary structures becomes increasingly important [4,5].

For these secondary structures, the choice of adhesive appears to be a critical issue given the wide range of conditions to which this material is subjected, including exposure to seawater, temperature fluctuations, and extended duty cycles [6,7]. This study will evaluate the performance of two adhesives specifically selected for their suitability in civil engineering and marine environments. The distinct performance differences observed between the two adhesives underscore their applicability within specific operational contexts, delineating their individual strengths and weaknesses under different conditions. For example, one adhesive's superior initial take-up or resistance to certain environmental stressors may make it more suitable for applications where these factors predominate. Conversely, the increased long-term stability or durability demonstrated by another adhesive may make it preferable for installations where extended service life is a priority. By recognizing these differences, engineers and designers can make informed decisions on adhesive selection tailored to the precise requirements of their projects, thereby optimizing performance and ensuring structural longevity in offshore wind environments. The expected results will provide concrete information for the integration of bonded assemblies while ensuring the structural integrity of offshore wind farms [8,9]. Recent works have focused on the durability of adhesive bonds in marine composite structures, providing a nuanced overview of the challenges and advances in this evolving field [10–12].

Other studies have investigated the aging behavior of epoxy based adhesives in marine environments, contributing to a better understanding of the long-term performance of adhesives subjected to these harsh conditions [3,12]. The effect of environmental influences on bonding in marine structures has been a major focus, providing crucial information on the longevity of adhesive joints subjected to the relentless forces of the marine environment [13].

Key references such as Andersen et al. [1], Brown et al. [14], and Gao et al. [9] provide valuable insights into the challenges and advances in adhesive technologies for offshore wind applications. Standards such as ASTM D638-14 [2,3] and ISO 3167:2013 [15] are highlights of the standardized test methods used in the field. Chen et al. [5] and Li et al. [16] investigated the effects of moisture absorption and the aging behavior of epoxy adhesives in marine environments. Kuang et al. [17] provided an overview of structural adhesives in offshore wind energy. In the studies of O'Connor et al. [18], Williams et al. [19], and Xu et al. [20], the focus was on the effects of the environment on bonding and assessing the durability of marine structures. The selection of these references is intended to provide a sound basis for this study, covering various aspects of bonding technology in the context of offshore wind turbines.

This study offers new insights into the behavior of Sikadur and Loctite adhesives under rigorous testing in different environmental conditions. In particular, the work of Pouyan Gand Noel M. H. [21] deals with the multi-scale modelling and life prediction of aged composites immersed in salt water. Their research provided a comprehensive understanding of how different composite materials age in such environments and valuable predictive insights into material degradation over time. In contrast, the study of Zike W and colleagues [22] investigated the long-term durability of BFRP/GFRP rods specifically in seawater and sea-sand concrete environments. This research provided insights into the specific performance of these reinforcing materials in marine environments, highlighting their durability and stability over long periods of time. In contrast, this study focuses on evaluating the performance of Sikadur and Loctite adhesives under various environmental conditions relevant to offshore wind environments. Through mechanical testing and accelerated aging simulations, including dry and wet conditions, the aim was to fully understand how these adhesives perform in real-world scenarios. This research contributes to the selection of materials for offshore wind turbines and advances the understanding of the role of adhesive technology in improving stability and durability in such environments.

2. Materials and Methods

The effectiveness of adhesive joints in offshore wind turbines is highly dependent on the choice of materials and the rigor of the testing methods. The materials studied are presented in detail. In addition, all the tests and methods used to characterize these adhesives are described.

2.1. Adhesive Materials

The adhesives studied in this work were Sikadur 30, manufactured by Sika Corporation, Bourget, France, and Loctite UK 1351 B25, manufactured by Henkel Adhesive Technologies, Düsseldorf, Germany. Both adhesives are intended to be used in bonded secondary structures for offshore wind turbines (balustrades, stairs, etc.).

2.1.1. Sikadur Adhesive

Sikadur 30 is a two-component, thixotropic structural adhesive, based on epoxy resin and special fillers, designed for use in a temperature range from +8 °C to +35 °C. It has a glass transition temperature (Tg) of +52 °C (Sika France SAS, Paris, France).

The resin contains epoxy groups, reactive compounds essential for forming strong chemical bonds during curing. The hardener contains amines that react with the epoxy groups, initiating polymerization and creating a solid matrix with a reputation for durability.

According to the manufacturer, in terms of mechanical properties (Table 1), this adhesive has notable characteristics in several key areas.

Table 1. Main characteristics of Sikadur 30 adhesive (SIKA France SAS).

Property	Symbol	Value (MPa)
Tensile modulus	E_t	9600
Tensile strength	σ_t	14–17 (at +15 °C) and 16–19 (at +35 °C)
Compressive modulus	E_c	11,200
Compressive strength	σ_c	24–27 (at +15 °C) and 26–31 (at +35 °C)
Shear strength	τ	14–17 (at +15 °C) and 16–19 (at +35 °C)
Tensile modulus	E_t	9600
Tensile strength	σ_t	14–17 (at +15 °C) and 16–19 (at +35 °C)

Sikadur adhesive has excellent tensile strength, enabling it to withstand high loads. It also has good compressive strength, making it suitable for applications with concentrated loads. Its proven adhesion to a wide range of substrates ensures strong, durable bonds. Finally, its chemical resistance and thermal stability allow it to be used in a variety of environments. The absence of volatile solvents reinforces its environmental credentials.

All of these properties make Sikadur adhesive a good choice for structural applications that require reliable, long-lasting performance.

2.1.2. Loctite Adhesive

The Loctite adhesive, UK 1351 B25, is a two-component, urethane-based adhesive which the manufacturer claims has excellent properties (Table 2).

Table 2. Characteristics of Loctite UK 1351 B25 (Henkel Adhesive Technologies, Düsseldorf, Germany).

Property	Symbol	Value (MPa)
Tensile strength	σ_t	26
Compressive modulus	E_c	4740
Compressive strength	σ_c	71

Germanischer Lloyd (GL®)-approved Loctite adhesive is specially formulated for bonding epoxy-based composites.

According to the manufacturer, it is characterized by its fatigue resistance and its ability to maintain a reliable bond, even under difficult conditions. In addition, the Loctite adhesive offers resistance to crack propagation, making it a good choice for applications where durability and structural integrity are paramount. Its adaptability to a wide range of substrates makes it a versatile adhesive, while its ability to withstand harsh environments seems to reinforce its reputation as a reliable option in a variety of applications.

2.2. Sample Preparation Method

The adhesive sample preparation process is an important step in the evaluation of adhesive performance. This section describes the methodology used to prepare adhesive samples to be tested under various conditions.

2.2.1. Mixing Preparation

In the case of Sikadur adhesive, the two components, epoxy resin and hardener, are accurately measured and then precisely mixed in the specified ratio. The resulting mixture is then applied evenly to the prepared surfaces of the mold (presented in the next section), taking care to maintain a constant thickness. Any air or vacuum trapped during application is carefully removed. The samples are then left in place for the specified polymerization time (7 days), to allow the adhesive to reach its maximum strength.

Similarly, with Loctite adhesive, the two components of the urethane-based adhesive are precisely proportioned and mixed. This mixture is applied to the designated areas of the mold, taking care to maintain an even thickness. Special attention is paid to the intimate contact between the adhesive and the mold through a controlled process. The adhesive is precisely placed in the mold and a heavy load is applied to exert considerable pressure. The samples are then cured under controlled conditions at room temperature to ensure optimum development of the adhesive's properties. For this adhesive, the polymerization time was set at 24 h.

Both adhesives undergo the same rigorous process to produce standardized test specimens, ensuring accurate and reliable results in subsequent mechanical testing. This meticulous approach to specimen preparation is essential for ensuring the accurate and comparative evaluation of the performance characteristics of Sikadur and Loctite adhesives.

2.2.2. Mold Fabrication

Ensuring accurate and reliable test results requires a precise and controlled environment for adhesive application. The manufacture of the mold is therefore of great importance in the preparation of adhesive samples for experimental testing. The mold was produced using a precision machining process. First, the design specifications were transferred to the machining software, which then guided the cutting tools to form the molds from blocks of Teflon material. This process ensured that the dimensions and features of the mold were accurately reproduced in accordance with the design requirements. The Teflon mold (Figure 1), measuring 600×600 mm^2 and 30 mm thick, was designed to produce uniform, well-defined adhesive plates. It consisted of two distinct parts: a lower plate and an upper plate, each 15 mm thick. Inside these plates were four precisely designed square cavities, measuring 200×200 mm^2 and 3 mm thick (Figure 1).

These cavities aid in the distribution of the adhesive, allowing the adhesive material to be evenly distributed. Once the cavities have been filled, the two mold plates were carefully closed to confine the adhesive to the specified areas. This process ensured the production of well-formed, uniform adhesive sheets (Figure 2), which were obtained by the molding process. These plates were then cut to obtain the adhesive samples for experimental testing.

Finally, the utilization of this Teflon mold guarantees the consistent production of test specimens, facilitating reliable and reproducible results across all tests. This allows reliable and reproducible results to be obtained in all tests. The careful manufacture of this mold is a key element in the preparation of adhesive samples in the form of plates for testing, as it ensures an accurate and reliable experimental results.

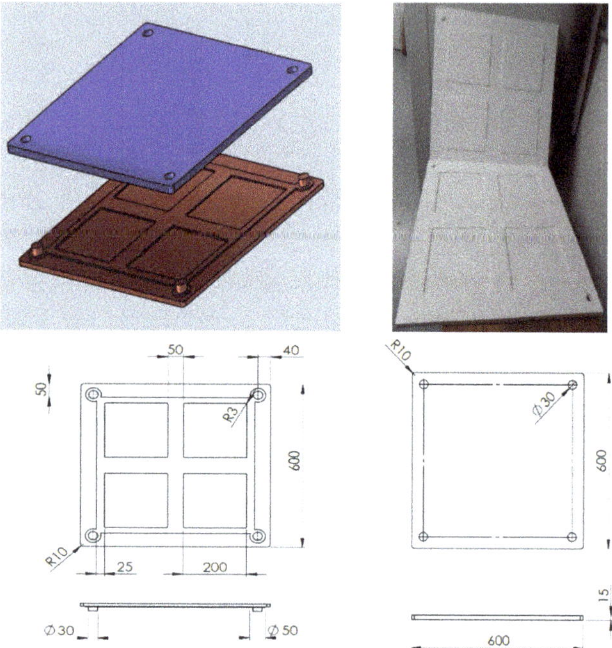

Figure 1. Geometry and dimensions of the Teflon mold.

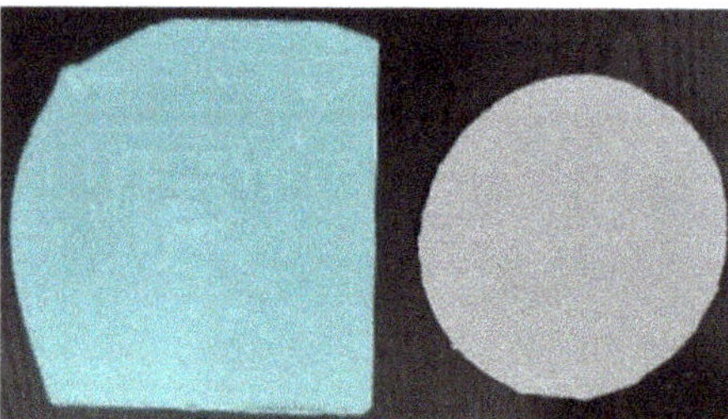

Figure 2. Examples of adhesive plates obtained through the molding process.

2.2.3. Specimen Preparation

In this study, it is important to clarify that we only molded the adhesives for our experimental tests. The adhesive materials were accurately measured, mixed, and then applied to the designated areas of the mold. After curing, the resulting adhesive sheets were carefully cut to provide standardized test specimens. This method ensured consistent specimen preparation for subsequent mechanical testing.

Two different methods were used to cut the adhesive plates and produce the dumbbell specimens: milling and water jet cutting. Initially, the milling technique was used, using a specialized milling tool to achieve the precise cutting of the adhesive plates and produce the dumbbell specimens. This method was chosen because of its ability to accurately create intricate contours. However, to speed up the process, a transition was made to waterjet cutting. Waterjet cutting was chosen for its ability to produce clean and accurate cuts without generating excessive heat or material deformation.

The use of these two complementary methods made it possible to produce adhesive samples to the required specifications. The adhesive samples obtained and subjected to tests have a specific geometry, shown in Figure 3.

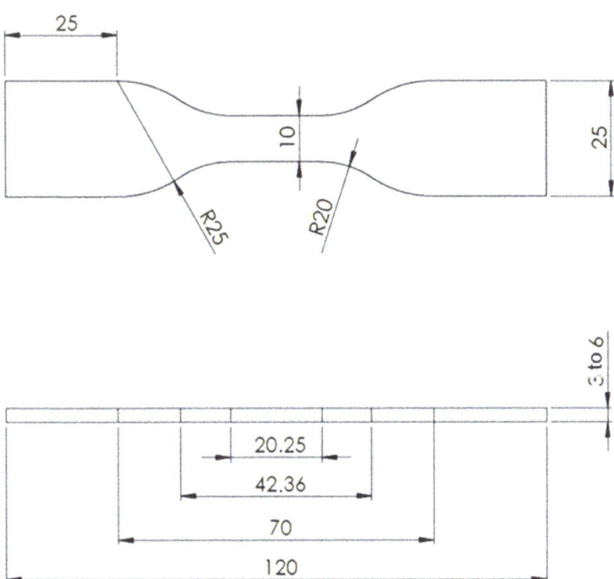

Figure 3. Geometry of the adhesive specimens used in the tensile tests (all units in mm).

2.3. Experimental Protocol of Aging

To investigate the durability performance of the two adhesives under different environmental conditions, accelerated aging tests were carried out by exposing the adhesive samples to different environments and temperatures. Special containers with covers were designed and used to hold the test specimens, providing a controlled test environment (Figure 4, left).

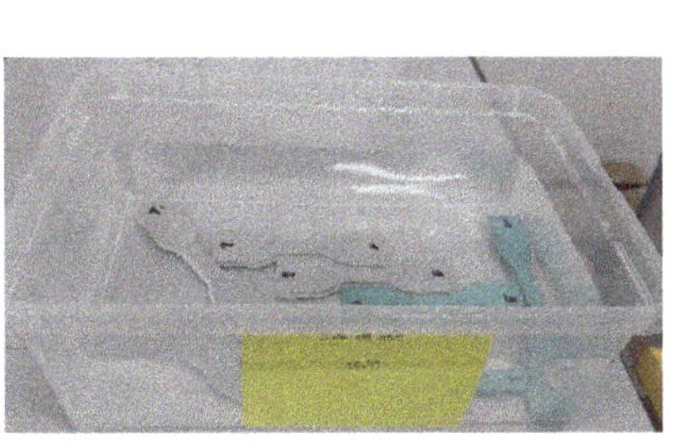

Figure 4. Testing container (on the (**left**)) and climate chamber/oven (on the (**right**)).

In addition, the adhesive samples were exposed to three different environments: a humid environment, distilled water, and seawater. Each of these environments simulated the realistic environmental conditions that adhesives may encounter in offshore wind turbines in a marine environment. The tests were therefore carried out at three different temperatures: ambient (22 °C), 35 °C, and 42 °C.

The adhesive samples were then placed in special ovens (Figure 4, right). This allowed the precise control of the environmental and thermal conditions throughout the aging test. This approach was important in accelerating the aging process and assessing the resistance of the adhesive materials under extreme conditions. This tool can also be effective in better understanding the long-term resistance of adhesive materials in marine environments.

The samples were aged for different lengths of time depending on the type of adhesive: the Sikadur samples were aged for up to 163 days, while the Loctite samples were aged for up to 231 days. These extended aging periods were chosen to thoroughly evaluate the long-term behavior and durability of the adhesives under different environmental conditions, providing valuable insight into their performance over time. In addition, by subjecting the samples to accelerated aging, we were able to speed up the prediction of their behavior over longer periods, making it easier to assess their durability and stability in marine environments.

2.3.1. Thermal Aging

As mentioned above, thermal aging was carried out at three temperatures, such as room temperature (22 °C), 35 °C, and 42 °C. The selection of accelerated aging temperatures was based on several considerations. Firstly, the aim was to ensure that the selected temperatures were within a range that would accelerate the aging process, while remaining below the glass transition temperature of the materials. This approach helps to simulate realistic conditions while avoiding extreme temperatures that could introduce artificial effects. In addition, the aim was to maintain above ambient temperatures to effectively accelerate aging. Consequently, temperatures of 35 °C and 42 °C were chosen to strike a balance between accelerating aging and maintaining relevance for practical applications in offshore wind turbine environments. In addition, the selected temperatures had to meet certain constraints to ensure the reliability of the results. In particular, they had to be within the range of feasible temperatures, with the proviso that they could not exceed the glass transition temperature of the material by more than 10 °C. The glass transition temperature of Sikadur is 50 °C and that of Loctite is 82 °C. This precaution is essential to avoid artificial effects that may not manifest themselves in real or exposed conditions. The purpose is to study the reaction and stability of the adhesive materials in environments to which offshore wind turbines may be exposed. At 22 °C, the adhesives were evaluated in conditions close to their normal use. At 35 °C, a slightly higher temperature, the samples were exposed to high thermal conditions. Finally, at 42 °C, the adhesive samples were tested in more extreme thermal conditions.

However, it is important to note that these tests are not designed to directly predict long-term performance. Instead, they establish a time–temperature equivalency to evaluate the long-term behavior of the adhesive under accelerated conditions. Accelerated testing allows adhesives to be observed more quickly at elevated temperatures, providing insight into their potential behavior over longer periods of time.

Finally, the effect of thermal aging on adhesive properties was analyzed to assess their long-term behavior. This will provide a better understanding of how adhesives maintain their structural integrity under severe environmental conditions. The results of these tests will play a crucial role in optimizing the selection of adhesives for these specific applications.

2.3.2. Multi-Environmental Aging

Multi-environmental aging is an important step in assessing the durability of adhesives. It involves exposing adhesive samples to different environments, including humidity, distilled water, and seawater, in order to simulate a variety of real-world conditions in offshore wind turbines. Distilled water typically has a very low salt content as it is purified by distillation to remove impurities and salts. On the other hand, the seawater used in the aging tests was taken from the Mediterranean Sea and has a conductivity of approximately 37.8 mS·cm^{-1} at 22 °C. In addition, multi-environmental aging will investigate how these different conditions affect the properties of the adhesives, such as mechanical strength, cohesion, resistance to degradation, and other key characteristics.

2.4. Gravimetric Measurements

The change in mass of the Sikadur and Loctite samples was measured periodically during aging at room temperature (22 °C) in distilled water, using a 10^{-5} g precision balance, with measurements taken from 1 day up to 1 week intervals.

The formula for the water absorption rate is expressed as a percentage and is given by Equation (1) [23]:

$$M(t) = \frac{m(t) - m_0}{m_0} \times 100, \tag{1}$$

where $M(t)$ represents the water absorption rate at time t; $m(t)$ represents the mass at time t; and m_0 represents the initial mass.

2.5. Tensile Tests

Tensile tests were used to evaluate the adhesive's response to an axial tensile force and to determine its mechanical properties such as tensile strength, tensile modulus, and elongation at break.

Tensile tests are performed on unaged adhesive materials to evaluate their as-received properties. They are also performed on aged adhesive samples. By subjecting the aged adhesive samples to tensile testing, the mechanical behavior of the material can be assessed for its response to exposure to environmental conditions.

Tensile tests were conducted on dumbbell-shaped adhesive specimens following the ASTM D638 standard [2], using a hydraulic testing machine equipped with a 100 kN load cell, at a crosshead displacement speed of 0.5 mm/min and at room temperature.

In addition, a GOM ARAMIS measurement system, with a remarkable resolution of 2448×2048 pixels at 15 Hz, was used to capture and analyze the deformation behavior of the specimens in real time (Figure 5). Finally, it should be noted that three tests were repeated for each configuration in order to obtain the average.

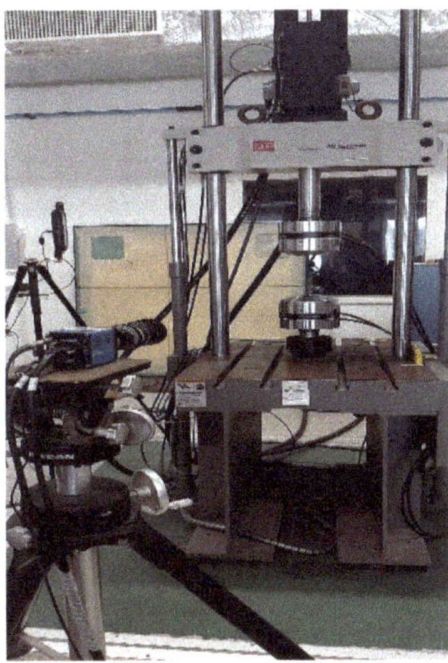

Figure 5. Hydraulic fatigue testing machines (tension/compression) 100 kN + GOM ARAMIS 5 M measurement System 2448×2048 pixels 15 Hz with workstation.

3. Results

This section presents the results of all the characterization series carried out on Sikadur and Loctite adhesives. All the data collected during the tensile and aging tests are analyzed in detail, providing valuable information on the performance of these adhesives. In addition, the focus will be on making comparisons between the two adhesives in order to determine their relative suitability for specific marine applications. The implications of these results for the industry and potential areas of research will also be discussed.

3.1. Adhesives Behavior before Aging

This section examines the initial mechanical behavior of the Sikadur and Loctite adhesives before they are subjected to the aging process. The results of tensile tests carried out on adhesive samples prior to aging are presented and analyzed in detail.

Figures 6 and 7 illustrate the tensile behavior of the adhesive samples under various conditions, focusing on the stresses recorded at the point of failure. Superimposed stress–strain curves for unaged Sikadur specimens are shown in Figure 6.

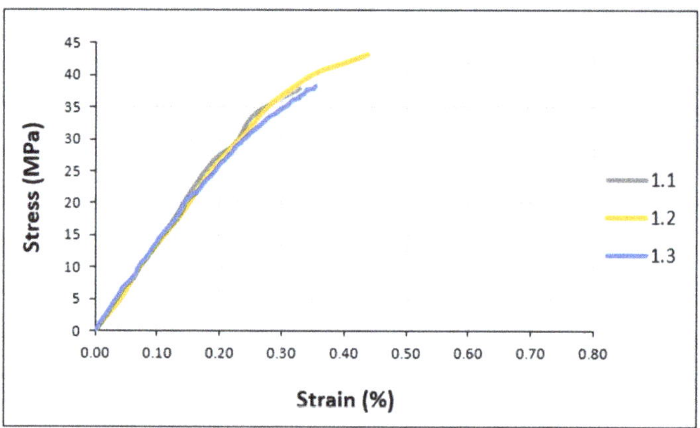

Figure 6. Typical tensile curve (stress–strain) for Sikadur adhesive.

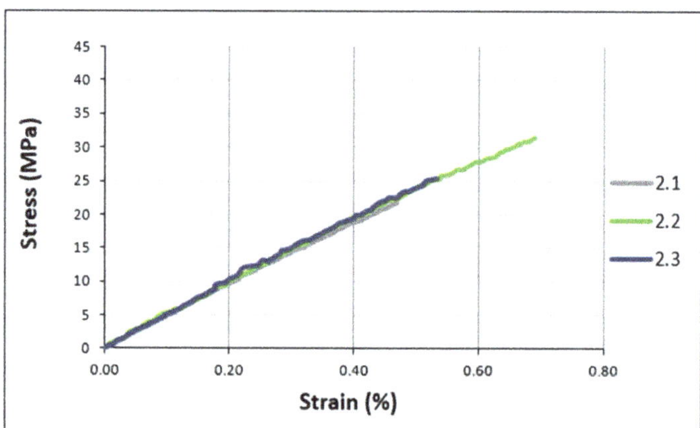

Figure 7. Typical tensile curve (stress–strain) for Loctite adhesive.

As can be seen from Figures 6 and 7, all the tensile curves show similar trends. This suggests a consistent mechanical behavior under different conditions. In fact, the results show that all the adhesives exhibit elastic behavior with a slight non-linearity observed up to brittle fracture. The assumed elastic character and the non-linearity of the elastic response could be interpreted as a discrete viscoplastic component.

The experimental tensile properties, extracted from Figures 6 and 7, are summarized in the histograms (Figure 8), including standard deviations for a better understanding.

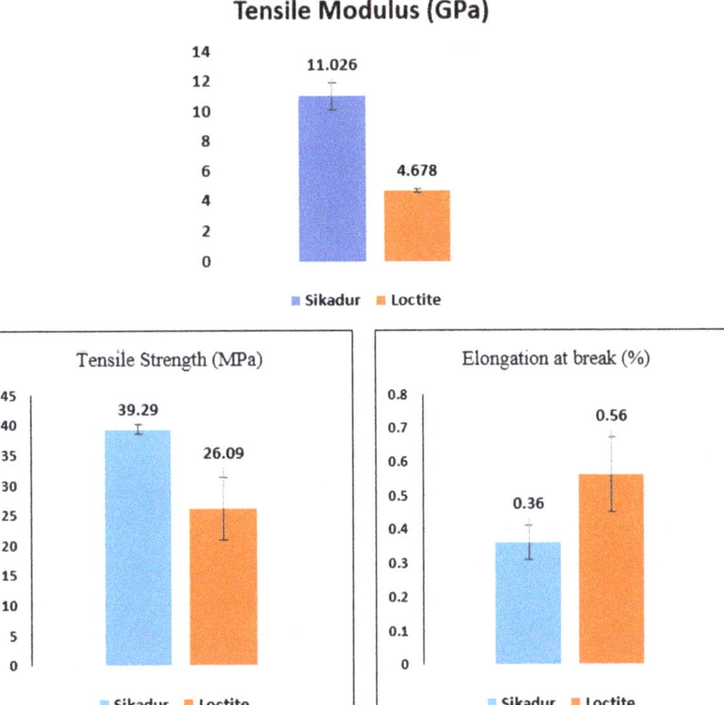

Figure 8. Results of tensile tests on Sikadur 30 and Loctite UK 1351 B25 adhesives.

Specifically, for the Sikadur adhesive, the average failure stress is 39.29 ± 0.84 MPa, and the elastic strain limit is $0.36 \pm 0.05\%$. On the other hand, for the Loctite adhesive, the average ultimate stress reached 26.06 ± 5.17 MPa, with an elastic deformation limit of $0.56 \pm 0.11\%$. These results are very close to the values quoted by the manufacturer.

Previous studies, such as those of Andersen et al. [1] and Gao et al. [9], have highlighted the importance of understanding the performance of adhesives in industrial applications. In addition, Chen et al. [5] investigated the effect of moisture absorption on the mechanical performance of structural adhesive bonds, providing insights that may contribute to the interpretation of the behaviors observed in this study.

These initial mechanical characteristics are used as a reference to evaluate the effect of the aging process on the adhesive materials, following the methodology used in the studies by O'Connor et al. [18] and Williams et al. [19]. This comparative approach provides a better understanding of the performance of Sikadur and Loctite adhesives, and provides information to optimize material selection and maintenance practices in an industrial context.

3.2. Water Sorption Kinetics

This section focuses on the behavior of the two adhesives under aging conditions. Figure 9 shows the mass uptake (in %) of both adhesives as a function of aging time in the specified environment and temperature. The simulations derived from the absorption model introduced by Bruneaux [4] are superimposed on the experimental results. It is of great importance to emphasize that this model characterizes the absorption of water from the material as a process that is governed by the superposition of two mechanisms: (i) firstly, the Fickian diffusion of water molecules through the free volume of the lattice;

(ii) secondly, the reorganization (or relaxation) of the macromolecular chains under the stresses of swelling. This allows the material to absorb more water than Fick's law alone predicts. This mechanism has been termed viscoelastic diffusion by Berens et al. [24].

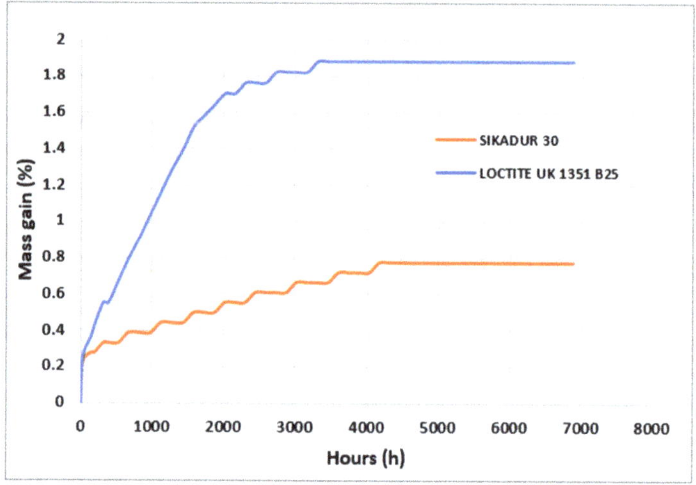

Figure 9. Comparison of experimental mass evolutions of Sikadur 30 and Loctite UK 1351 B25 adhesives during aging at 22 °C in distilled water.

The general shape of the mass evolution curves indicates the existence of two apparent regimes in the sorption kinetics.

In the short term and in the early stages, the kinetics of water absorption are very fast. In particular, the Loctite adhesive shows a significant absorption rate in the initial phase and it appears to have a higher free volume occupied by water. This initial phase therefore corresponds mainly to the filling of the free volume by the diffusion of water molecules, which is a thermally activated mechanism [7,25].

In the long term (over longer periods), the increase in water uptake is slow but continuous. No equilibrium is reached, but the evolution becomes asymptotic. This behavior cannot be explained by classical diffusion phenomena and would indicate either the presence of irreversible phenomena within the material or an internal reorganization of the macromolecular structure (chain relaxation) under the effect of swelling stresses [4,24]. Whatever the exact origin of the phenomenon, the observed effect appears to be the long-term non-saturation of the adhesive material.

According to Figure 9, the comparison in the sorption kinetics for the two adhesives immersed in distilled water at 22 °C shows that the initial water uptake is faster for Loctite adhesive. This could be related to a larger free volume accessible to water molecules by diffusion. In the long-term, however, the asymptotic plateau appears to be less pronounced for Loctite than for Sikadur. This result seems to be related to the extraction of the plasticizer during the aging time.

From a quantitative point of view, when examining the slope of the curve between 1000 and 30,000 h, a slope value of 6.5×10^{-4} (mass per hour) is observed for the Loctite adhesive and a slope value of 2.5×10^{-4} (mass per hour) for the Sikadur adhesive. This means that the Loctite adhesive has much faster absorption kinetics and it absorbs water at a slightly faster rate than the Sikadur adhesive.

Figure 9 also shows an asymptotic plateau, which indicates the point at which water absorption slows down and the material appears to have absorbed as much water as it can under the specific conditions. For Sikadur, this plateau appears to be reached at around 2016 h, whereas for Loctite it appears to be reached at around 2736 h.

Similarly, Figure 9 clearly shows the maximum absorption rates for the two adhesive samples. In fact, the Sikadur adhesive reaches a plateau or equilibrium value at around 0.7%, while for the Loctite adhesive this plateau is reached at around 1.8% by mass.

Furthermore, expressing the changes in mass as a percentage of the initial mass of the samples allows for a comparison of the relative performance of the materials in terms of water absorption. For example, after 2880 h, Sikadur had absorbed approximately 60.77% of its initial mass, while Loctite had absorbed approximately 182.59% of its initial mass.

Finally, taking into account all the results obtained in this section, it can be concluded that there are important implications for the long-term durability and performance of adhesive materials in environments where water absorption is a critical factor.

3.3. Adhesive Behavior after Thermal and Hygrothermal Aging (after 46 Days)

As mentioned above, in the dynamic and demanding environment of offshore wind turbines, adhesives play a vital role in ensuring the structural integrity and longevity of critical joints. However, as these installations are subject to a wide range of environmental stressors, including temperature fluctuations and exposure to moisture, the durability of adhesives is of paramount importance. Understanding how adhesives react to prolonged exposure to these conditions is essential to ensure their reliability and performance over time.

This section examines the behavior of adhesives following periods of thermal and hygrothermal aging. This study offers insights into how the two adhesives adapt to the challenges posed by aging. Accelerated aging is employed to identify any alterations in the mechanical properties and structural integrity of the adhesives, with the aim of predicting their long-term behavior.

The histograms in Figures 10–12 show the tensile behavior of the two adhesives aged at different temperatures and in different environments.

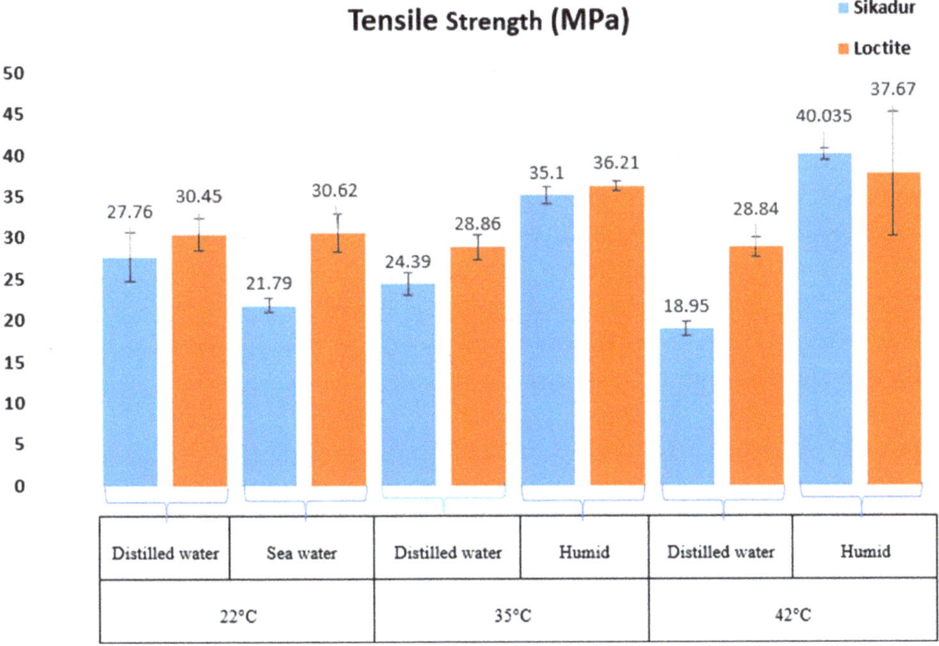

Figure 10. Results of tensile strength of Sikadur 30 and Loctite UK 1351 B25 adhesives after aging for 46 days.

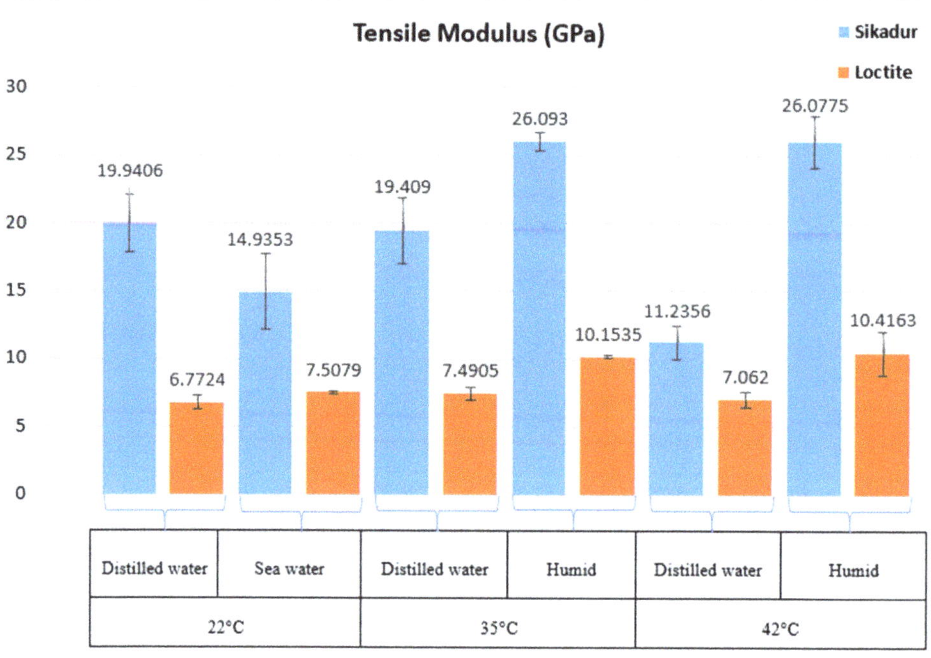

Figure 11. Results of tensile modulus of Sikadur 30 and Loctite UK 1351 B25 adhesives after aging for 46 days.

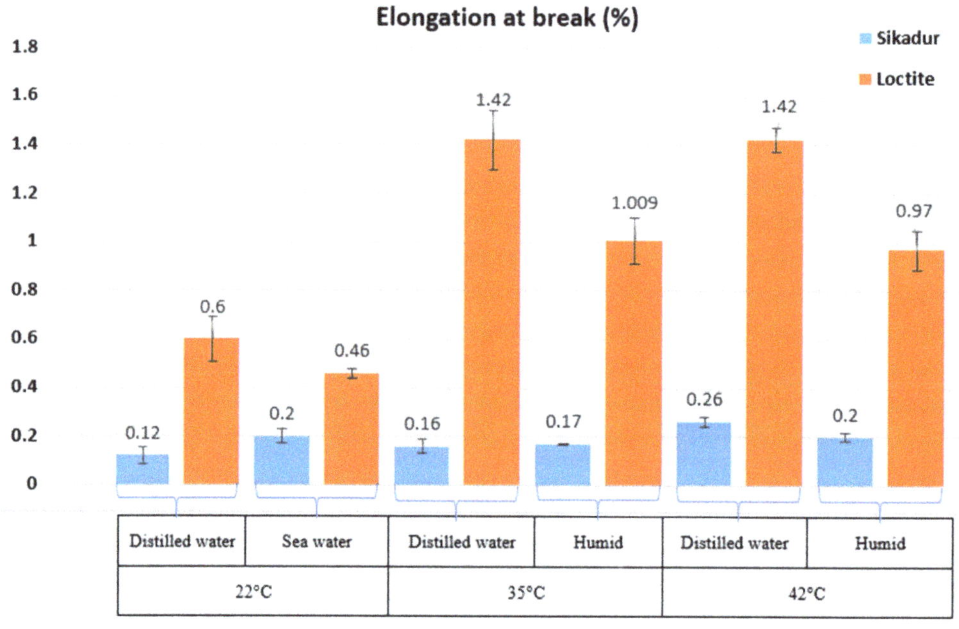

Figure 12. Results of elongation at break of Sikadur 30 and Loctite UK 1351 B25 adhesives after aging for 46 days.

This section focuses on the effect of aging temperature and environment on the tensile behavior of the adhesives studied, as published in many papers on marine adhesives [6,10,13,26]. For this purpose, the adhesive samples were exposed to different conditions—humidity, distilled water, and seawater—coupled with different temperatures (22 °C, 35 °C, and 42 °C) for 46 days. The main results obtained are shown in Figures 10–12. The initial parameters, consistent with the initial pre-aging tests, were used, and each temperature condition was evaluated with three samples, in accordance with established standards [8,15,27].

With respect to the temperature range maintained, various changes in the behavior of the adhesives studied were observed, as previously reported in the literature [3,14]. From Figure 10, it can be seen that in a humid environment, Sikadur showed a 9% decrease in yield stress at 35 °C, while at 42 °C it showed a 3% increase. The same tendency was observed for Loctite adhesive, with a 22% increase in stress at 35 °C and a 25% increase at 42 °C. These variations highlight the role of recrystallisation in modifying the stiffness of the polymer matrix [10,16].

A comparison of the parameters in distilled water at different temperatures revealed some interesting information. For the Sikadur adhesive, the stress decreased by 12% between 22 and 35 °C and by 31% between 22 and 42 °C. On the other hand, Loctite adhesive showed an almost stable decrease of 5% between 22 and 35 °C and remained constant between 22 and 35 °C and between 22 and 42 °C. These comparative observations are consistent with those reported in published work on the effect of temperature on the mechanical properties of adhesives [7,17].

In the case of Sikadur adhesive, the main observations correlate well with thermomechanical principles [16,28]. An increase in temperature resulted in an increase in deformation (Figure 12), indicating the increased ductility of the material under elevated thermal conditions. Sikadur adhesive exhibited its highest tensile strength (40.035 MPa) at 35 °C in a humid environment, highlighting its suitability for demanding conditions [1].

In terms of material stiffness, Figure 11 shows that the tensile modulus of the material reaches its maximum at 35 °C (26,093 MPa), indicating that the material effectively maintains its stiffness. The superior performance in distilled water compared to seawater is consistent with Refs. [11,29]. It also appears that the presence of moisture improves the performance, particularly at higher temperatures, suggesting a beneficial interaction with moisture. A similar trend was also observed for Loctite adhesive. In fact, as the temperature increases, the material deformation (ductility) becomes more important, which is the case for Sikadur. Loctite adhesive reached its maximum tensile strength (37.67 MPa) at 42 °C in a humid environment, indicating remarkable strength, particularly at elevated temperatures. The tensile modulus reached its maximum at 35 °C (10,153.5 MPa), demonstrating excellent stiffness.

It should be noted that when seawater is combined with the positive influence of humidity on performance, the behavior of Sikadur is well illustrated.

In terms of comparative evaluation, both Loctite and Sikadur adhesives showed excellent mechanical properties under various conditions, confirming the findings of recent work in similar areas of research [6,30,31].

3.4. Adhesives Behavior at Longer Periods

According to previous published work [9,32], the comparison between the two adhesives should consider all aspects in order to make the right decisions. For this purpose, in this section, the variation in the ultimate tensile strength (UTS) for both adhesives at different aging times and at different temperatures (22, 35, and 42 °C) has been followed. Then, in a second phase, this section deals with the detailed description of the evolution of the mechanical behavior of the two adhesives, especially at longer aging times. The results obtained are shown in Figures 13 and 14 for Sikadur and Loctite adhesives, respectively.

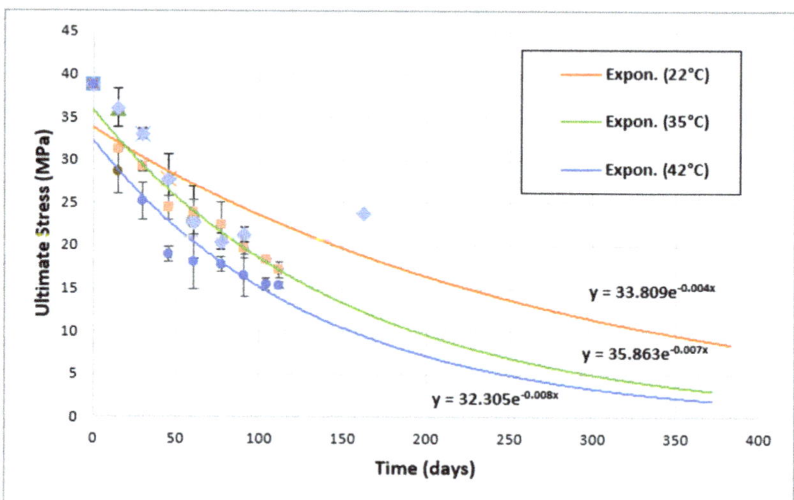

Figure 13. Evolution of ultimate tensile strength (UTS) over time for Sikadur adhesive.

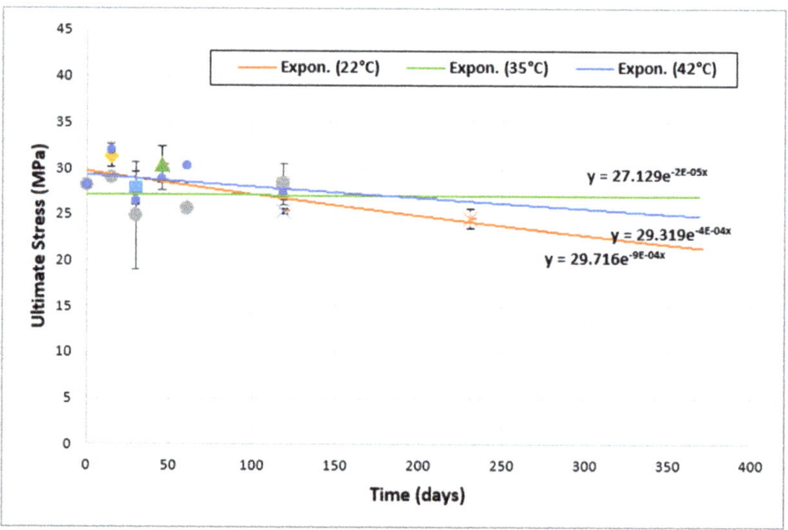

Figure 14. Evolution of ultimate tensile strength (UTS) over time for Loctite adhesive.

As can be seen in Figures 13 and 14, this study of the mechanical properties of aged Sikadur (up to 163 days) and Loctite (up to 231 days) shows a clear trend towards a progressive decrease in ultimate tensile strength (UTS) with the exposure time. Such a trend can take the form of an exponential function, as follows:

$$\sigma = ae^{-kt},\tag{2}$$

where σ represents the ultimate stress; t is the aging time; and a and k are parameters. The correlation between our observations and those reported in other publications on the aging of adhesives utilized in marine environments is noteworthy [12,19,33].

For Sikadur adhesive exposed in distilled water at 22 °C, the UTS decreased from an initial value of 38.75 MPa to 23.67 MPa after 163 days. A similar trend was observed for exposure at 35 °C and 42 °C. This progressive degradation is consistent with the expected effects of aging. This also highlights the importance of regular maintenance for adhesive materials during their service life [18,20].

Loctite adhesive also shows a similar aging behavior. In fact, at 22 °C, the UTS of Loctite adhesive decreased from 28.12 MPa to 24.56 MPa after 231 days. Similar trends were observed at 35 °C and 42 °C, indicating the evolution of the mechanical properties of the adhesive material over time [34–36].

Based on Figures 13 and 14, all parameters describing the aging behavior of the two adhesives were extracted and then summarized in Table 3.

Table 3. Parameters of exponential decrease in ultimate tensile strength (UTS) as function of time.

	T (°C)	a	k
	22	32.305	0.008
Sikadur adhesive	35	35.863	0.007
	42	33.809	0.004
	22	29.716	9.00×10^{-4}
Loctite adhesive	35	27.129	0.20×10^{-4}
	42	29.319	4.00×10^{-4}

Comparing Sikadur and Loctite, it appears that Loctite generally retains a higher tensile strength under similar conditions, indicating potential resistance to the effects of aging. However, both adhesives show significant changes in UTS over time. In fact, both Sikadur and Loctite adhesives undergo some degree of degradation over time. Furthermore, this degradation is more pronounced at elevated temperatures which accelerate the aging process [2,27].

The exponential trends also highlight the need for a better understanding of the aging behavior of adhesives [20,36]. This fact underlines the essential role of regular evaluation and maintenance in order to ensure the long-term reliability of adhesives [17,20,32].

On the other hand, in addition to the tensile strength, the stiffness of the aged materials was also evaluated during the aging period. The evolution of the tensile modulus over time for Sikadur and Loctite adhesives is shown in Figures 15 and 16, respectively.

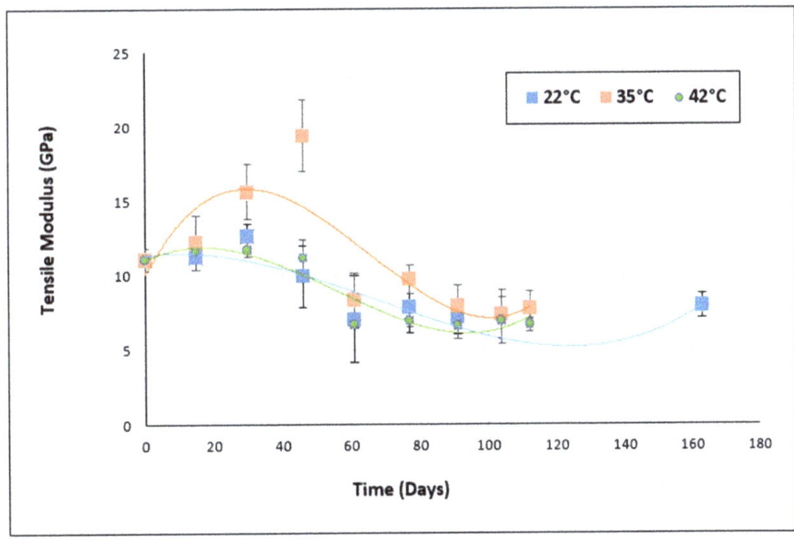

Figure 15. Evolution of tensile modulus over time for Sikadur adhesive.

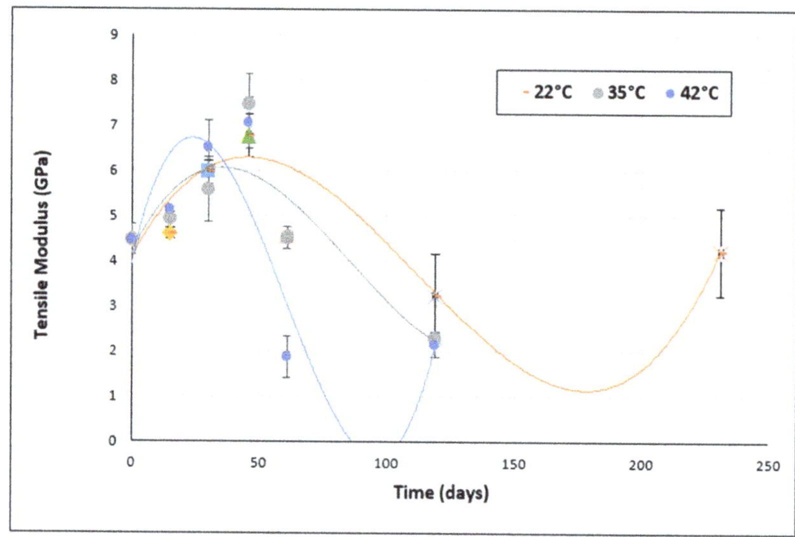

Figure 16. Evolution of tensile modulus over time for Loctite adhesive.

From Figures 15 and 16, it can be seen that for both Sikadur and Loctite adhesives, the variation in tensile modulus over time shows several trends.

Firstly, for the Sikadur adhesive exposed at 22 °C, an initial increase in tensile modulus was observed during the first 30 days of aging, followed by a gradual decrease up to 77 days. Thereafter, the tensile modulus appears to stabilize. This trend may indicate an initial phase of aging during which chemical bonds are strengthened, followed by a possible gradual deterioration of mechanical properties.

At 35 °C, there is a significant increase in the tensile modulus during the first 30 days of aging, suggesting a more rapid response at higher temperatures. However, after this period, there is a significant decrease in the modulus, which could indicate an early deterioration of mechanical properties at elevated temperatures.

For samples exposed at 42 °C, a progressive decrease in tensile modulus was observed throughout the aging process. This again confirms the continuous degradation of mechanical properties with increasing temperature.

In the case of Loctite adhesive, an initial increase in tensile modulus was observed during the first 30 days (at 22 °C), followed by a gradual decrease. This suggests a period of initial cure followed by subsequent degradation. At 35 °C, an initial increase is observed, but this is followed by a marked decrease, indicating an early degradation of mechanical properties at higher temperatures. Finally, at 42 °C, a similar trend of gradual decrease in tensile modulus is observed, suggesting continuous degradation at elevated temperatures.

In general, when comparing the performance of Sikadur and Loctite adhesives, Loctite appears to be more resistant to mechanical degradation at higher temperatures. In fact, Loctite adhesive maintains relatively higher tensile moduli than Sikadur adhesive under the same conditions.

However, it is worth noting that adhesive performance can be influenced by various factors, such as chemical composition, molecular structure, and environmental conditions. Further analysis would therefore be required to fully understand the results obtained.

The following section focuses on the limitations of the Arrhenius law in the case of the two adhesives when exposed to an aging environment. To achieve this, the logarithm of the aging times (t) is plotted against the reciprocal of the temperature ($1/T$) in Figures 17 and 18.

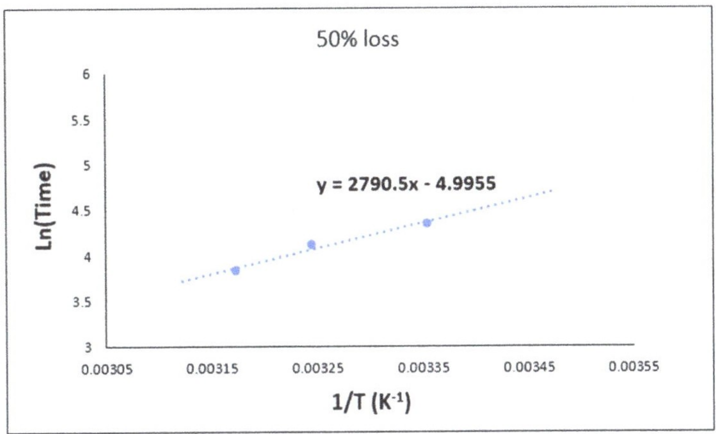

Figure 17. Validation of the Arrhenius law for Sikadur adhesive in distilled water.

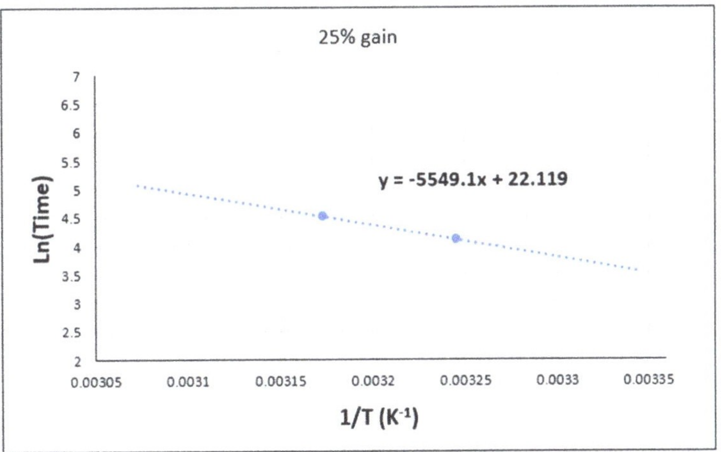

Figure 18. Validation of the Arrhenius Law for Sikadur adhesive in humid environment.

As can be seen in Figure 17, the aging of Sikadur adhesive shows a remarkable validation of the Arrhenius law, as it shows a linear relationship between the logarithm of time (in days) and the reciprocal of the absolute temperature (in Kelvin), as in Equation (3):

$$Ln(t) = -\frac{E_a}{R} \times \frac{1}{T} + \ln(A)$$

(3)

where t is the aging time; E_a is the activation energy; R is the perfect gas constant; T is the absolute temperature; and A is the pre-exponential factor.

The results obtained are consistent with recent studies on adhesives used in marine environments [11,14,26]. In particular, the linear correlation between Ln(time) and $1/T$ (Figure 17) indicates that the aging process of Sikadur adhesive follows Arrhenius kinetics [37]. This result provides a mathematical model for predicting the behavior of the adhesive over time under different temperature conditions.

In addition, the curve obtained in Figure 17 also shows a 50% loss of properties, which represents a significant reduction in the mechanical performance of the adhesive [28]. This can give an indication of the lifetime of the material and its ability to resist.

On the other hand, the curve obtained in Figure 18 shows an increase in the mechanical performance of the adhesive as the wet temperature increases. In fact, at higher temperatures, the chemical reactions responsible for polymerization and curing of the adhesive can take place more rapidly. This makes it possible to improve its structure and mechanical properties.

It is, however, essential to acknowledge that this interpretation is contingent upon an empirical relationship. Other factors, such as water, humidity, and environmental composition, may also influence the mechanical performance of adhesive materials.

By applying the Arrhenius law (Equation (3)) and utilizing the provided data, the long-term behavior of the Sikadur adhesive under these specific conditions can be estimated. Indeed, by extrapolating the Ln(days) values for additional values of $(1/T)$, the evolution of the aging time in these environments can be predicted.

However, it should be noted that long-term extrapolation may be subjected to uncertainties and limitations. Indeed, the extrapolation is based on limited experimental data and on the assumption of continuity of the observed behavior in the available data. In addition, other environmental and aging factors could potentially affect adhesive performance over time. As a result, long-term projections may be subject to potential variation.

4. Discussion

The results of the extensive experimental program shed light on the performance of Sikadur and Loctite adhesives in offshore wind applications, and provided important insights that can be interpreted in the context of previous studies and working hypotheses.

Comparing the results of this study with existing research in the field of adhesive behavior, it is clear that both Sikadur and Loctite adhesives exhibit important mechanical properties. These results are consistent with previous studies that have highlighted the importance of durable and resilient adhesives to withstand the harsh conditions encountered in offshore environments.

In addition, the differences observed in water absorption kinetics between the Sikadur and Loctite adhesives add to the body of knowledge on adhesive behavior under varying environmental conditions. By understanding these differences, engineers and researchers can make more informed decisions when selecting adhesives for specific offshore wind turbines projects, taking into account factors such as initial absorption rates and maximum absorption capacities.

The implications of the findings extend beyond the specific materials tested in this study, providing important insights into the broader challenges and opportunities in offshore wind applications. By considering the implications of these results in the broadest possible context, researchers can identify areas for further investigation and improvement in adhesive technology for offshore wind structures.

Future research directions may include refining predictive models for assessing the long-term performance of adhesives, incorporating additional environmental factors such as UV exposure and mechanical loading. In addition, studies focusing on the development of novel adhesive formulations optimized for offshore wind turbines applications could further advance the field and contribute to the sustainability and reliability of renewable energy sources.

5. Conclusions

In this work, an extensive experimental program was carried out at the LMA laboratory in Marseille in order to evaluate the durability performance of two adhesives and to determine their suitability for offshore wind applications. The investigation included a series of mechanical tests and accelerated aging simulations (in terms of temperature, relative humidity (RH), aging times).

The following main conclusions can be drawn from the studies carried out on Sikadur and Loctite adhesives under different aging conditions:

- Both adhesives showed significant mechanical properties, demonstrating their potential for this industrial use.
- Loctite adhesive has a slightly faster initial absorption rate than Sikadur adhesive, but the latter reaches an asymptotic plateau at a lower maximum absorption rate than Loctite adhesive.
- When subjected to aging conditions, Sikadur adhesive showed good tensile strength, particularly at elevated temperatures and in humid environments. Loctite adhesive also showed good mechanical properties, with notable resistance observed at higher temperatures and in wet conditions.
- When both adhesives were exposed to longer periods of time, the aging results showed a progressive deterioration in the mechanical properties as a function of aging time. This appears to follow an exponential function.
- The aging results show a clear correlation with the Arrhenius law, providing a predictive tool for the aging process. The aging process thus follows Arrhenius kinetics.

In addition to confirming the results obtained during this investigation, it is imperative to emphasize the innovative aspects of the methodology employed in comparison to existing research paradigms. This study integrates a comprehensive experimental program encompassing mechanical tests and accelerated aging simulations to assess the durability performance of two adhesives for offshore wind applications. The incorporation of these diverse approaches yields valuable insights into the mechanical properties and aging behavior of the adhesives, as well as a predictive framework for evaluating their service life. Of particular significance is the correlation with the Arrhenius law, which represents a significant advance in comprehending the long-term performance of adhesives in offshore wind turbine applications. Further research could be conducted with the objective of refining predictive models to encompass additional environmental factors and operational conditions. This would enhance the understanding of material behavior and optimize the design and maintenance of offshore wind structures.

Author Contributions: All co-authors agreed with the content of this submission and they are in agreement with the submission. They have also obtained consent from the responsible authorities at the institutes/organizations where the work has been carried out, before the work was submitted. In terms of contributions to this work, all the mentioned authors (K.I., A.M.-P., F.L. and N.G.) have contributed to the conception of the work; or the acquisition, analysis, or interpretation of data. Also, they have approved the version to be published; and they agree to be accountable for all aspects of the work in ensuring that questions are related. All authors have read and agreed to the published version of the manuscript.

Funding: This research received no external funding.

Institutional Review Board Statement: Not applicable.

Informed Consent Statement: Not applicable.

Data Availability Statement: Data are contained within the article.

Conflicts of Interest: The authors declare no conflicts of interest.

References

1. Andersen; Sivert, O.; Trond, O.; Hagen; Berg, J.K. Enhancing Adhesive Performance in Offshore Wind Turbine Applications. *J. Wind Eng. Ind. Aerodyn.* **2016**, *149*, 27–38.
2. *ASTM D638-14*; Standard Test Method for Tensile Properties of Plastics. ASTM International: West Conshohocken, PA, USA, 2014.
3. Harper, P.; Smith, M.; Johnson, R.; Lee, S. Evaluating Adhesive Performance in Marine Environments: A Comprehensive Review. *Ocean Eng.* **2015**, *104*, 25–35.
4. Bruneaux, M.-A. Durability of Edhesively Bonded Structures: Development of a Predictive Mechanical Modelling Taking into Account Physico-Chemical Characteristics of the Adhesive. Ph.D. Thesis, University of Rome II, Rome, Italy, 2004.
5. Chen, J.; Li, W.; Zhang, H. Effect of Moisture Absorption on Mechanical Performance of Structural Adhesive Joints. *J. Adhes. Sci. Technol.* **2019**, *33*, 1137–1150.

6. Chen, Y.; .Liu, W.Z.; Zhang, Q. Advanced Adhesive Solutions for Offshore Wind Farms: A Review. *Renew. Sustain. Energy Rev.* **2020**, *130*, 109934.
7. Chin, J.W.; Aouadi, K.; Haight, M.R.; Hugues, W.L.; Nguyen, T. Effect of water, salt solution and simulated concrete pore solution on the properties of composite matrix resins used in civil engineering applications. *Polym. Compos.* **2001**, *22*, 282–297. [CrossRef]
8. *EN 17024:2019*; Adhesives—Determination of Tensile Strength of Joints. European Committee for Standardization (CEN): Brussels, Belgium, 2019.
9. Gao, L.; Zhang, H.; Wang, Z.; Li, X.; Chen, Q. Adhesive Bonding Technologies for Offshore Wind Turbine Structures: A Comprehensive Review. *Renew. Energy* **2020**, *145*, 2008–2025.
10. Gao, Z.; Zhao, X. Experimental Study on Mechanical Properties of Composite Adhesives in Marine Environment. *Ocean Eng.* **2019**, *192*, 106523.
11. Gonzalez, M.; Smith, J.; Johnson, K.; Martinez, L.; Patel, S. Challenges and Solutions in Adhesive Bonding for Marine Applications. *J. Adhes. Sci. Technol.* **2017**, *22*, 245–262.
12. Green, R.E.; Brown, T.K.; Carter, M.J.; Davis, H.R.; Evans, G.P. Long-Term Durability of Structural Adhesives in Marine Environments. *Polym. Degrad. Stab.* **2018**, *147*, 10–18.
13. Kim, S.H.; Lee, H. Characterization of Epoxy Adhesives for Offshore Wind Turbine Blades. *Int. J. Adhes. Adhes.* **2019**, *90*, 1–9.
14. Brown, L.E.; Miller, J.D.; Taylor, S.R.; Wilson, K.P.; Harris, A.B. Advancements in Adhesive Technologies for Harsh Marine Environments. *Renew. Energy* **2020**, *45*, 123–136.
15. *ISO 3167:2013*; Adhesives—Determination of Tensile Lap-Shear Strength of Rigid-to-Rigid Bonded Assemblies. ISO: Geneva, Switzerland, 2013.
16. Li, Y.; Zhang, H. Aging Behavior of Epoxy Resin Adhesive in Marine Environment. *Mater. Today Proc.* **2020**, *27*, 1354–1359.
17. Kuang, Y.; Zhang, Q.; Lee, M.; Yang, S.; Zhao, H. Structural Adhesives in Offshore Wind Energy: A Comprehensive Review. *Renew. Sustain. Energy Rev.* **2016**, *56*, 745–758.
18. O'Connor, A.; Smith, B.; Jones, C.; Patel, R.; Thomas, D. Environmental Effects on Adhesive Bonding in Marine Structures. *Mar. Struct.* **2021**, *58*, 102982.
19. Williams, P.A.; Brown, L.E.; Garcia, M.; Johnson, K.; Wilson, S. Durability Assessment of Structural Adhesives in Offshore Wind Turbine Environments. *J. Renew. Mater.* **2019**, *15*, 387–402.
20. Xu, W.; Li, H.; Wang, J.; Zhang, L.; Chen, Y. Adhesive Bonding for Structural Applications in Marine Environments: A Review. *J. Mar. Sci. Eng.* **2017**, *5*, 50.
21. Ghabezi, P.; Harrison, N.M. Multi-scale modeling and life prediction of aged composite materials in saltwater. *J. Reinf. Plast. Compos.* **2023**, *43*, 205–219. [CrossRef]
22. Zike, W.; Johnson, R.; Smith, L.; Brown, M.; Taylor, D. Long-term durability of basalt- and glass-fiber reinforced polymer (BFRP/GFRP) bars in seawater and sea sand concrete environment. *Constr. Build. Mater.* **2017**, *139*, 467–489.
23. Niu, R.; Yang, Y.; Liu, Z.; Ding, Z.; Peng, H.; Fan, Y. Durability of Two Epoxy Adhesive BFRP Joints Dipped in Seawater under High Temperature Environment. *Polymers* **2023**, *15*, 3232. [CrossRef]
24. Berens, A.R.; Hopfenberg, H.B. Diffusion and relaxation in glassy polymer powders: 2. Separation of diffusion and relaxation parameters. *Polymer* **1978**, *19*, 489–496. [CrossRef]
25. Nogueira, P.; Ramirez, C.; Torres, A. Effect of water sorption on the structure and mechanical properties of an epoxy resin system. *J. Appl. Polym. Sci.* **2001**, *80*, 71–80. [CrossRef]
26. Molenaar, J.; Johnson, A.; Williams, E.; Brown, M.; Thomas, R. Durability of Adhesive Bonds in Composite Marine Structures: A State-of-the-Art Review. *J. Compos. Sci.* **2019**, *3*, 35.
27. *ISO 527-2:2012*; Plastics—Determination of Tensile Properties—Part 2: Test Conditions for Moulding and Extrusion Plastics. ISO: Geneva, Switzerland, 2012.
28. Wang, X.; Zhang, Q.; Lee, H.; Chen, Y.; Yang, S. Recent Advances in Marine Adhesives for Structural Bonding Applications. *Int. J. Adhes. Adhes.* **2018**, *81*, 28–41.
29. Zhang, L.; Lee, H.; Johnson, R.; Brown, M.; Taylor, D. Adhesive Technologies for Composite Materials in Offshore Wind Turbines: A Review. *Compos. Part B Eng.* **2017**, *111*, 135–146.
30. Soutis, C.; Garcia, M.; Wilson, K.; Patel, R.; Thomas, D. Recent Advances in Adhesive Bonding: Metal Adhesives. *J. Adhes.* **2019**, *95*, 1151–1191.
31. Wang, Y.; Li, H.; Chen, J.; Zhang, L.; Xu, W. Advanced Adhesive Technologies for Structural Bonding in Marine Environments. *Front. Mater.* **2021**, *8*, 641002.
32. Smith, J.K.; Johnson, A.R. Adhesive Selection and Testing for Offshore Wind Applications. *J. Mater. Eng. Perform.* **2018**, *27*, 3396–3405.
33. Ma, Y.; Li, W. Effects of Environmental Factors on the Mechanical Properties of Marine Adhesive. *Mater. Struct.* **2017**, *50*, 188.
34. *ISO 19030-1:2016*; Guidelines for Measurement of Changes in Hull and Propeller Performance (Part 1: General Principles). International Organization for Standardization (ISO): Geneva, Switzerland, 2021.
35. Stutz, B.; Miller, J.D.; Brown, L.E.; Wilson, K.P.; Thomas, R. Aging of High-Performance Structural Adhesives: A Review. *Int. J. Adhes. Adhes.* **2017**, *76*, 70–88.

36. Zhang, H.; Song, Y.; Song, Y. Aging Behavior and Durability Evaluation of Epoxy Adhesives in Marine Environment. *J. Adhes. Sci. Technol.* **2017**, *31*, 1975–1990.
37. Lomov, S.V.; Brown, T.K.; Miller, J.D.; Wilson, S.R.; Taylor, D.H. Elastic Constants and Short Beam Shear Strength of Adhesively Bonded Composite Joints. *Compos. Struct.* **2016**, *157*, 1–12.

Disclaimer/Publisher's Note: The statements, opinions and data contained in all publications are solely those of the individual author(s) and contributor(s) and not of MDPI and/or the editor(s). MDPI and/or the editor(s) disclaim responsibility for any injury to people or property resulting from any ideas, methods, instructions or products referred to in the content.

Article

Nonlinear Analytical Procedure for Predicting Debonding of Laminate from Substrate Subjected to Monotonic or Cyclic Load

Marco Lamberti [1,2,*], **Francesco Ascione** [3], **Annalisa Napoli** [3], **Ghani Razaqpur** [4,5,*] and **Roberto Realfonzo** [3]

[1] ENEA, Research Center Brasimone, 40032 Camugnano, Italy
[2] CNRS, Centrale Marseille, LMA, Aix-Marseille University, 13453 Marseille, France
[3] Department of Civil Engineering, University of Salerno, 84084 Fisciano, Italy
[4] Sino-Canada Joint R&D Center on Water and Environmental Safety, College of Environmental Science and Engineering, Nankai University, Tianjin 300071, China
[5] Department of Civil Engineering, McMaster University, Hamilton, ON L8S 4L8, Canada
*   Correspondence: lamberti@lma.cnrs-mrs.fr (M.L.); razaqpu@mcmaster.ca (G.R.)

Abstract: The bonding of steel/fiber-reinforced polymer (SRP/FRP) laminate strips to concrete/masonry elements has been found to be an effective and efficient technology for improving the elements' strength and stiffness. However, premature laminate–substrate debonding is commonly observed in laboratory tests, which prevents the laminate from reaching its ultimate strength, and this creates uncertainty with respect to the level of strengthening that can be achieved. Therefore, for the safe and effective application of this technology, a close estimate of the debonding load is necessary. Towards this end, in this paper, a new, relatively simple, semi-analytic model is presented to determine the debonding load and the laminate stress and deformation, as well as the interfacial slip, for concrete substrates bonded to SRP/FRP and subjected to monotonic or cyclic loading. In the model, a bond-slip law with a linearly softening branch is combined with an elasto-plastic stress-strain relationship for SRP. The model results are compared with available experimental data from single-lap shear tests, with good agreement between them.

Keywords: SRP; semi-analytic model; cyclic loading

Citation: Lamberti, M.; Ascione, F.; Napoli, A.; Razaqpur, G.; Realfonzo, R. Nonlinear Analytical Procedure for Predicting Debonding of Laminate from Substrate Subjected to Monotonic or Cyclic Load. *Materials* **2022**, *15*, 8690. https://doi.org/10.3390/ma15238690

Academic Editor: Konstantinos Tserpes

Received: 3 November 2022
Accepted: 2 December 2022
Published: 6 December 2022

Publisher's Note: MDPI stays neutral with regard to jurisdictional claims in published maps and institutional affiliations.

Copyright: © 2022 by the authors. Licensee MDPI, Basel, Switzerland. This article is an open access article distributed under the terms and conditions of the Creative Commons Attribution (CC BY) license (https://creativecommons.org/licenses/by/4.0/).

1. Introduction

After decades of service, concrete structures may exhibit insufficient strength or stiffness due to either environmental degradation or changes in loading requirements, thus creating the need for their repair/rehabilitation [1–3]. Relatively recently, the repair and retrofit of deficient concrete members by C (SRP) textiles or laminates have been found structurally and economically feasible. Such textiles are made of high-tensile-strength steel micro wires, twisted into small diameter cords or strands, which are unidirectionally laid in a polymer matrix to form a composite fabric. Similar to the more familiar fiber-reinforced polymer (FRP) laminates, the SRP can be externally bonded to a substrate via wet lay-up, using either epoxy or polyester resin as adhesive. The tensile behavior of steel textiles has been investigated in a number of studies, either specifically devoted to their mechanical characterization or in the context of assessing their suitability as external reinforcement for structural elements [4,5].

As demonstrated in the above investigations, differently from similar FRP strengthened members, SRP-retrofitted elements can exhibit relatively ductile behavior and higher energy dissipation at failure. However, the bond between the laminate and the concrete substrate is a crucial factor that affects the efficacy of this retrofit method. Consequently, analytical/numerical assessment of the retrofit requires, among other things, a suitable constitutive law for the laminate–substrate interface.

Modeling the global behavior of existing members strengthened in bending by means of bonded FRP laminates is a major topic of interest in practice and it has been extensively researched [6–14], but unlike FRP, which behaves linear-elastically until failure, SRP is an elasto-plastic material with strain-hardening characteristics. Consequently, the analysis of SRP retrofitted elements can be relatively more complex than that of FRP-retrofitted elements. To date, some mainly experimental studies have been conducted to investigate the physical and mechanical properties of SRP and the behavior of SRP-retrofitted concrete elements via single-lap shear tests. Furthermore, some analytical/numerical studies involving fracture mechanics and finite element analyses have been conducted with the aim of developing some empirical models [6], following the models intended for FRP-retrofitted elements as described below.

Teng et al. [7] studied the behavior of FRP–concrete interface between two adjacent cracks, assuming all forces applied at the two ends of the FRP strengthening plate and concrete member to remain proportional during the entire loading process. Although their study was aimed at describing the behavior of retrofitted concrete members, they applied a simplified relationship to describe the complex interfacial shear stress-slip relationship. Similarly, Chen et al. [8] developed a simple analytical solution for determining the strength of FRP–concrete bonded joints based on a linear softening bond-slip curve without allowing any slip before the maximum shear at the interface reaching or exceeding the interfacial shear resistance. Another detailed treatment of this problem was undertaken by Quiao and Chen [9], who presented a solution using a linear law to describe the bond-slip behavior of the interface.

To characterize the debonding mechanism, numerous other models have been proposed based on linear or nonlinear fracture mechanics, regression analysis and some semi-empirical methods. The available models focus on the analysis of the stress transfer and fracture propagation in different kinds of adhesive joints by adopting different shear stress-slip models, with or without consideration of a softening branch. For example, cohesive debonding of a bonded strip from an elastic substrate has been studied by Franco. and Royer Carfagni [10] using a model representing a finite stiffener bonded to the boundary of a semi-infinite plate to determine the effective bond length of the stiffener. Cornetti and Carpinteri [11] obtained an analytical solution by assuming an exponential decaying softening branch for the interfacial bond-slip law. Accordingly, expressions for the interfacial shear stress distribution and the load–displacement response were derived for the different loading levels. In [12] the effect of different bonded lengths on FRP–concrete interfacial debonding behavior was investigated. Finally, Liu et al. [13] presented an analytical model to simulate the debonding process at the FRP–concrete interface in a single-lap shear test, but the accuracy of the model was gauged by only comparing its results with the corresponding finite element analysis results.

Using a suitable bond-slip law or relationship, analytical solutions for determining the tensile stress and strain in the FRP laminate, the failure load of the retrofitted assemblage, the shear stress distribution and the associated slip along the FRP–concrete interface, and the degree of interfacial damage due to a certain applied load can be obtained. The bond-slip relationship depends on the material properties of the adherents and the geometrical dimensions of the elements involved [14]. Since SRP as an elasto-plastic material has different properties than linear elastic FRP, the analytical models developed for FRP–concrete interfaces may not directly apply to SRP–concrete interfaces. Consequently, there is a need for simple analytical models that can be used to predict the behavior and strength of SRP–concrete interfaces.

Another important factor that may govern interfacial failure or debonding is the type of load to which the FRP is subjected. To date, the majority of studies have focused on monotonic loading while some structures are retrofitted to increase their seismic resistance or, in the case of bridges, to increase their live load capacity. In both cases, the interface will be subjected to cyclic loading. Consequently, it is necessary to investigate the FRP/SRP–concrete interfacial behavior and ultimate strength under cyclic loads. In particular, the effect of those cyclic loadings that induce very high bond stresses must be determined to avoid premature debonding.

Some researchers have investigated the fatigue performance of the FRP–concrete interface [15–31]. Other studies [15,16] have shown that debonding failure due to fatigue loads is often smaller than the ultimate capacity under monotonic loading. Furthermore, the combined effects of cyclic loading and hygrothermal environment accelerate the degradation process of the FRP-to-concrete bonded interface and thus reduce the fatigue life of the interface [17,18]

Despite numerous studies dealing with the fatigue behavior of strengthened structures [19,20] and the advantages of the FRP strengthening systems to delay the crack propagation and limiting the crack width, there are still some unresolved issues with regard to the fatigue performance of the FRP-to-concrete bond [21–26], partly due to the difficulties of conducting sufficiently robust tests to comprehensively capture the fatigue behavior of the FRP/SRP-concrete interface under direct shear and ensuring the pure shear stress state of the bonded interface [22,29]. Among the various bond tests available in the literature, single and double shear tests are probably the most common.

Zheng et al. [18] performed double shear test on specimens with carbon fiber laminate by subjecting them to fatigue load under a series of constant temperature and relative humidity (RH) conditions. The results showed that the temperature and RH negatively affected the bond behavior of the carbon–concrete interface. Other experimental double-shear tests using an improved test set-up were performed by Yun et al. [21], demonstrating the effect of different anchoring systems on the fatigue behavior of the FRP-to-concrete bond; the influence of the load amplitude and number of cycles on the bond performance. The test results showed the better fatigue performance of the hybrid-bonded FRP (HB-FRP) system compared to the other FRP strengthening systems that were examined.

Ferrier et al. [22] studied the fatigue behavior of the bond interface using a standard double-lap shear test. The experimental results suggested a linear relationship between the maximum strength of the interface and the logarithm of the number of load cycles. Particulary, Mazzotti et al. [26] presented the results of an experimental program involving the cyclic behavior of FRP–concrete interface while Nigro et al. [27] reported the results of fatigue tests on single-lap shear specimens. Ko et al. [29] proposed a bond-slip model intended to simulate the observed response of a series of specimens retrofitted with aramid, carbon or polyacetal FRP strips bonded to concrete blocks and subjected to monotonic or cyclic load. In the model a Popovics-like constitutive law was assumed and it involved seven mechanical parameters that must be calibrated experimentally. Finally, Martinelli et al. [30], formulated a model using two different expressions to represent the bond-slip behavior of FRP strips bonded to concrete substrate.

Considering the need for a robust, yet simple, analytical model that can be used to analyze the interfacial behavior and obtain the ultimate capacity of FRP- (Carbon, glass, aramid, etc.) or SRP-retrofitted concrete elements, in the current investigation such a model is proposed. The proposed model is intended for application to elements subjected to monotonic or cyclic load. Essentially, a closed-form analytical solution is presented to rapidly analyze the interfacial response of linear elastic or elasto-plastic laminates bonded to concrete or other similar substrates subjected to monotonic or cyclic load. The accuracy and robustness of the model are demonstrated by analyzing many test specimens and comparing the analytical results with companion test results reported by others. Besides the ultimate capacity of the retrofitted element, the model can provide interfacial shear stress distribution as well as the laminate stress and strain along its bonded length.

2. Governing Equations

The purpose of the present investigation is to provide an analytical model that can be used to study the pull-out behavior and interfacial debonding of a FRP or SRP laminate bonded to a concrete substrate as encountered in single lap shear tests. The geometry of a typical single-lap shear test is represented in Figure 1, where the symbol L_f denotes the bonded length, bf the width of the reinforcing strip, and b_c and h_c the width and the height of the concrete prism, respectively. The pull-out specimen illustrated in Figure 1 is composed of a FRP or steel strip and a concrete prism, which can be treated as two adherents subjected to axial deformation only, while the adhesive layer that bonds the two can be assumed to be subjected only to shear deformation. Bending effects are neglected, while the shear stress and the axial deformation are assumed to be constant across and through the thickness of the adhesive layer and across the width of the laminate.

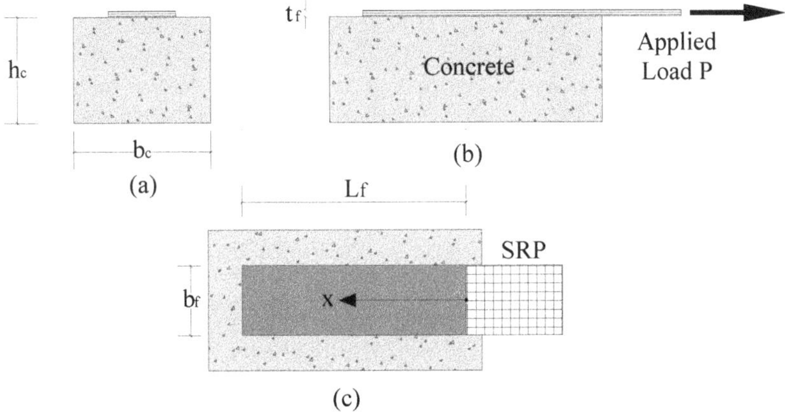

Figure 1. Single lap shear test specimen geometry: (**a**) front view; (**b**) side view; (**c**) top view.

To derive the basic equations of the analytical model, the equilibrium of the forces acting on an infinitesimal element of length dx along the strengthened prism is shown in Figure 2.

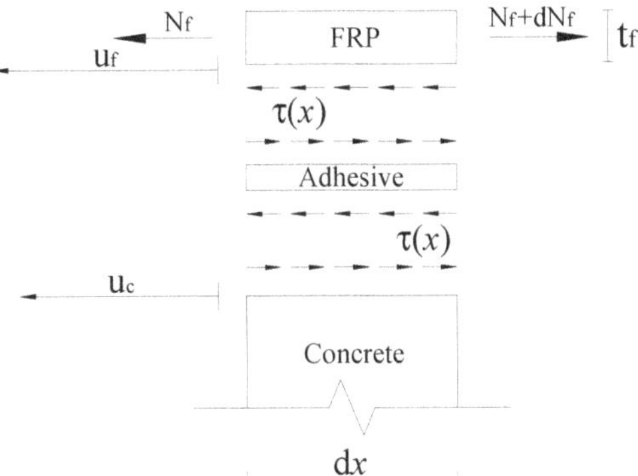

Figure 2. Equilibrium of an infinitesimal element of length dx along the specimen.

Based on the horizontal equilibrium of the forces in Figure 2,

$$\frac{dN_f}{dx} - \tau b_f = 0 \tag{1}$$

where $N_f = \sigma_f b_f t_f$ is the axial force resisted by the strengthening strip or laminate, σ_f the axial stress of strip or laminate, $\tau(x)$ the shear stress of adhesive layer.

The interfacial slip is the difference between the horizontal displacement of FRP laminates, u_f and the concrete surface, u_c:

$$\lambda = u_f - u_c \tag{2}$$

The evaluation of u_c requires the introduction of the equation of equilibrium of the axial forces acting on the specimen cross section as follows

$$N_c + N_f = \sigma_c h_c b_c + \sigma_f t_f b_f = 0 \tag{3}$$

It is important to point out that Equation (3) allows consideration not only of the geometry of the concrete prism section, but also its mechanical properties without disregarding substrate axial strain.

The SRP-Concrete Interface and the Pertinent Materials Constitutive Laws

The tensile behavior of steel textiles has been experimentally investigated by De Santis et al. [4,5] through traditional tensile tests, some of whose results are recapped in Figure 3. In this figure, it can be observed that the tensile stress-strain relationship of SRP is initially linear elastic up to 60–70% of its tensile strength, followed by a nonlinear segment that exhibits gradual stiffness reduction until the specimen rupture. This nonlinear pre-rupture behaviour is due to the intrinsic ductility of the steel cords and the partial unwinding of the twisted wires that form the cords. Cords with relatively steep twist angles are reported to exhibit more ductility. The latter behavior is not exhibited by typical FRP laminates; therefore, for analytical purposes, the SRP textiles behavior can be approximated by a bi-linear strain-hardening stress-strain relationship as shown in Figure 4 and expressed by Equation (4a,b). Such a relationship is adopted in the proposed analysis.

$$\sigma_f = E_{f,1}\varepsilon \text{ for } \varepsilon < \varepsilon_{fy} \tag{4a}$$

$$\sigma_f = \sigma_{fy} + E_{f,2}\left(\varepsilon - \varepsilon_{fy}\right) \text{ for } \varepsilon_{fy} < \varepsilon < \varepsilon_{fu} \tag{4b}$$

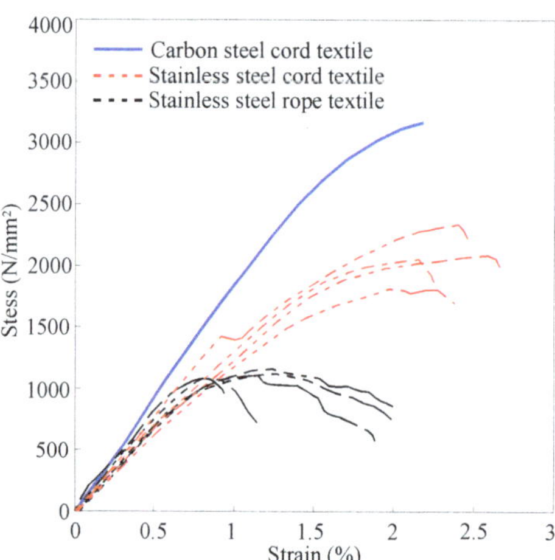

Figure 3. Tensile stress-strain relationship different SRP textiles.

The adhesive layer bonding the SRP to the substrate is mainly subjected to planar shear stress, which results in Mode II fracture. Fracture in Mode II is commonly represented by a bi-linear law composed of an ascending branch until the maximum shear stress is reached, followed by a descending or softening branch until complete loss of strength [32–34]. After the ultimate deformation or slip, λ_u, is reached or exceeded, full delamination is assumed to have occurred. Mathematically, the bond-slip relationship can be expressed as

$$\tau = k_s\lambda \text{ for } \lambda < \lambda_0 \tag{5a}$$

$$\tau = k_b(\lambda_u - \lambda) \text{ for } \lambda_0 < \lambda < \lambda_u \tag{5b}$$

$$\tau = 0 \text{ for } \lambda > \lambda_u \tag{5c}$$

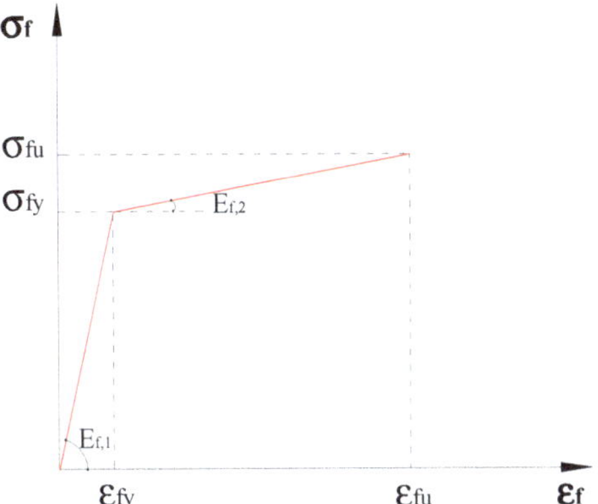

Figure 4. Steel textile stress-strain relationship.

With reference to Figure 5, in Equation (5a–c), λ_0 represents the slip corresponding to the maximum shear resistance τ_{max} while k_s and k_b are the shear stiffnesses of the interface corresponding, respectively, to the slope of the ascending and descending branches in the interfacial shear-slip relationship.

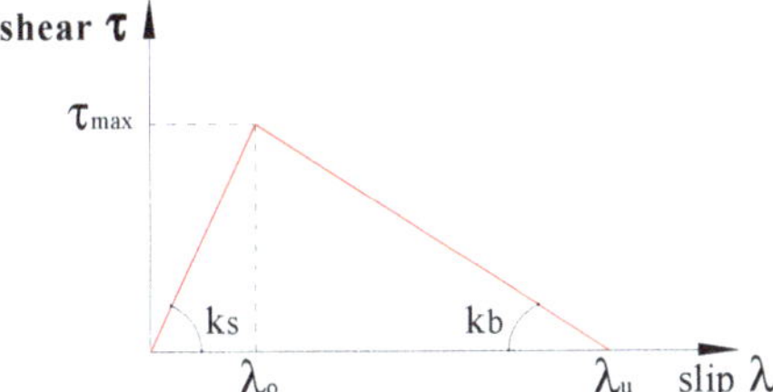

Figure 5. Shear stress-slip relationship of the adhesive layer at the interface.

Finally, in conformity with previous studies [14], concrete in tension is assumed to be a brittle linear elastic material whose stress-strain relationship can be written as

$$\sigma_c = E_c \frac{du_c}{dx} \tag{6}$$

where E_c represents the concrete elastic modulus. Although concrete is known to exhibit tension stiffening, this characteristic is not germane to the current analysis because failure in the cases with which the present analysis is concerned is almost always initiated by interfacial debonding.

3. Proposed Analytical Model

3.1. Monotonic Load Case

In this section, the complete debonding process in a single lap shear test subjected to a monotonically increased load P_i will be analyzed using a stage-by-stage approach. In particular, the proposed model permits the investigation of the axial strain and shear stress distribution, the interfacial slip along the bonded length, and the load-deformation curve of the bonded portion of SRP.

3.1.1. Elastic or Ascending Stage

For low load levels, the shear stress along the bonded length will be in the elastic or ascending stage. In this case, based on the constitutive relationship of the SRP textile Equation (7) and its substitution in Equations (1) and (4a), one can write

$$\frac{d\sigma_f}{dx} = E_{f,1}\frac{d^2 u_f}{dx^2} \tag{7}$$

$$E_{f,1}t_f\frac{d^3 u_f}{dx^3} - k_S\left(\frac{du_f}{dx} - \frac{du_c}{dx}\right) = 0 \tag{8}$$

Based on Equation (3), the axial stress at the concrete substrate level can be written as

$$\sigma_c = -\sigma_f\frac{l_f b_f}{t_c b_c} \tag{9}$$

Considering the linear elastic behaviour of concrete per Equation (6), the strain in the concrete can be expressed as

$$\frac{du_c}{dx} = -\frac{\sigma_f A_f}{E_c A_c} \tag{10}$$

Finally, substituting Equation (10) into Equation (8) results in the following second-order homogeneous differential equation

$$\varepsilon_f'' - \omega_{I,1}^2\varepsilon_f = 0 \text{ with } \omega_{I,1}^2 = k_s\left(\frac{1}{E_{f,1}t_f} + \frac{b_f}{E_c A_c}\right) \tag{11}$$

where the relation $\varepsilon_f = \frac{du_f}{dz}$ represents the axial strain of the SRP textile.

The general solution, $\varepsilon_f(x)$, of the above differential equation is

$$\varepsilon_{fI,1}(x) = A_{I,1}e^{\omega_{I,1}x} + B_{I,1}e^{-\omega_{I,1}x} \tag{12}$$

In which the two unknown constants can be determined by insertion into Equation (12) the relevant boundary conditions, expressed as Equation (13a,b), that is

$$N_f\left(L_f\right) = P \tag{13a}$$

$$\varepsilon_{fI,1}(0) = 0 \tag{13b}$$

3.1.2. Softening or Descending Stage

After the interfacial slip exceeds λ_0, the softening stage commences along the bonded length, starting from the loaded end and propagating towards the unloaded end. To model this stage, Equation (4b) is inserted in Equation (1), resulting in

$$E_{f,1}t_f\frac{d^3 u_f}{dx^3} - k_b\left(\frac{du_c}{dx} - \frac{du_f}{dx}\right) = 0 \tag{14}$$

Inserting Equation (10) in Equation (14) leads to the following governing differential equation in the latter case

$$\varepsilon_f'' + \omega_{1,II}^2\varepsilon_f = 0 \text{ with } \omega_{II,1}^2 = k_b\left(\frac{1}{E_{f,1}t_f} + \frac{b_f}{E_c A_c}\right) \tag{15}$$

The solution of Equation (15) can be written as

$$\varepsilon_{f\,II,1}(x) = A_{II,1}\cos(\omega_{II,1}x) + B_{II,1}\sin(\omega_{II,1}x) \tag{16}$$

In this stage, the bonded length must be divided into two regions. The first is the region from the unloaded end of the SRP to the point where $\lambda \leq \lambda_0$, while the second region comprises the remaining length. The bond-slip characteristics of the two regions are governed by the ascending and descending shear stress-slip relationship in Figure 4.

The general solution of the governing equation is given by Equation (12) for the elastic region and by Equation (16) for the softening region. The solutions of these two equations produce four constants of integration, which can be solved by enforcing the four boundary conditions given by Equation (17a–d):

$$N_f\left(L_f\right) = P \tag{17a}$$

$$\varepsilon_{fI,1}(x_a) = \varepsilon_{fII,1}(x_a) \text{ (due to the continuity)} \tag{17b}$$

$$\varepsilon'_{f\,I,1}(x_a) = \varepsilon'_{f\,II,1}(x_a) \text{ (due to the continuity)} \tag{17c}$$

$$\varepsilon_{fI,1}(0) = 0 \tag{17d}$$

Note, the parameter distinguishing region one from two is x_a, which can be determined by enforcing either of the following conditions:

$$\lambda(x_a) = \lambda_0 \tag{18a}$$

$$\tau(x_a) = \tau_{\max} \tag{18b}$$

3.1.3. Debonding with the SRP Remaining Elastic

Besides considering the interfacial conditions leading to the above stage, two scenarios must be considered depending on the stress level in the SRP textile, i.e., whether it is in the elastic or plastic state before the initiation of debonding. In the following analysis, it is first assumed that the SRP stress is below its yield stress and is consequently behaving elastically.

In this scenario, the applied load can be increased until ultimate slip s_u is reached, and the debonding process begins at the loaded end of the SRP textile. The pertinent debonding stage can be represented by the following equation:

$$\varepsilon'_{f\,deb}(x) = 0 \tag{19}$$

The solution of Equation (19) is a constant as given in Equation (20):

$$\varepsilon_{f\,deb}(x) = A_{deb} \tag{20}$$

Integrating Equation (20) one can find the axial displacement of the textile in the debonding zone as

$$u_{f deb}(x) = A_{deb}x + B_{deb} \tag{21}$$

This stage is characterized by the presence of three regions along the bonded length: (a) the region governed by elastic shear stress-slip response, (b) the region governed by the softening branch of the shear stress-slip relationship, and (c) the SRP textile being completely separated from the concrete substrate.

The relevant solutions for the above three cases are given by Equations (12), (16) and (20). Due to the presence of three zones, there will be five integration constants that need to be determined using the relevant boundary conditions as follows:

$$N_f\left(L_f\right) = P \tag{22a}$$

$$\varepsilon_{fII,1}(x_p) = \varepsilon_{f\,deb}(x_p) \tag{22b}$$

$$\varepsilon_{fI,1}(x_a) = \varepsilon_{fII,1}(x_a) \text{ (due to the continuity)} \tag{22c}$$

$$\varepsilon'_{f\,I,1}(x_a) = \varepsilon'_{f\,II,1}(x_a) \text{ (due to the continuity)} \tag{22d}$$

$$\varepsilon_{f\,I,1}(0) = 0 \tag{22e}$$

It can be observed in Equation (22a–e) that the three regions can be identified by determining the two position parameters x_a and x_p, where x_a represents the point along the interface where the interfacial shear stress equals the maximum shear resistance, $\tau_{\max}$, while x_p represents the point where resistance is exhausted and drops to zero, thus initiating the FRP/SRP separation.

The four additional conditions that need to be satisfied to enforce the above scenario are:

$$\lambda(x_a) = \lambda_0 \tag{23a}$$

$$\tau(x_a) = \tau_{\max} \tag{23b}$$

$$\lambda(x_p) = \lambda_u \tag{23c}$$

$$\tau(x_p) = 0 \tag{23d}$$

3.1.4. Debonding after the SRP Entering the Plastic State

In the previous section, the equations governing SRP textile debonding under the scenario of the textile remaining elastic until complete debonding were developed. In the following, the alternative scenario is considered where the SRP enters the plastic state before the incidence of total interfacial separation. In this case, in view of Equation (4), one can write

$$\frac{d\sigma_f}{dx} = E_{f,2}\frac{d^2 u_f}{dx^2} \tag{24}$$

Substituting Equation (24) in Equation (1) and assuming the softening part of the shear stress-slip relationship (Equation (4b)), one can write

$$E_{f,2}t_f\frac{d^3 u_f}{dx^3} - k_b\left(\frac{du_c}{dx} - \frac{du_f}{dx}\right) = 0 \tag{25}$$

Equation (25) can be recast as follows, which is the governing equation of the scenario under consideration.

$$\varepsilon_f'' + \omega_{II,2}^2 \varepsilon_f + \eta_{II,2} = 0 \tag{26}$$

The solution of Equation (26) is given as

$$\varepsilon_{f\,II,2}(x) = A_{II,2}\cos(\omega_{II,2}x) + B_{II,2}\sin(\omega_{II,2}x) - \frac{\eta_{II,2}}{\omega_{II,2}^2} \tag{27}$$

With

$$\omega_{II,2}^2 = k_b\left(\frac{1}{E_{f,2}t_f} + \frac{b_f}{E_c A_c}\right) \tag{28a}$$

and

$$\eta_{II,2} = k_b\left(\frac{b_f\sigma_{f_y}}{E_c A_c E_{f,2}} - \frac{b_f\varepsilon_{f_y}}{E_c A_c}\right) \tag{28b}$$

The preceding set of equations can be used to obtain the stress and deformation of the concrete substrate, the SRP textile and the SRP-concrete interface for various failure modes that are observed in single-lap shear tests.

3.1.5. Effective Bond Length

The effective bond length represents a key parameter in the study of the delamination of reinforcing laminates from their substrate. It is well known that the effective bond length may be shorter than the provided bond length. Hence, knowledge of the former length is necessary for determining the debonding load or ultimate capacity.

In the proposed model, the effective bond length, L_{eff}, is given by

$$L_{eff} = L_{eff,I} + L_{eff,II} = \frac{1}{\omega_{I,1}} + \frac{\pi}{2\omega_{II,1}} \tag{29}$$

As can be observed in Equation (29), the effective bond length is composed of the sum of the length of the elastic and softening regions, denoted as $L_{eff,I}$ and $L_{eff,II}$, respectively. Furthermore, it is important to underline that the effective bond length is a function of several parameters, including the thickness and stiffness of the reinforcing textile, the concrete prism mechanical properties, and the stiffness of the elastic and softening branches of the interface.

3.2. Cyclic Load Case

In practice, a retrofitted concrete element may be subjected to cyclic loads as in the case of bridges under traffic load or in the case of structures subjected to earthquake, wind or other fluctuating dynamic loads. The response of the retrofitted element, as idealized in the single-lap shear test in which the laminate is subjected to cyclic tensile load, as in Figure 6, is modelled analytically here and the relevant governing equations and their solutions are presented below.

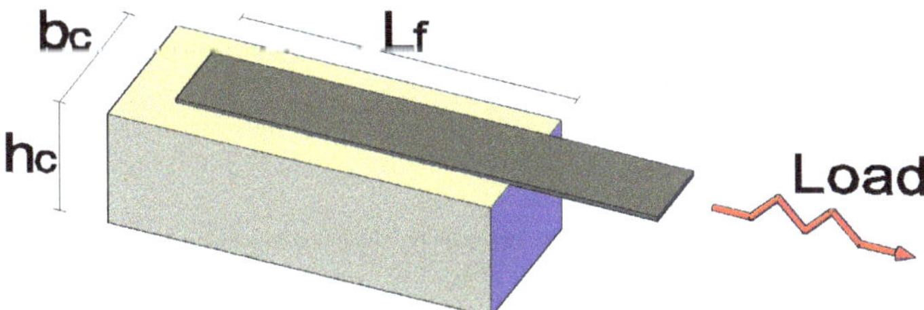

Figure 6. Single-lap shear test geometry under cyclic load.

The cohesive stress-separation law is defined by Equation (5) as presented earlier. The total fracture energy is given by

$$G_{II} = \int_0^{\lambda_u} \tau dx \tag{30}$$

where G_{II} represents the total energy and is equal to the sum of the elastic and inelastic energies represented by area under the shear stress-slip curve, and denoted by $G_{E,II}$ and $G_{S,II}$, respectively, as in Equation (31):

$$G_{II} = G_{E,II} + G_{S,II} \tag{31}$$

García-Collado et al. [35] named $G_{E,II}$ the forward region and $G_{S,II}$ the wake zone on the basis of the concepts of Ritchie [36], who essentially defined two possible classes of fatigue mechanisms: intrinsic and extrinsic.

Under the unloading cycle, two possible scenarios can be envisaged along the bonded length: (a) the interfacial shear stress being less than the maximum shear resistance, (b) the slip exceeding the slip corresponding to the maximum shear resistance, λ_0. The first case involves only elastic energy, less than or equal to $G_{E,II}$, which is fully released/recovered after complete unloading and no damage or permanent deformation occurs under this scenario. The second scenario, on the other hand, involves both $G_{E,II}$ and $G_{S,II}$ where the former energy is completely recoverable as under scenario one, while the latter is fully dissipated. In the proposed model, in scenario two, loading and unloading are assumed to occur along a line parallel to the ascending part of the shear stress-slip curve as shown in Figure 7. This means that unloading from a certain stress level to zero and reloading the specimen back to the same stress level does not cause any additional damage beyond that incurred at the start of the unloading process. In other words, hysteresis is assumed to be negligible.

The preceding unloading/reloading process can be expressed as follows

$$k_C(x) = k_S \text{ for } \lambda \leq \lambda_0 \tag{32a}$$

$$k_C(x) = k_S \left[\frac{\lambda_u - \lambda(x)}{\lambda_u - \lambda_e} \right]^2 \text{ for } \lambda > \lambda_e \tag{32b}$$

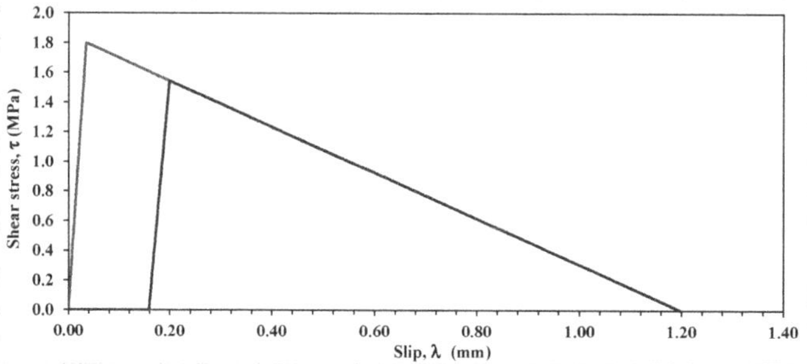

Figure 7. Representation of the cohesive shear stress-slip law for the loading-unloading-reloading process.

The expression Equation (32) can be obtained after some manipulations considering Equation (5a,b), the elastic energy $G_{E,II}$, and if the elastic area does not change under any unloading/reloading process.

The shear stress-slip law can be evaluated in the following way:

$$\tau = k_C \lambda \tag{33}$$

It can be observed in Figure 7 that upon full unloading, permanent damage occurs, which is represented by the permanent slip. In the post-elastic state, the slip at any point located at distance x from the laminate unloaded end can be determined using

$$\lambda(x) = Ae^{(w_C x)} + Be^{(-w_C x)} \tag{34a}$$

$$\lambda(x) = A + Bx + Cx^2 \tag{34b}$$

where A, B and C are constants of integration and the coefficient w_x is given by

$$w_C^2 = k_C b_f \left(\frac{A_f E_f + A_c E_c}{A_f E_f A_c E_c} \right) \tag{35}$$

Knowing the interface deformation function, $\lambda(x)$, the total strain can be obtained as

$$\varepsilon(x) = \frac{d\lambda}{dx} \tag{36}$$

The total strains can be expressed as sum of its elastic, ε_e, and plastic ε_p parts as

$$\varepsilon_t(x) = \varepsilon_e(x) + \varepsilon_p(x) \tag{37}$$

The tensile force resisted by the laminate–concrete ensemble can be evaluated by integrating the shear stresses acting on the interface as follows

$$F = b_f \Delta x \sum \tau(x) \tag{38}$$

with Δx being the distance between two points along the interface with average shear stress $\tau(x)$.

4. FEM Model

In this study, a finite element model was made and analyzed to simulate the aforementioned single-lap shear test. Due to symmetry, shown in Figure 8, only half of the specimen was discretized by an assemblage of 3D finite elements as shown in Figure 9. Using the software Abaqus [37], a fully implicit integration scheme was adopted in the analysis

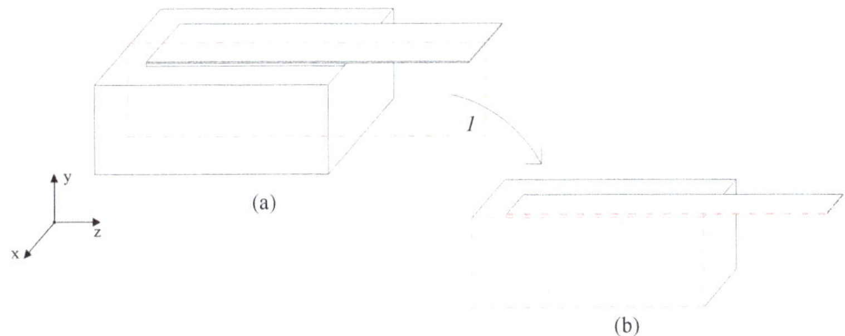

Figure 8. Symmetry plane: (**a**) single-lap shear specimen, (**b**) symmetrical part.

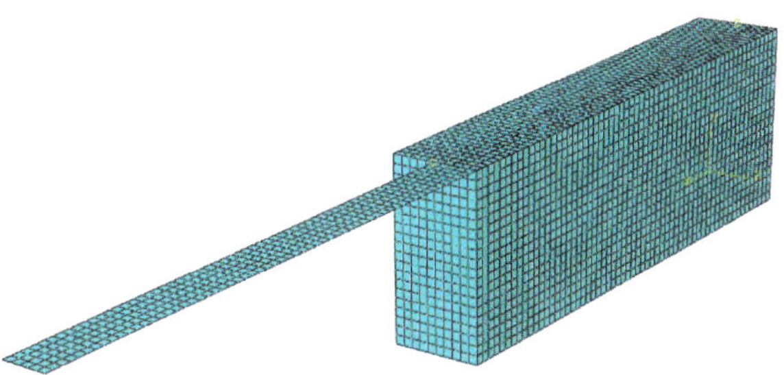

Figure 9. 3D FEM model of single-lap shear test.

Eight-node quadratic brick elements (C3D8), each having side lengths of 5 mm, were used to discretize the concrete block, while four-node 2D shell elements (S4) with side lengths of 5 mm were used to discretize the SRP strip. Further details of the adopted mesh are provided in Table 1. The constitutive models adopted for the concrete elements and FRP sheet were isotropic and their mechanical characterizing parameters are shown in Table 2a.

Table 1. Checking FE solution convergence through mesh refinement.

Trial	Element Size	Concrete	FRP Sheets	Force
	[mm]	Number of Finite Elements		[kN]
1	25	6000	600	28.210
2	20	7500	750	27.124
3	15	100,000	1000	26.900
4	10	150,000	1500	26.794
5	5	300,000	3000	26.785

Table 2. (**a**) Mechanical properties of Concrete and FRP sheets. (**b**) Mechanical properties of the adhesive.

(a)			
		Unit	Value
Concrete Young's Modulus	E_c	MPa	34,000
FRP Young's Modulus	E_f	MPa	216,000

(b)			
		Unit	Value
Elastic Stiffness	k_S	N/mm	100.0
Tensile Strength	τ_{II}	MPa	5.0
Fracture Energy	G_{II}	MPa·mm	1.0
Ultimate displacement	s_u	mm	0.4

To simulate the adhesive layer, a cohesive law, representing the damage between the laminate and concrete, was adopted, which is characterized by linear ascending and descending parts as illustrated in Figure 5 in relation to the law used to represent Mode II fracture. The values of the relevant parameters of this law, as used in the current analysis, are summarized in Table 2b.

For checking convergence of the solution, the force control criterion was selected while the analysis was performed via displacement control. The Newton–Raphson method was adopted to solve the non-linear system of equations, with nonlinearity being caused by yielding of the SRP laminate and/or by the phenomena associated with the fracture process.

In Figure 10, the results of the finite element analysis and analytical model are compared in terms of the load-slip relationship. The laminate–concrete system was also subjected to cyclic load where unloading was allowed when the slip first reached 0.1 mm and then 0.2 mm.

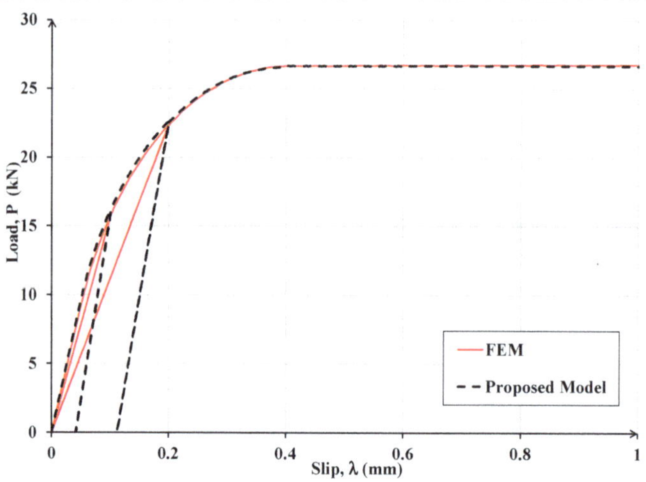

Figure 10. Comparison of the FEM and the proposed analytical model results for simulating the single-lap shear test under cyclic load.

One can observe in Figure 10 that under monotonic load, the FEM and the proposed model results are practically identical, but the unloading responses under cyclic load are different. The difference is due to different damage definitions used in the FEM and the proposed analytical model. For clarity, the loading–unloading law used in FEM analysis is shown in Figure 11. In the latter law, damage is defined in terms of the rate of stiffness degradation after damage initiation, triggered by the maximum interfacial shear stress exceeding the interface maximum shear resistance. A scalar damage index, D, is applied, which is assumed to represent the overall damage in the material and captures the combined effects of all the active damage mechanisms. It initially has a value of zero, but as damage is accumulated, D is assumed to monotonically evolve from 0 to 1. The post-damage stress component, τ_s, is calculated as

$$\tau_s = (1 - D)\overline{\tau}_s \tag{39}$$

where $\overline{\tau}_s$ is the shear stress component predicted by the elastic stress-slip behavior without damage. This law may correctly predict the post-peak stress and the concomitant damage level, but the associated deformation is not correctly captured while the proposed model is expected to accurately predict all three quantities. It should be noted that if the evolution of interfacial damage is associated with stress or load redistribution, application of Equation (12) may not predict the correct failure load.

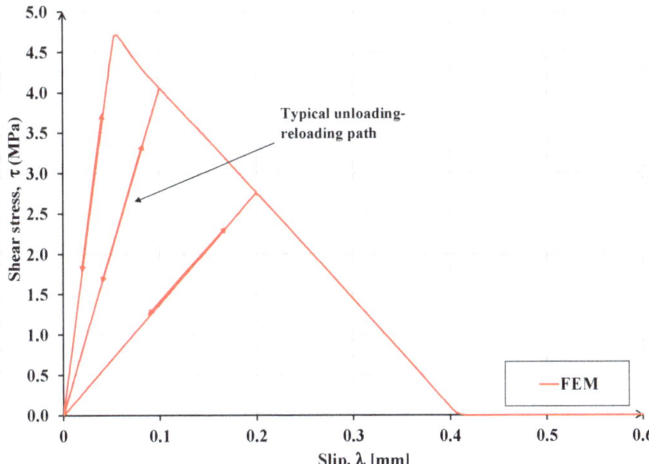

Figure 11. Cohesive stress-separation law applied in the FEM model for representing loading-unloading-reloading sequences.

It can be noticed in Figure 11 that the interface constitutive law adopted in the FEM analysis does not comprise any permanent slip upon unloading, which does not comport with empirical evidence while the proposed model, as shown in Figure 7, furnishes the extent of permanent slip upon unloading.

5. Comparison with Experimental Results

5.1. Monotonic Loading Tests

To ascertain the validity of the assumptions in the proposed model as well as the model's accuracy and robustness, several analyses are made, and the results are compared with the companion experimental data obtained from the literature. The cases investigated involve (a) CFRP plate-concrete specimen with CFRP having a thickness of 1.016 mm [38]; (b) SRP textile-concrete specimens, with SRP having a thickness of 0.084 mm, 0.254 mm or 0.381 mm, and designated as low-density (LD), medium-density (MD) and high-density (HD) SRP, respectively [1]; (c) MD textile bonded to concrete block (CB) or MD textile bonded to tuff (TU) block specimens [39]. In all three cases, in the physical experiments, single-lap shear tests were performed under monotonically increasing load until failure.

Relevant properties to the concrete–laminate system are given in Table 3. As the values of certain parameters in the experiments were not reported here, for the purposes of performing the analysis, they are assumed based on suggested values in the literature or using engineering judgment. The assumed values are shown in italics in Table 3.

Figure 12 shows the experimental data and the presently computed load-slip curves for the CFRP plate bonded to the concrete substrate in Case (a). On the other hand, for the same case, Figure 13 shows for the CFRP laminate the experimental and presently computed strain distribution along its bonded length.

In Figure 14, the results for Case (b) are shown in terms of applied load versus slip. It is important to remark that in Case (b) the parameters of the bond-slip law were calibrated by Ascione et al. [40] using the experimental results in [1].

In Figure 15, analytical results for Case (c) are compared with the companion experimental data. It can be observed that, in all cases the analytical results are in good agreement with the corresponding experimental data. The observed differences are primarily due to the natural variability in the properties of the substrate and the quality of the workmanship when bonding the laminate to the substrate. In practical applications, such variabilities can be accounted for by reliability-based resistance factors.

Table 3. Geometrical and material properties of specimens analyzed.

Property	Unit	Specimen		
		Case (a) [38]	Case (b) [1]	Case (c) [39]
Specimen width, b_c	mm	100	200	120
Specimen height, h	mm	100	150	120
Specimen strength, f_{cs}	MPa	15 -	22.5 -	4.4 14.8
Specimen elastic modulus, E_{cs}	GPa	25 -	28 -	7.8 19.5
FRP laminate thickness, t_f	mm	1.016 - -	0.084 0.254 0.381	0.254 - -
FRP laminate length, L_f	mm	203.2	300	200
FRP laminate width, b_f	mm	25.4	100	50
FRP laminate elastic modulus	GPa	110.4	-	-
Steel textile yield stress, f_y	MPa	-	2410	2410
Steel textile ultimate stress, f_u	MPa	-	3191	3191
Steel yield strain, ε_y	-	-	0.013	0.013
Steel strain corresponding to its ultimate strength, ε_{su}	-	-	0.021	0.021
Maximum shear stress, τ_{max}	MPa	5.75	2.6	3.5
Slip corresponding to τ_{max}, λ_0	mm	0.05	0.05	0.04
Ultimate slip, λ_u	mm	0.32	0.40	0.13

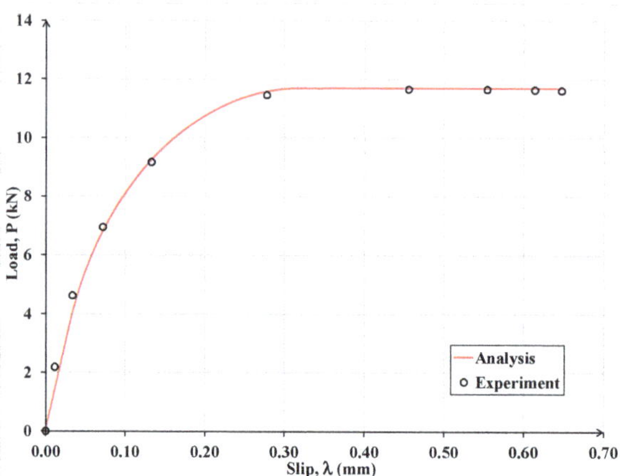

Figure 12. Current analytical versus experimental load-slip curve of the CFRP for Case (a).

Finally, the proposed Equation (29) is used to compute the effective bonded length of the specimen's strength with LD, MD, and HD SRP textiles, and the computed values are compared with the corresponding experimental values reported by Ascione et al. [1]. They reported the effective length for the specimens with LD textile in the range of 60–90 mm, for those with MD textiles in the range of 120–150 mm and for HD textiles in the range of 150–200 mm. Here, the textiles' mechanical properties given in [1] and the parameters of the interface bond-slip law given in [40] are used to compute effective bond lengths of 89 mm, 154 mm and 188 mm for the specimens made with LD, MD and HD textile, respectively. It can be observed that the computed values are practically all within the relevant observed ranges.

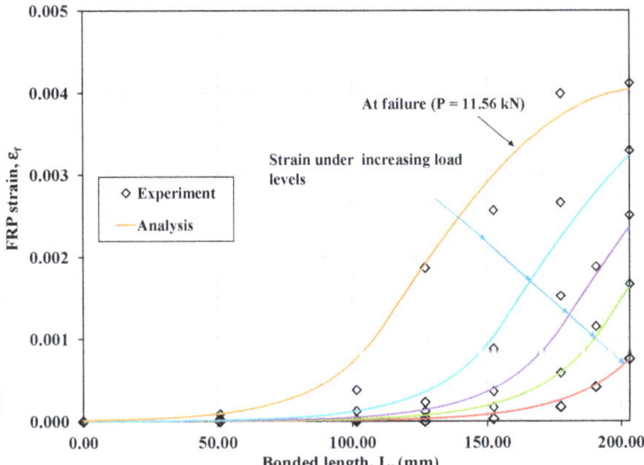

Figure 13. Analytical versus experimental CFRP strain distribution along the bonded length in Case (a).

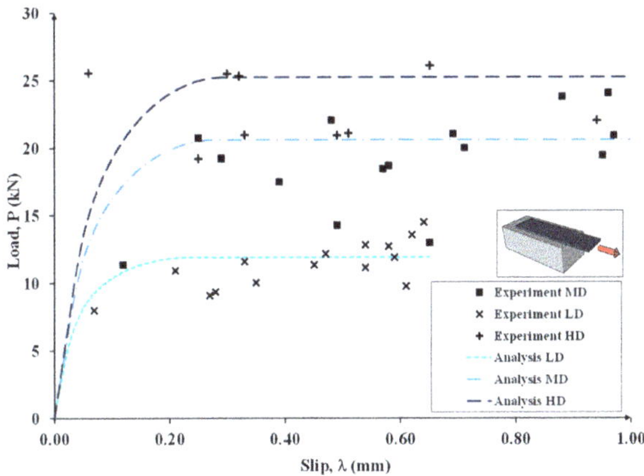

Figure 14. Experimental and analytical load-slip curves for the test specimens with low-density, medium-density and high-density steel textiles in Case (b).

For comparison, Italian guideline CNR-DT 200 R1/2013 [41] provides a semi-empirical expression for calculating the required bonded length of FRP laminates attached to concrete substrate by a suitable adhesive. Based on that expression, for the above test specimens, the computed values are 77 mm, 133 mm and 163 mm for the specimens made with LD, MD and HD textile, respectively. Based on these values, the expression in the Italian guidelines also gives reasonable values for the effective bonded length of SRP textile-concrete assemblages.

5.2. Cyclic Load Case

To validate the proposed analytical model and to evaluate its performance in the case of test specimens subjected to cyclic loading, in the following, some physical tests reported in the literature are analyzed and the analytical results are compared with the corresponding experimental data.

Nigro et al. [27] Tests

These investigators reported the results of monotonic and cyclic load tests on concrete prisms strengthened with CFRP sheets. In particular, the prisms were 150 mm wide,

200 mm high and 500 mm long. The concrete cylinder mean strength, f_{cm}, was 22.5 MPa. The CFRP strip was 100 mm wide and it had a bonded length of 400 mm, with an elastic modulus and strength of 216 GPa and 3240 MPa, respectively. The authors tested duplicate specimens to gauge the consistency of the interface performance.

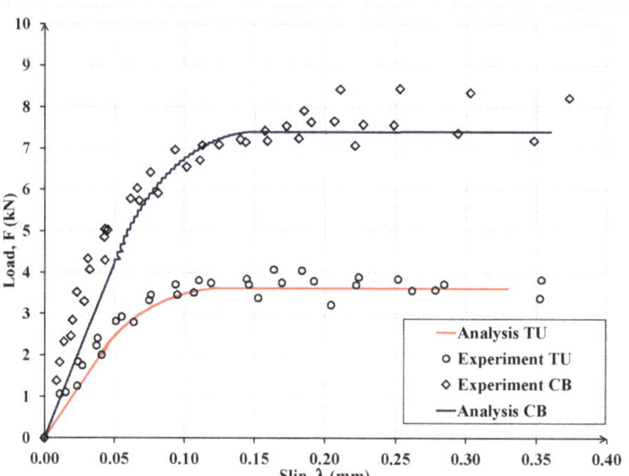

Figure 15. Experimental versus analytical load-slip curves for the test specimens in Case (c) involving MD steel textile bonded to concrete and tuff substrates.

In Figure 16, the experimental load-slip curves for three repeat specimens, designated as SM_1, SM_2 and SM_3, subject to monotonic load, are plotted, together with the companion computed curve obtained by using the proposed model. In the analytical model, based on the results in [27], the maximum slip, corresponding to total debonding, was taken as 0.2 mm for monotonic loading and 0.21 for cyclic loading.

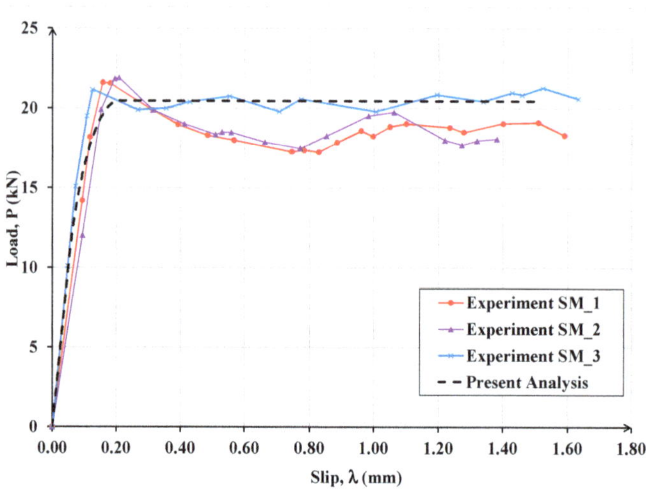

Figure 16. Comparison between experimental and analytical results under monotonic load.

It can be observed that the computed results are in good agreement with the corresponding experimental data and the differences among them can be mainly ascribed to the random variations in material properties and workmanship in specimen preparation.

The latter investigators tested two nominally identical specimens, designated as SC_4 and SC_5, under cyclic load. These specimens were analyzed using the proposed model and the results are shown in Figures 17 and 18, respectively.

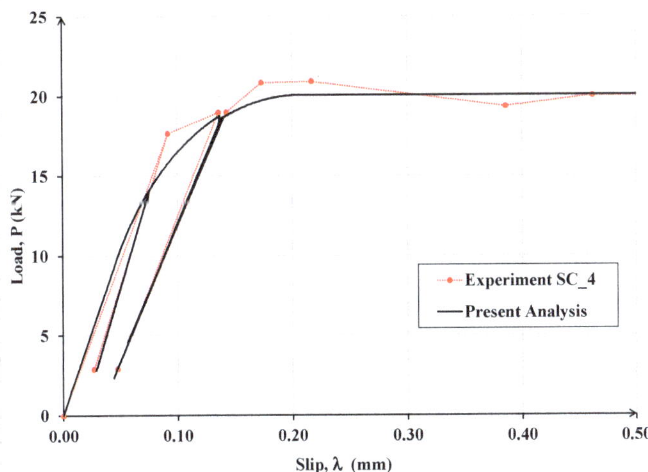

Figure 17. Comparison of experimental and analytical load-slip curves for specimen SC_4 subjected to cyclic load.

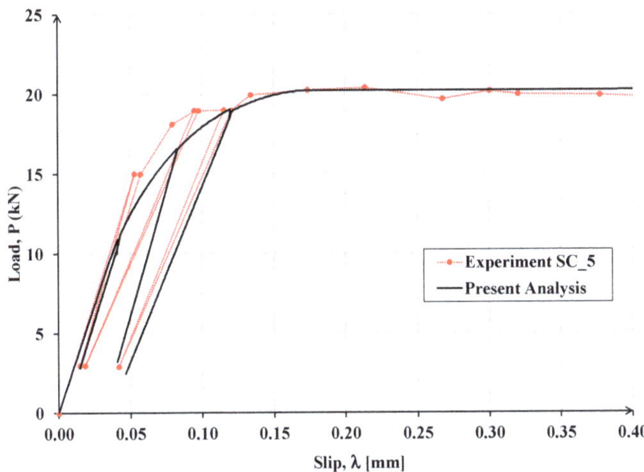

Figure 18. Comparison of experimental and analytical load-slip curves for specimen SC_5 subjected to cyclic load.

It can be observed that the results of the proposed analytical model agree well with the corresponding experimental results throughout the loading, unloading and reloading process. Consequently, the model seems sufficiently robust and can provide a reasonable estimate of the debonding load for specimens in single-lap shear tests subjected to monotonic or cyclic load.

6. Parametric Analysis Using the Proposed Model

6.1. Static Load Case

Using the proposed model, in this section, debonding of the steel textile from the concrete substrate is further analyzed by investigating the effect of some key parameters on the debonding load and deformations. To avoid premature debonding due to insufficient bonded length, a bonded length equal to 300 mm is chosen, which is guided by experimental data in 0. The concrete prism dimensions are held fixed at $200 \times 150 \times 400$ mm (width $\times$ height $\times$ length) for all the analyzed specimens. The thickness of steel textile is set equal to 0.381 mm, its width to 100 mm and the maximum interfacial shear resistance, τ_{max}, equal to 3.5 MPa.

Figures 19–22 show the textile axial strain, its tensile stresses along the bonded length, and the interfacial shear stress and slip evolution, respectively, under increasing applied load. In Figure 19, it can be noticed that the curvature of the axial strain curve changes from concave to convex as the interfacial shear stress begins to exceed the associated shear resistance and the shear resistance enters the descending or softening branch of the shear-slip constitutive law. As the load is further increased beyond that eliciting the maximum shear stress, the strain over a large part of the bonded length becomes constant, which signifies debonding over that part. It can be also noticed that the debonding initiates at approximately 90% of the maximum or failure load; hence, the debonding process is relatively sudden. It is also important to mention that up to 90% of the failure load, the effective bonded length is about 150 mm, which is only half of the provided bonded length.

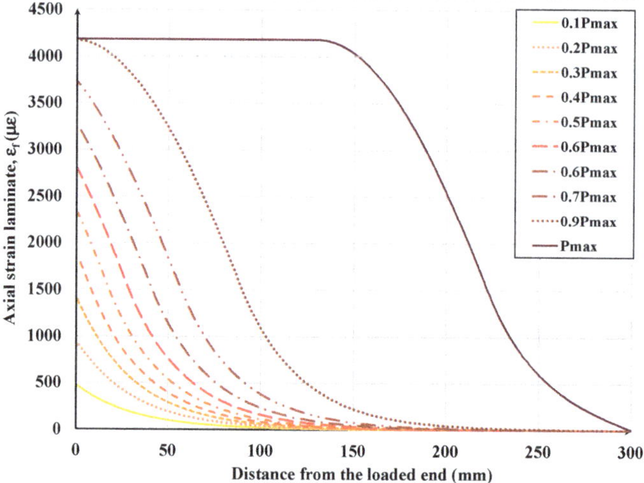

Figure 19. Development of steel textile axial strain along the bonded length under increasing load.

Figure 20 shows the evolution of axial stress in the textile along its bonded length under increasing load. As expected, the stress follows a practically identical pattern as the associated strain due to the textile stress being less than the SRP yield stress. Since in this analysis the textile was assumed to be HD, such high-strength textile will not generally yield because failure will be initiated by loss of interfacial shear resistance and debonding before the initiation of yielding.

Figure 21 shows the variation of slip along the bonded length under various load levels. This figure again demonstrates that up to 90% of the failure load, the slip is relatively small, and it slowly increases as the load is increased, but as the load is increased beyond the above level, the slip sharply increases and soon after, failure occurs. Consequently, in practical applications, the proposed model can be used to determine the design service and ultimate load for a certain combination of FRP/SRP, adhesive and substrate while avoiding sudden or unexpected failure.

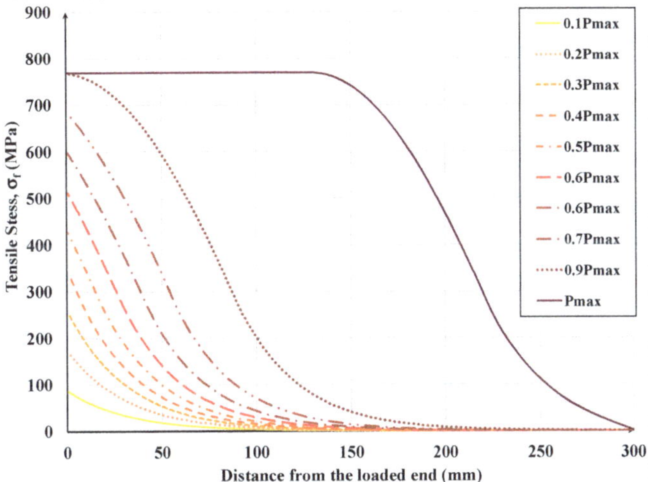

Figure 20. Development of steel textile tensile stress along its bonded length.

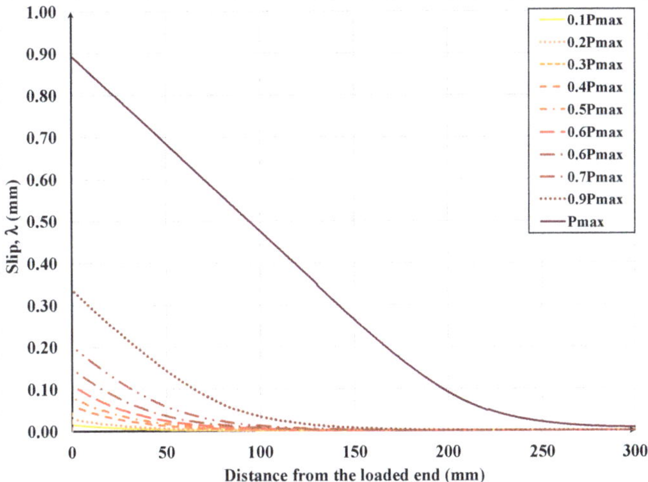

Figure 21. Slip variation along the bonded length under increasing load.

Figure 22 shows the distribution of interface shear stress along the bonded length under increasing applied load. Notice how the location where maximum shear stress occurs begins to shift along the bonded length. In addition, it can be observed that maximum shear stress is first reached at the laminate loaded end at only 40% of the failure load of the assemblage; hence, the FRP/SRP assemblage has substantial load redistribution capacity. For design purposes, shear stress distribution graphs, similar to the ones in Figure 22 can be plotted for various SRP/substrate combinations to obtain the average interfacial shear at failure, a quantity that is normally obtained from physical tests and used in design.

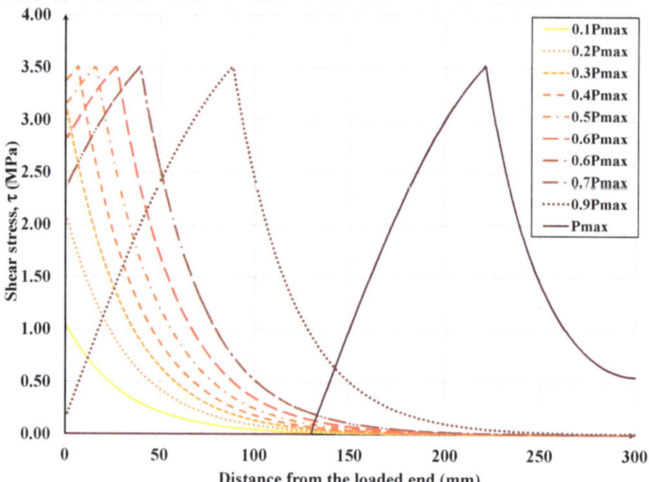

Figure 22. Interfacial shear stress variation along the bonded length under increasing load.

Figure 23 shows the variation of the normalized axial stress in the steel textiles of different densities along the bonded length. Note that the stress is expressed as fraction of the ultimate strength of the relevant textile. Since the textile density is a function of its thickness and since its ultimate load capacity is also a function of its thickness, the higher the textile density, the higher its ultimate load capacity. For this reason, the LD textile achieved the greatest fraction of its ultimate strength compared to the MD and HD textiles. Despite this higher stress in LD, the highest axial force was resisted by HD, MD and LD, respectively. Finally, it would appear that the debonded interfacial length is practically independent of the textile density.

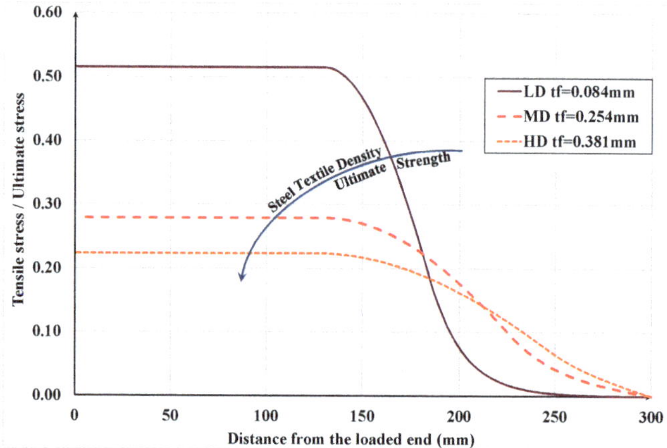

Figure 23. Normal strain behavior for different steel textile density.

6.2. Cyclic Load Case

In this section, some analyses, using the proposed model, are performed to gain greater insight into the behavior and strength of FRP/SRP retrofitted concrete elements subjected to cyclic load. It must be emphasized that the analyzed cases are supposed to represent physical single-lap shear tests. The maximum cyclic load to be applied was established by considering the maximum or failure load reached in analogous specimens under monotonic loading. Specifically, the load cycles involved the following:

1° Cycle: increase the load from zero to 50% of the maximum load, then begin unloading to zero

2° Cycle: reload from zero to 70% of the maximum load, then begin unloading to zero

3° Cycle: reload from zero to 85% of the maximum load, then begin unloading to zero

The analytical results in terms of force and displacement are depicted in Figure 24.

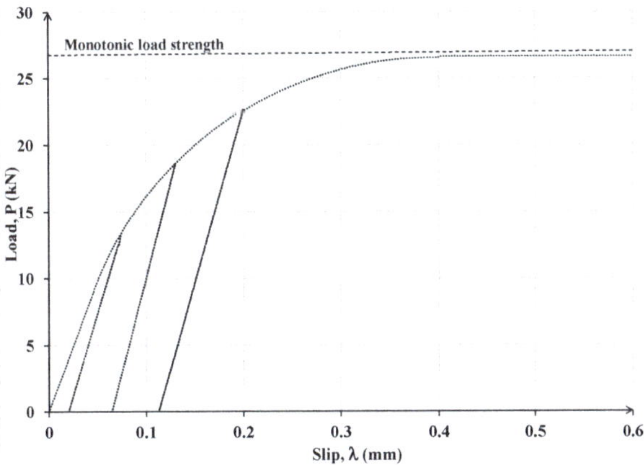

Figure 24. Load-slip curve under cyclic load.

In Figures 25 and 26 the permanent slip and the plastic strain distributions are plotted along the bonded length at the end of each of the above-described three load cycles. Notice that with each cycle, as the load at which unloading occurs is increased, the permanent slip and plastic strain in the textile increase accordingly, but the increase is not linearly proportional. The last two figures also show that as unloading is carried out from higher load levels, the permanent slip and plastic strain propagate to a larger portion of the bonded length. Although these responses are expected, the current model provides a quantitative measure of the extent of plastic deformations in the RP/SRP laminate and its interface with the concrete substrate. In practice, to avoid unexpected failure, the computed values by the current model can be compared to the recommended limits for these quantities specified in design guidelines and standards.

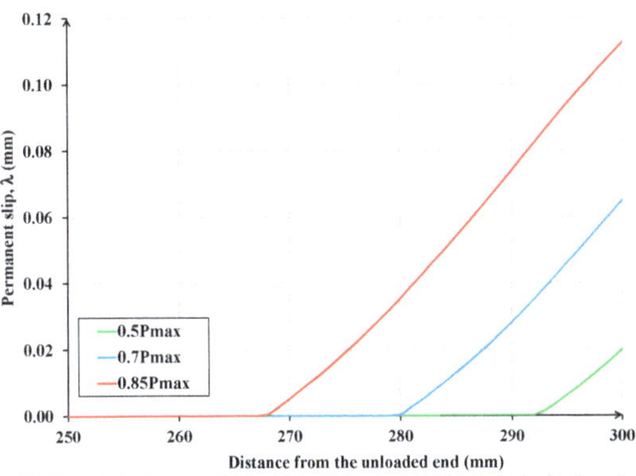

Figure 25. Permanent slip distribution along the bonded length under different load cycles.

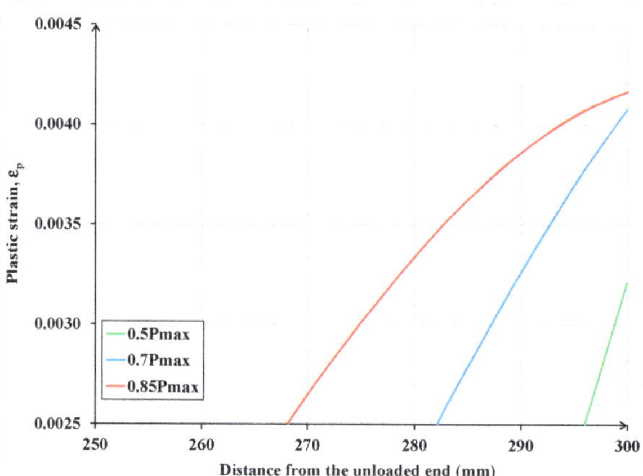

Figure 26. Plastic strains distribution along the bonded length.

7. Conclusions

In this paper, a novel and relatively simple analytical model is proposed to analyze the interfacial behavior and strength of FRP/SRP laminates or textiles bonded by an adhesive to concrete or other similar substrate. The model is applicable to elastic or elasto-plastic strain-hardening laminates subjected to monotonic or cyclic loading. The model reliability and accuracy are verified by comparing its results with the corresponding experimental data and with the results of three-dimensional finite element analysis. These comparisons are made by analyzing concrete prisms retrofitted with FRP/SRP laminates subjected to monotonic or cyclic loading in single-lap shear tests. Based on the comparisons and subsequent parametric studies, the following principal conclusions are reached:

The proposed analytical model and its associated bond-slip law can predict the interfacial response of FRP/SRP laminates bonded to concrete, or other similar, substrates subjected to monotonic or cyclic load.

The model can accurately predict the ultimate load and the associated slip of FRP/SRP retrofitted prisms tested in single-lap shear.

The model can trace the evolution of the debonding zone along the bonded length of laminate and can provide the corresponding interfacial shear stress distribution and the laminate axial stress and strain along the entire bonded length at any load level up to failure.

The model can be used to assess, under a given monotonic or cyclic load, the extent of interfacial damage incurred by the retrofitted element. In assessing the damage, a damage index based on permanent interfacial slip can be applied.

Author Contributions: Conceptualization, M.L.; methodology, M.L.; software, M.L.; formal analysis, M.L.; validation M.L., G.R., F.A., A.N. and R.R.; investigation, M.L., G.R., F.A., A.N. and R.R.; data curation, M.L., G.R., F.A., A.N. and R.R.; writing—original draft preparation, M.L.; writing—review and editing, M.L., G.R., F.A., A.N. and R.R.; supervision, M.L., G.R., F.A., A.N. and R.R. All authors have read and agreed to the published version of the manuscript.

Funding: This work was financially supported by the research grants of the Ministry of Science and Technology of China [C021801601], and the Tianjin City Government [BE044741] to Professor A. Ghani Razaqpur at Nankai University, which are gratefully acknowledged.

Institutional Review Board Statement: Not applicable.

Informed Consent Statement: Not applicable.

Data Availability Statement: The data required to reproduce these findings are all reported in the paper.

Acknowledgments: The authors are grateful to the State Administration of Foreign Experts of China, the Tianjin Municipal Government and Nankai University for their financial support of this research.

Conflicts of Interest: The authors declare no conflict of interest.

References

1. Ascione, F.; Lamberti, M.; Napoli, A.; Razaqpur, A.G.; Realfonzo, R. An experimental investigation on the bond behavior of steel reinforced polymers on concrete substrate. *Compos. Struct.* **2017**, *181*, 58–72. [CrossRef]
2. Ascione, F.; Lamberti, M.; Napoli, A.; Realfonzo, R. *SRP/SRG Strips Bonded to Concrete Substrate: Experimental Characterization*; American Concrete Institute, ACI Special Publication: Farmington Hills, MI, USA, 2018; p. 326.
3. Carloni, C.; Ascione, F.; Camata, G.; de Felice, G.; De Santis, S.; Lamberti, M.; Napoli, A.; Realfonzo, R.; Santandrea, M.; Stievanin, E.; et al. *An Overview of the Design Approach to Strengthen Existing Reinforced Concrete Structures with SRG*; American Concrete Institute, ACI Special Publication: Farmington Hills, MI, USA, 2018; p. 326.
4. De Santis, S.; de Felice, G.; Napoli, A.; Realfonzo, R. Strengthening of structures with Steel Reinforced Polymers: A state-of-the-art review. *Compos. Part B Eng.* **2016**, *104*, 87–110. [CrossRef]
5. De Santis, S.; Ceroni, F.; de Felice, G.; Fagone, M.; Ghiassi, B.; Kwiecień, A.; Lignola, G.P.; Morganti, M.; Santandrea, M.; Valluzzi, M.R.; et al. Round Robin Test on tensile and bond behaviour of Steel Reinforced Grout systems. *Compos. Part B Eng.* **2017**, *127*, 100–120. [CrossRef]
6. Yao, J.; Teng, J.G.; Chen, J.F. Experimental study on FRP-to-concrete bonded joints. *Compos. Part B Eng.* **2004**, *36*, 99–113. [CrossRef]
7. Teng, J.; Yuan, H.; Chen, J. Frp-to-concrete interfaces between two adjacent crack: Theoretical model for debonding faiulure. *Int. J. Solids Struct.* **2006**, *43*, 5750–5778. [CrossRef]
8. Chen, J.F.; Yuan, H.; Teng, J.G. Debonding failure along a softening FRP-to-concrete interface between two adjacent cracks in concrete members. *Eng. Struct.* **2006**, *29*, 259–270. [CrossRef]
9. Qiao, P.; Chen, F. An improved adhesively bonded bi-material beam model for plated beams. *Eng. Struct.* **2008**, *30*, 1949–1957. [CrossRef]
10. Franco, A.; Royer-Carfagni, G. Cohesive debonding of a stiffener from an elastic substrate. *Compos. Struct.* **2014**, *111*, 401–414. [CrossRef]
11. Cornetti, P.; Carpinteri, A. Modelling the FRP-concrete delamination by means of an exponential softening law. *Eng. Struct.* **2011**, *33*, 1988–2001. [CrossRef]
12. Caggiano, A.; Martinelli, E.; Faella, C. A fully-analytical approach for modelling the response of FRP plates bonded to a brittle substrate. *Int. J. Solids Struct.* **2012**, *49*, 2291–2300. [CrossRef]
13. Liu, S.; Yuan, H.; Jiayu, W. Full-range mechanical behaviour of FRP-to-concrete interface for pull-pull bonded joints. *Compos. Part B Eng.* **2019**, *164*, 333–334. [CrossRef]
14. Jinlong, P.; Wu, Y.F. Analytic modeling of bond behavior between FRP plate and concrete. *Compos. Part B Eng.* **2014**, *61*, 17–25.
15. Harries, K.A.; Aidoo, J. Debonding- and Fatigue-Related Strain Limits for Externally Bonded FRP. *J. Compos. Constr.* **2006**, *10*, 87–90. [CrossRef]
16. Ferrier, E.; Bigaud, D.; Hamelin, P.; Bizindavyi, L.; Neale, K.W. Fatigue of CFRPs externally bonded to concrete. *Mater. Struct.* **2005**, *38*, 39–46. [CrossRef]
17. Diab, H.M.; Wu, Z.; Iwashita, K. Theoretical Solution for Fatigue Debonding Growth and Fatigue Life Prediction of FRP-Concrete Interfaces. *Adv. Struct. Eng.* **2009**, *12*, 781–792. [CrossRef]
18. Zheng, X.H.; Huang, P.Y.; Chen, H.M.; Tan, X.M. Fatigue behavior of FRP–concrete bond under hygrothermal environment. *Constr. Build. Mater.* **2015**, *95*, 898–909. [CrossRef]
19. Kim, Y.J.; Heffernan, P.J. Fatigue behavior of externally strengthened concrete beams with fiber-reinforced polymers: State of the SRT. *J. Compos. Constr.* **2008**, *12*, 246–256. [CrossRef]
20. Dong, J.F.; Wang, Q.Y.; Guan, Z.W. Structural behaviour of RC beams externally strengthened with FRP sheets under fatigue and monotonic loading. *Eng. Struct.* **2012**, *41*, 24–33. [CrossRef]
21. Yun, Y.; Wu, Y.-F.; Tang, W.C. Performance of FRP bonding systems under fatigue loading. *Eng. Struct.* **2008**, *30*, 3129–3140. [CrossRef]
22. Ferrier, E.; Bigaud, D.; Clément, J.C.; Hamelin, P. Fatigue-loading effect on RC beams strengthened with externally bonded FRP. *Constr. Build. Mater.* **2011**, *25*, 539–546. [CrossRef]
23. Bizindavyi, L.; Neale, K.W.; Erki, M.A. Experimental Investigation of Bonded Fiber Reinforced Polymer-Concrete Joints under Cyclic Loading. *J. Compos. Constr.* **2003**, *7*, 127–134. [CrossRef]
24. Peng, H.; Wang, B.; Zhang, J.; Li, S. Test study on fatigue behavior of bonded FRP–concrete joint. *J. Exp. Mech.* **2014**, *29*, 189–199.
25. Zheng, X.H.; Huang, P.Y.; Han, Q.; Chen, G.M. Bond behavior of interface between CFL and concrete under static and fatigue load. *Constr. Build. Mater.* **2014**, *52*, 33–41. [CrossRef]
26. Mazzotti, C.; Savoia, M. FRP-Concrete Bond Behaviour under Cyclic Debonding Force. *Adv. Struct. Eng.* **2009**, *12*, 771–780. [CrossRef]

27. Nigro, E.; Di Ludovico, M.; Bilotta, A. Experimental Investigation of FRP-Concrete Debonding under Cyclic Actions. *J. Mater. Civ. Eng.* **2011**, *23*, 360–371. [CrossRef]
28. Carloni, C.; Subramaniam, K.V.; Savoia, M.; Mazzotti, C. Experimental determination of FRP–concrete cohesive interface properties under fatigue loading. *Compos. Struct.* **2012**, *94*, 1288–1296. [CrossRef]
29. Ko, H.; Sato, Y. Bond Stress–Slip Relationship between FRP Sheet and Concrete under Cyclic Load. *J. Compos. Constr.* **2007**, *11*, 419–426. [CrossRef]
30. Martinelli, E.; Caggiano, A. A Unified Theoretical Model for the Monotonic and Cyclic Response of FRP Strips Glued to Concrete. *Polymers* **2014**, *6*, 370–381. [CrossRef]
31. Dai, J.G.; Sato, Y.; Ueda, T.; Sato, Y. Static and fatigue bond characteristics of interfaces between CFRP sheets and frost damage experienced concrete. In Proceedings of the 7th International Symposium on Fiber Reinforced Polymer Reinforcement for Reinforced Concrete Structures, Kansas City, MO, USA, 6–9 November 2005; ACI: Farmington Hills, MI, USA, 2005; pp. 1515–1530.
32. Monti, M.; Renzelli, M.; Luciani, P. FRP adhesion in uncracked and cracked concrete zones. In Proceedings of the 6th International Symposium on FRP Reinforcement for Concrete Structures, Singapore, 8–10 July 2003; World Scientific Publications: Singapore, 2003; pp. 183–192.
33. Lu, X.Z.; Teng, J.G.; Ye, L.P.; Jiang, J.J. Bond–slip models for FRP sheets/plates bonded to concrete. *Eng. Struct.* **2005**, *27*, 920–937. [CrossRef]
34. Faella, C.; Martinelli, E.; Nigro, E. Direct versus Indirect Method for Identifying FRP-to-Concrete Interface Relationships. *J. Compos. Constr.* **2009**, *13*, 226–233. [CrossRef]
35. García-Collado, A.; Vasco Olmo, J.M.; Díaz, F.A. Numerical analysis of plasticity induced crack closure based on a irreversible cohesive zone model. *Theor. Appl. Fract. Mech.* **2017**, *89*, 56–62. [CrossRef]
36. Ritchie, R.O. Mechanisms of fatigue-crack propagation in ductile and brittle solids. *Int. J. Fract.* **1999**, *100*, 55–83. [CrossRef]
37. Smith, M. *ABAQUS/Standard User's Manual*, version 6.9; Dassault Systèmes Simulia Corp.: Providence, RI, USA, 2009.
38. Chajes, M.; Finch, W.; Januska, T.; Thomson, T. Bond and force transfer of composite material plates bonded to concrete. *ACI Struct. J.* **1996**, *93*, 208–217.
39. Napoli, A.; de Felice, G.; De Santis, S.; Realfonzo, R. Bond behaviour of Steel Reinforced Polymer strengthening systems. *Compos. Struct.* **2016**, *152*, 499–515. [CrossRef]
40. Ascione, F.; Lamberti, M.; Napoli, A.; Razaqpur, A.G.; Realfonzo, R. Modeling SRP-concrete interfacial bond behaviour and strength. *Eng. Struct.* **2019**, *187*, 220–230. [CrossRef]
41. Advisory Committee on Technical Recommendations for Constructions, Italian National Research Council CNR DT200 R1/2013. *Istruzioni per la Progettazione, l'Esecuzione ed il Controllo di Interventi di Consolidamento Statico Mediante l'utilizzo di Compositi Fibrorinforzati*; Consiglio Nazionale Delle Ricerche: Napoli, Italy, 2013.

Article

Adhesive Joint Degradation Due to Hardener-to-Epoxy Ratio Inaccuracy under Varying Curing and Thermal Operating Conditions

Jakub Szabelski [1,2,*], Robert Karpiński [3,*], Józef Jonak [3] and Mariaenrica Frigione [2]

1. Department of Computerisation and Production Robotisation, Faculty of Mechanical Engineering, Lubli University of Technology, Nadbystrzycka 36, 20-618 Lublin, Poland
2. Department of Innovation Engineering, University of Salento, Provinciale Lecce-Monteroni, 73100 Lecce, Italy
3. Department of Machine Design and Mechatronics, Faculty of Mechanical Engineering, Lublin University of Technology, Nadbystrzycka 36, 20-618 Lublin, Poland
* Correspondence: j.szabelski@pollub.pl (J.S.); r.karpinski@pollub.pl (R.K.)

Abstract: This paper presents the results of an experimental study of adhesive joint strength with consideration of the inaccuracy of the hardener dosage, in the context of evaluating the degradation of joints when used either at ambient or elevated temperatures. The butt joint strength characteristics were assessed for two types of adhesives—rigid and flexible—and two curing scenarios—with and without heat curing. An excess hardener was shown to be significantly more unfavourable than its deficiency, which can ultimately be considered as a recommendation for forming epoxy adhesive joint assemblies. In order to fully understand the relationship between the analysed mechanical properties of the material and the influence of component ratio excesses and heating, a process of fitting basic mathematical models to the obtained experimental data was also performed.

Keywords: epoxy adhesives; adhesive joint; degradation; heat curing; thermal stability; resin hardener mix ratio; mathematical modelling

Citation: Szabelski, J.; Karpiński, R.; Jonak, J.; Frigione, M. Adhesive Joint Degradation Due to Hardener-to-Epoxy Ratio Inaccuracy under Varying Curing and Thermal Operating Conditions. *Materials* **2022**, *15*, 7765. https://doi.org/10.3390/ma15217765

Academic Editors: Marco Lamberti and Aurelien Maurel-Pantel

Received: 3 October 2022
Accepted: 31 October 2022
Published: 3 November 2022

Publisher's Note: MDPI stays neutral with regard to jurisdictional claims in published maps and institutional affiliations.

Copyright: © 2022 by the authors. Licensee MDPI, Basel, Switzerland. This article is an open access article distributed under the terms and conditions of the Creative Commons Attribution (CC BY) license (https://creativecommons.org/licenses/by/4.0/).

1. Introduction

Adhesives have been known to man for thousands of years. Initially, natural materials based, for example, on birch tar, bituminous substances (asphalts) or animal collagen derivatives were used, but the constant advance of knowledge in the fields of chemistry, physics, materials engineering and mechanics itself has led to the development of this technique as well [1]. Today, thousands of adhesive compositions are commercially available. There are adhesives dedicated to specific substrates, specific operating conditions and joint loading conditions, as well as universal adhesives. We can find adhesives for high-strength joints, requiring precise steps of the bonding operation (preparation of the surface, the adhesive, the joint curing method), and adhesives for detachable joints, e.g., light paper adhesives, adhesives neutralised by solvents or heat. Today, a wide variety of materials are bonded using this technique, from tissues (adhesives used in biomedical engineering) and simple construction materials to even the most complex manufacture of honeycomb composites used in the aerospace industry (Figure 1 [2]). This shows how important the adhesive bonding operation is today. Technologists are using it more frequently today not only as a complementary solution for joining materials, as in certain cases, adhesive bonding may be the only possible joint forming method [3]. These joint types are successively substituting traditional joining methods such as welding, soldering and bolted joints. The reason for this is that adhesive joints have several significant advantages over classic material joining techniques. These include the simplicity of manufacture, no increase in the weight of the joined elements, universality (many materials that are difficult to join using other methods can be joined) and cost (it is usually the cheapest method of making connections. Wherever the typical disadvantages of joining are not disqualifying for the use of this method, it can be a serious competitor to existing techniques. Among the biggest disadvantages are: the joint strength, which is difficult to calculate and often requires destructive experimental trials, the bonding time, the thermal resistance and the need for a proper and precise execution of the entire bonding process. Therefore, much attention is paid to the analysis of the adhesive joint strength and the testing of factors that can ensure an increase in the quality and functional properties of the joint.

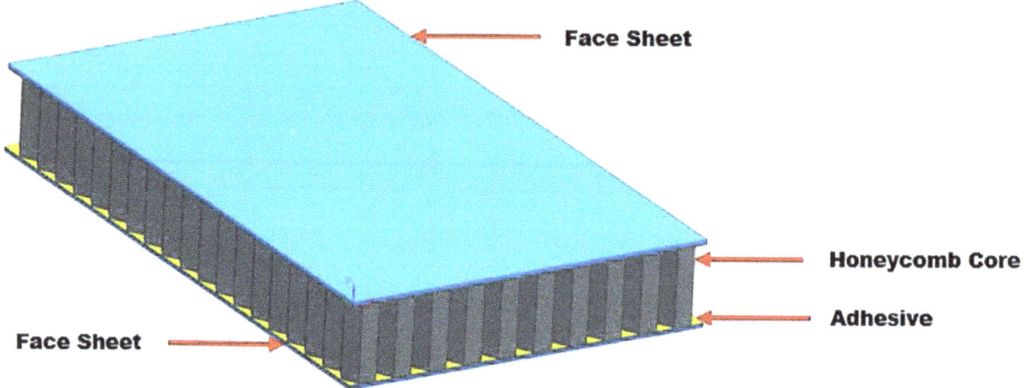

Figure 1. 3D model of honeycomb sandwich structure.

The final strength of structural adhesive joints depends on many factors, both technological and structural [4,5]. The most important ones, the optimisation of which can significantly increase the strength of structural adhesive joints, include the method of preparing the surfaces of the elements to be joined (achieving an optimal surface condition through mechanical, electrochemical or chemical treatment operations) [6,7]. Another important issue is the selection of a proper adhesive for joining specific surfaces and the modification of the adhesive composition (introducing fillers, plasticisers, etc.) [8–12]. The conditions of preparing the joints also play a significant role in obtaining strong and reliable joints [13,14], as well as the manufacturing technology, taking into account the geometry of the joint, the method of loading the joint and the conditions of curing the adhesive joint [15–17]. The strength of the adhesive bond is also affected by various types of defects in the form of non-joined areas. They can be caused, for example, by air bubbles in the adhesive, which are the result of mixing too much of the two-component adhesive too quickly, or by physical impurities in the adhesive. Such an adhesive will have a porous structure after cross-linking, with the risk of local voids at the interface between the bonding material and the adhesive, which will impair adhesion, and voids in the internal structure of the adhesive, which will weaken the cohesion of the adhesive material [18,19].

Typical defects in adhesive bonds include damage [20]:

(1) adherend

    a.   failure of one or both adherends—substrate failure,
    b.   failure of an adherend—cohesive substrate failure,
    c.   failure through lamination—delamination failure,

(2) adhesive

    a.   Cohesion failure

        i.   Cohesion failure
        ii.   Special cohesion failure
        iii.   Failure with stress whitening of adhesive

    b.   Adhesion failure

  c. Adhesion and cohesion failure with peel

(3) Corrosion at the interface—debonding due to bondline corrosion.

Taking into consideration the wide variety of adhesive compositions, the large quantitative parameters and the vast range of parameters that need to be taken into account when producing the strongest possible joints, any tests carried out on this subject are complex and require considerable time and financial expenditure, as, very often, the final strength of the finished joint is verified by destructive testing. The analysis of changes in mechanical parameters is extremely important for the estimation of the service life of adhesive bonds and in the design of modern materials. Conducting a large number of destructive experimental tests is often not possible due to time and economic constraints. The methods for solving such a problem can include the implementation of mathematical modelling [21,22], computer methods such as the finite element method (FEM) [23–28], the boundary element method (BEM) [29,30], the application of predictive models [31–33], machine learning methods [33–36] and analytical data analysis [37,38]. The models developed by this approach allow the most important relationships between individual parameters and mechanical properties to be determined. Using previously acquired experimental data, they can also indicate the most promising research direction and help to minimise the number of physical tests performed, thereby significantly reducing the cost and time of research.

The research analysed the problem of joints made with adhesives of a very popular type—epoxies, produced by the chemical reaction of the polymerisation of an epoxy resin combined with a hardener. Adhesives of this type, as they require two different components, are usually supplied as commercial compositions in a separated form, although there also exist single-component solutions, where the components are already mixed but require activation, e.g., by temperature or UV light. It is also possible to develop custom compositions by reacting one of the many available epoxy resins with selected hardeners (e.g., amine). When preparing adhesives (as well as other, two-component polymeric materials [39–42]), the correct stoichiometric ratio of the curing resin and hardener is crucial. This can determine not only the strength of the adhesive bond but many other performance parameters such as the resistance to heat, harmful environmental influences, oils, etc. The qualitative impact of the inaccuracy of the ratio of the components of an epoxy adhesive on these characteristics does not seem to have been fully analysed, especially when combined with consideration of the other two factors [43], and additional research expanding the range of tested adhesives by adhesives of different properties seems to be justified. The heat role mentioned earlier, as an initiator of the polymerisation reaction of single-component epoxy adhesives, can also be used in the reaction of typical two-component adhesives not only as a reaction accelerator but also as something that can allow a higher degree of polymerisation to be achieved and thus create stronger bonds.

In view of the above-described problems of bonding defects related to the methods of their prevention, attention was paid to the aspect of the potential degradation of adhesive bonds as a result of the combination of harmful effects of temperature during bond operation and, at the same time, inaccuracy in the technological production of the adhesive (epoxy–resin–amine–hardener ratio), taking into account the technology of bond curing (cold/heat curing). Although some degradation of bond strength with increasing temperatures and some degradation related to the inaccuracy of the above-mentioned ratio can be assumed, it is not entirely certain what effect the curing conditions will have on the degree of degradation by the above-mentioned factors. Therefore, this study set out to determine the degree of adhesive joint strength degradation with an inaccurate component ratio of the hardener in the epoxy adhesive and to evaluate this degradation when heat-cured joints are operated at ambient and elevated temperatures for rigid and elastic adhesives. This will allow general recommendations to be formulated for the use of heat curing operations when there is uncertainty in the adhesive composition and may help to ensure greater adhesive joint assembly durability at elevated temperatures.

2. Materials and Methods

2.1. Materials and Sample Preparation

The research involved testing adhesive butt joints prepared using two adhesives that differed in their parameters: a commercial Loctite Hysol 9492 composition and an adhesive based on Epidian 57 epoxy resin and a PAC hardener. Both are two-component, chemically curing epoxy formulations with similar curing times and strengths of the joints formed. The most important difference between joints made with the two above-mentioned adhesives is that they will differ significantly in stiffness after curing. Hysol 9492 is more rigid, with the Young's modulus being approximately $5\times$ that of the PAC-crosslinked Epidian 57.

Hysol 9492 is an epoxy adhesive belonging to the wide range of Loctite-brand adhesives. It is supplied in double cartridges with an applicator (mixer) at the tip and is designed to mix the components, in the recommended volumetric proportion, as soon as it leaves the applicator. The volumetric ratio of parts A to B (resin to hardener) is 2:1. It is a multi-purpose adhesive and can be used in many different applications in addition to its function as a traditional adhesive—e.g., as a sealant or for repairing materials (e.g., removing pores and irregularities from the surface of castings or forgings).

Epidian 57 epoxy resin is a clear, yellow, viscous liquid that is produced by the reaction of bisphenol A and epichlorohydrin, modified with unsaturated polyester resin. In addition to its traditional application, i.e., making adhesive bonds, the resin is also used to produce, in combination with glass fibre, high-strength glass-epoxy laminates. For the curing of Epidian 57, the hardener used in the study was PAC, an amber-coloured viscous liquid consisting of approx. 8–12% triethylenetetramine and enriched with unsaturated fatty acids. The ratio range recommended by the manufacturer for the curing of Epidian 57 is 100:50–80.

The choice of adhesive materials was driven by an attempt to test materials with differing properties. Typically, flexible adhesives are more resistant to elongation but can withstand less stress. The effect of the other parameters examined in the paper on the final strength of the joint may be also different for rigid adhesives and different for flexible ones. Modifying the amount of the hardener, the research analysed joints produced by adhesives made with a deficiency and with an excess of the hardener, in the combinations of −50%, −30%, −10%, +10%, +30% and in the ratio that is considered appropriate (2:1 for both adhesives, i.e., 0% inaccuracy). Also included in the test plan were the curing conditions of the bond, the effect of which was also studied in this work, i.e., according to the manufacturers' recommendations: long-term curing at ambient temperature (25 °C: 7 days for E57/PAC and 3 days for Hysol 9492) and heat curing (100 °C: 2 h for E57/PAC and 1 h for Hysol 9492).

Using the above adhesives, cylindrical steel specimens (S235JR/USt37-2/A283 Gr.C) with dimensions of ⌀20 × 100 mm were butt jointed. The faces of the specimens were prepared for bonding in accordance with PN-EN 13887:2005 (Adhesives for structural joints—Guidelines for preparing the surfaces of metals and plastics before bonding) in a turning operation with the surface roughness. The average of the profile height deviations from the mean line Ra = 3.4 ± 0.6 μm, and the ISO roughness grade: N8-N9 [30]. To minimise the impact of corrosion processes on the properties of the joint, the time from finishing the surface treatment to bonding was no more than 2 h. Maintaining a precise fit and parallelity of the bonded surfaces required the use of a custom-built bonding stand and special U-joint fixtures. The series for which the joints were planned to be heat cured were placed in an oven controlled by a Shimaden FP93 digital controller connected to a PC (Figure 2).

Figure 2. Positioning brackets for the correct orientation of the bonded specimens in the heating chamber.

2.2. Mechanical Testing

The adhesive joints were tested for axial tensile strength, which corresponds to pulling straight, in-plane, and away from the adhesive bond. For this purpose, the joints were fixed in the clamps of the testing machine (Figure 3) using U-joint grips and then pulled at a constant speed of 4 mm/min while recording the time, deformation and value of the tensile force. The tests were carried out in two series for each of the prepared combinations, i.e., at an ambient temperature of 20 °C and, using the heating chamber of the testing machine, at an elevated temperature (50 °C for the joints made with E57/PAC adhesive and 70 °C for Hysol 9492). At the end of each test run, the failure mode of the individual joints was recorded according to the EN ISO 10365:2022 standard (Adhesives—Designation of main failure patterns) [20].

Figure 3. The adhesive specimen in the jointed fixture of the testing machine.

2.3. Statistical Analysis

Statistical analysis is necessary to answer questions regarding actual batch-to-batch comparison, and it is only in this way that it is possible to estimate changes in strength due to variations in adhesive preparation accuracy and other parameters. If multiple averages need to be compared, simple statistical tests (Student's *t*-tests) do not fulfil their purpose, as, the higher the number of averages being compared, the greater the risk of error for such comparisons. The correct method, the results of which are described below, involves isolating homogeneous groups of results, between which the differences are not statistically significant (at the assumed significance level of $\alpha = 0.05$). A detailed description of the mathematical operations carried out to test the comparison of multiple batches is given, among others, in: [44–46]. The assumptions preceding the applicability of the method (normal distribution of results within the series, equality of variance within groups) were tested and confirmed. Tukey's statistic tests were used in a version for different numbers of samples. This approach was chosen due to the unequal numbers of samples within the individual series, resulting, for example, from the exclusion of samples whose values were characterised by coarse errors, i.e., they differed significantly from other measurement results of the same quantity. The aforementioned test is one of several available in the software used in the analyses described in this paper (Tibco Statistica 13). Other tests of this type include Scheffé's test, Newman and Keuls, Duncan's test and Fisher's NIR [44].

2.4. Mathematical Modelling

In order to gain a better understanding of the relationship between the degree of quantitative inaccuracy of the resin/hardener ratio, heat curing and the temperature of the working joint and its strength, further analysis was carried out to match typical mathematical models to the results obtained. A detailed description of the technique of regression modelling carried out in the paper is presented in: [47,48]. Given the nature, process and arrangement of the results obtained, simple models were tested: a linear model (which can perform well within the range of inaccuracy of the ratios studied in this paper) and a polynomial model of the 2nd, or possibly 4th, degree (which, as polynomial models of an even degree, will correspond to the course of a function whose extremes will tend towards zero from above, as is to be expected in a complete approach to the problem). For each fit, the R^2 value was estimated, i.e., the coefficient of determination, which is a description of the quality of the fit of the model under test to the data obtained. The degree of fit of the selected models was examined.

3. Results

The results of the strength tests are presented in Figure 4 (samples joined with Epidian 57/PAC adhesive) and Figure 5 (Hysol 9492). The results of the joints, grouped according to the same curing conditions (heat- and non-heat cured) and divided into two series—depending on the test temperature (ambient and elevated temperature) and taking into account the degree of the under/over-hardener in the adhesive formulation—are summarised there. The error bars indicating the standard deviation of the results for the two series as well as the failure pattern of the joint according to ISO 10365:2022 (Adhesives—Designation of main failure patterns) [20] are included in both diagrams: AF—adhesive failure, CF—cohesive failure, AC + AF—adhesive-cohesive failure, ACFP—adhesive failure with adhesive layer detachment. An example of the specimens after the destruction of the adhesive bond is shown in (Figure 6 [20]). In the case of the mixed-failure character, the percentage of each type within the reported series was additionally given. Below the plots are tables indicating the average strength values of each series.

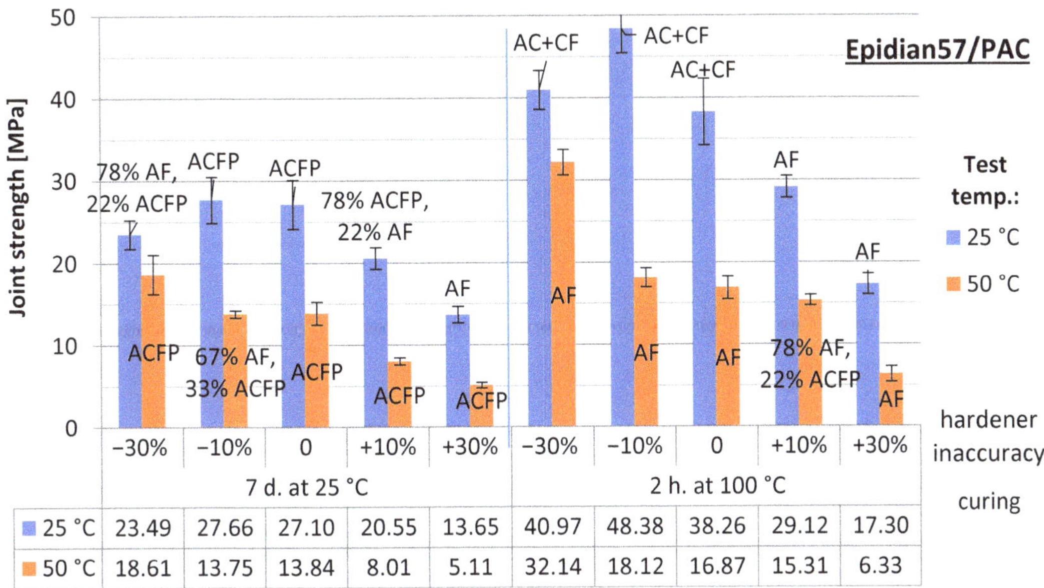

	−30%	−10%	0	+10%	+30%	−30%	−10%	0	+10%	+30%
25 °C	23.49	27.66	27.10	20.55	13.65	40.97	48.38	38.26	29.12	17.30
50 °C	18.61	13.75	13.84	8.01	5.11	32.14	18.12	16.87	15.31	6.33

Figure 4. Test results for the adhesive joint strength made with Epidian57/PAC.

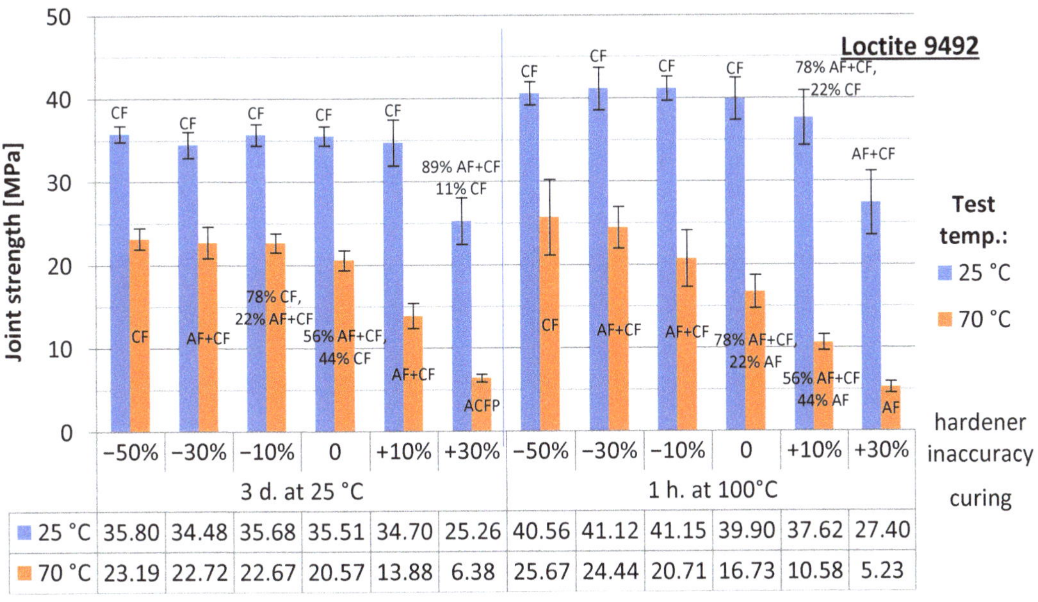

	−50%	−30%	−10%	0	+10%	+30%	−50%	−30%	−10%	0	+10%	+30%
25 °C	35.80	34.48	35.68	35.51	34.70	25.26	40.56	41.12	41.15	39.90	37.62	27.40
70 °C	23.19	22.72	22.67	20.57	13.88	6.38	25.67	24.44	20.71	16.73	10.58	5.23

Figure 5. Test results for adhesive joint strength made with Hysol 9492.

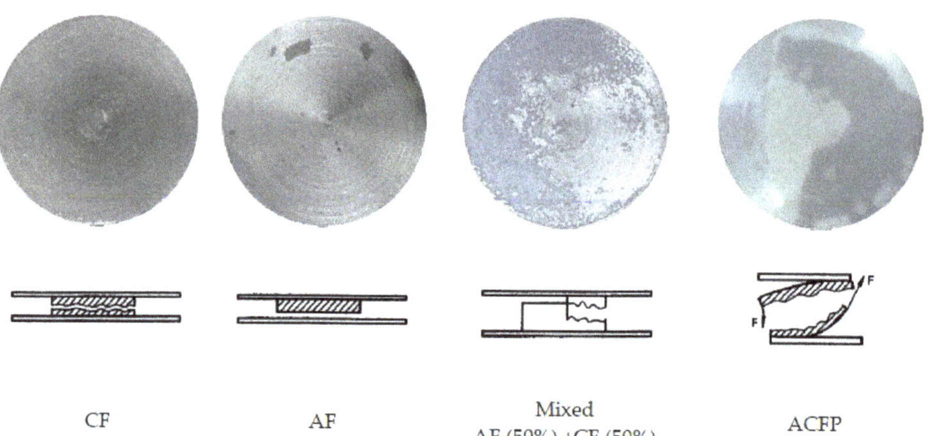

CF   AF   Mixed
AF (50%) +CF (50%)   ACFP

Figure 6. Main failure patterns of the tested adhesive joints [20].

From all the results obtained, the trend towards joint degradation with an excess of the hardener is most evident. The deficiency is considerably less unfavourable, even at high hardener deficiencies of 30–50%. The hardener, as a crosslinking reaction agent, allows it to continue even at lower dosage levels. The crosslinking reaction continues, producing an adhesive material with reasonably good performance as an adhesive. The analysis of failure models for joints made with Epidian 57/PAC adhesive shows that, often, the lesser the hardener is, the more important the cohesive nature of the failure becomes, especially for joints tested at ambient temperature. Operating the joints at elevated temperatures leads to a decrease in adhesion at the interface between the adhesive phases and the bonded material, so the joints are destroyed more often due to the detachment of the adhesive from the sample surface. Joints made with Hysol 9492 exhibit very strong adhesion at room temperature. Most samples show a cohesive failure pattern. Only a large excess of the hardener leads to a weakening of the adhesive structure and an increase in the proportion of cohesive failure (regardless of whether the joints were heat cured or cured at room temperature during the crosslinking stage).

Tables 1 and 2 present the results of the statistical analysis, i.e., the obtained groups of results of homogeneous joint strengths made with Hysol 9492 adhesive, grouped according to the joint curing conditions and taking into account the joint test temperature. The groups include series marked in the tables with the symbol X, for which, with the adopted significance level $\alpha = 0.05$, there are no statistically significant differences.

Table 1. Homogeneous groups (1-2, 1-2-3) of strength at 20 °C for joints made with Hysol 9492.

Curing Conditions	3 Days in 20 °C			1 h in 100 °C			
Hardener Inaccuracy	Mean Strength (MPa) at 20 °C	1	2	Mean Strength (MPa) at 20 °C	1	2	3
−50%	35.81	X		41.06	X		
−30%	33.46	X		41.12	X		
−10%	35.68	X		39.89	X	X	
0%	**35.51**	**X**		**39.90**	**X**	**X**	
10%	33.82	X		35.71		X	
30%	25.44		X	28.50			X

Table 2. Homogeneous groups (1-2-3-4, 1-2-3-4-5) of strength at 70 °C for joints made with Hysol 9492.

Curing Conditions	3 Days in 20 °C					1 h in 100 °C					
Hardener Inaccuracy	Mean Strength (MPa) at 70 °C	1	2	3	4	Mean Strength (MPa) at 70 °C	1	2	3	4	5
−50%	23.18	X				25.67	X				
−30%	21.60	X	X			23.09	X	X			
−10%	22.67	X	X			19.93		X			
0%	**20.12**		X			**14.77**			X		
10%	13.49			X		10.10				X	
30%	6.64				X	5.25					X

The statistical analysis clearly shows and confirms what the analysis of the average strength results initially indicated. An excess hardener almost always leads to a degradation of the strength characteristics of the joints. This is particularly noticeable when the joint is used at elevated temperatures. However, a deficiency of the hardener, in terms of the strength of the joint, can slightly, but in some cases significantly, increase the strength.

Tables 3 and 4 show similar results for joints made with Epidian 57/PAC.

Table 3. Homogeneous groups (1-2-3-4, 1-2-3-4) of strength at 20 °C for joints made with Epidian 57/PAC.

Curing Conditions	7 Days in 20 °C					2 h in 100 °C				
Hardener Inaccuracy	Mean Strength (MPa) at 20 °C	1	2	3	4	Mean Strength (MPa) at 20 °C	1	2	3	4
−30%	22.88		X	X		39.50		X		
−10%	27.39	X				47.21	X			
0%	**25.95**	X	X			**37.29**		X		
10%	19.97			X		29.12			X	
30%	13.65				X	17.30				X

Table 4. Homogeneous groups (1-2-3-4, 1-2-3-4) of strength at 50 °C for joints made with Epidian 57/PAC.

Curing Conditions	7 Days in 20 °C					2 h in 100 °C				
Hardener Inaccuracy	Mean Strength (MPa) at 50 °C	1	2	3	4	Mean Strength (MPa) at 50 °C	1	2	3	4
−30%	18.10	X				31.41	X			
−10%	13.71		X			18.13		X		
0%	**13.84**		X			**16.87**		X	X	
10%	9.09			X		15.55			X	
30%	4.79				X	6.39				X

In the case of joints made with Epidian 57/PAC, the strength changes are of a similar nature to those observed for joints made with the first adhesive analysed. An excess of the hardener leads to strength degradation, especially in joints tested at elevated temperatures. At the same time, similarly, a shortage of the hardener allows stronger joints to be obtained.

On the basis of the data obtained, an attempt was made to fit them into simple mathematical models. Given the nature of the problem under analysis, it appears that the best fit will be a second-order polynomial model. This choice is due to the fact that there is a closed bounding interval for the hypothetical joint inaccuracy: from -100% (i.e., resin without hardener) to $+100\%$ (hardener alone without resin), with an extreme (maximum) somewhere in the middle. Of course, the range investigated in this work is limited on the practical side, from $-50\%/-30\%$ to $+30\%$. For this reason, an additional attempt was made to fit a linear model within this range. Tables 5 and 6 present the results of the mathematical analyses of the regression models, together with the values of the individual coefficients of determination that indicate the quality of a given model's fit to the data (R^2).

Table 5. Fitting the strength variation curve against the inaccuracy of the adhesive preparation to the linear model: Strength = a + b · (inaccuracy).

Adhesive	Test Temperature	Curing Conditions	a	b	R^2
H9492	20 °C	1 h at 100 °C	36.43	−14.50	0.473
H9492	20 °C	3 d at 20 °C	32.53	−9.28	0.291
H9492	70 °C	1 h at 100 °C	14.22	−26.73	0.804
H9492	70 °C	3 d at 20 °C	16.36	−19.28	0.673
E57 + PAC	20 °C	2 h at 100 °C	33.61	−41.16	0.611
E57 + PAC	20 °C	7 d at 20 °C	22.13	−17.17	0.380
E57 + PAC	50 °C	2 h at 100 °C	17.67	−38.82	0.909
E57 + PAC	50 °C	7 d at 20 °C	11.97	−23.03	0.888

Table 6. Fitting the strength variation curve with respect to the inaccuracy of the adhesive preparation to a second-order polynomial model: strength = a + b·(inaccuracy) + c·(inaccuracy)^2.

Adhesive	Test Temperature	Curing Conditions	a	b	c	R^2
H9492	20 °C	1 h at 100 °C	38.48	−21.99	−35.16	0.651
H9492	20 °C	3 d at 20 °C	34.34	16.44	−31.94	0.514
H9492	70 °C	1 h at 100 °C	15.28	−30.71	−18.37	0.822
H9492	70 °C	3 d at 20 °C	18.69	−28.51	−41.56	0.874
E57 + PAC	20 °C	2 h at 100 °C	37.76	−41.16	−103.68	0.774
E57 + PAC	20 °C	7 d at 20 °C	24.94	−17.96	−75.51	0.697
E57 + PAC	50 °C	2 h at 100 °C	16.70	−38.82	24.35	0.924
E57 + PAC	50 °C	7 d at 20 °C	12.29	−22.98	−7.74	0.892

An analysis of the results above shows that, in most cases, the linear model does not very accurately describe the variation in strength from the adhesive inaccuracy. The fit coefficients R2, depending on the combination of joint curing conditions and test temperatures, varied between 0.29 and, in some cases, even up to 0.91. However, the linear model more accurately described only joints tested at elevated temperatures (0.67–0.91), while the same joints tested at ambient temperature obtained values in the range of 0.29–0.61. In these cases, second-order (quadratic) polynomial models often proved to be significantly more accurate (0.51–0.77), representing a coefficient gain in the range of 27 to 77%. Attempts were made to model with higher-order polynomials due to the physical nature of the experiment (even-order), but the increase in fitting accuracy was no longer so significant, as, for example, when analysing a 4th degree polynomial model, the increments averaged about 7% (9.5% for joints tested at 20 °C and 4.4% for those tested at an elevated temperature). However, it should be kept in mind that increasing the polynomial degree of the model, although it may lead to a better fit to the local data, will result in a high sensitivity of the model outside the range of the data used and is not necessarily more beneficial in a general sense.

Finally, an attempt was made to model the area of change in the strength of joints made with the analysed adhesives, depending on the use of heat at the stage of crosslinking the adhesive material (forming the adhesive joint) and taking into account the thermal conditions of joint operation. Figure 7 shows "wafer" diagrams modelling the course of changes in joint strength depending on the inaccuracy of the share of the hardener in the adhesive composition and the joint test temperature, grouped according to the type of adhesive used and its curing conditions.

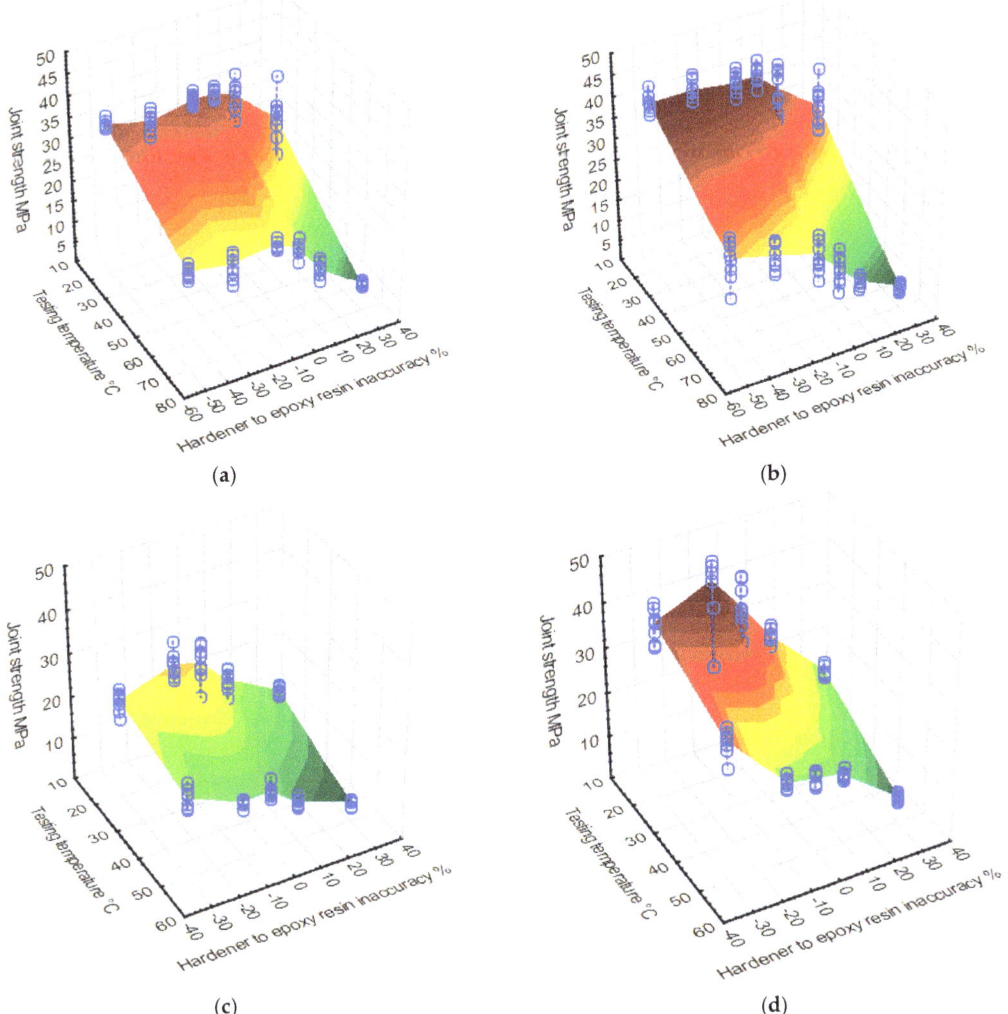

Figure 7. Strength of adhesive joints prepared using the presented adhesives crosslinked under different thermal conditions: (**a**) Hysol 9492 cured for 3 days at 20 °C, (**b**) Hysol 9492 cured for 1 h at 100 °C, (**c**) Epidian 57/PAC cured for 7 days at 20 °C, (**d**) Epidian 57/PAC cured for 2 h at 100 °C.

The above modelling results, in the form of the presented graphs, provide a comprehensive overview of the strength variations and the influence of two analysed factors: the degree of accuracy of the epoxy formulation and the joint test temperature. The three-dimensional structure of the presentation of the results obtained from destructive experimental tests, together with the modelled plane, also allow for the observation of the character of the transition of the area of the estimated joint strength at intermediate values of the operating temperature, which may be important when there is a need to estimate the strength at parameters with values in between those used for modelling.

4. Discussion

Adhesion engineering is directly related to chemistry, physics and mechanics; therefore, the analysis of adhesive bonding issues often involves topics from more than one of the aforementioned disciplines (multidisciplinary approach). In this paper, the focus is on the mechanical strength approach to the analysis of adhesive joints and the adhesives themselves, without touching on the chemical aspects in too much detail, which, in the longer term, may contribute interesting observations to the results of the paper. As described above, the final strength of structural adhesive joints has been shown to be relative to several factors, such as the state of preparation of the surfaces to be bonded [49,50], the type of adhesive used and its modifications [51], the curing conditions of the joint [52] and the operating conditions, among others.

Deviation from the stoichiometric amount of a curing agent has an impact on resin characteristics at the processing stage. Differences in the melt flow time and glass transition temperature (processability) of aerospace-grade resins and changes in ageing behaviour due to non-ideal DDS isomers, amine-to-epoxy (a/e) stoichiometric ratio material and process conditions have been found [53]. The predictive models were developed with the potential to provide the adaptive manufacturing of high-quality parts. The change in the characteristics of the cured products, e.g., among other things, the flexibility of materials based on the standard epoxy resin and the modified epoxy resin with different glycol molar ratios, was also analysed [54]. It has also been shown that samples with a low stoichiometric ratio have a predominantly branched molecular structure, whereas samples with a high stoichiometric ratio have a predominantly chain extension type structure. As a result, samples with a higher stoichiometric ratio increase the ductility of the materials, and the elongation at break increases, but the Young's modulus decreases as the stoichiometric ratio increases [55]. The effect of the epoxy–amine ratio on the thermal properties, cryogenic mechanical properties and liquid oxygen compatibility of a phosphorus-containing epoxy resin was investigated, suggesting that reduced stiffness (resulting from a change in the epoxy–amine ratio) may be beneficial for improving the compatibility of the polymer with liquid oxygen [56]. A different study [57] analysed the mechanical properties as a function of the stoichiometry and chain structure of the components used. A good correlation was found between the toughness and ductility of these materials. The hardness behaviour was explained as a function of the homogeneity of the network and the extensivity of the segments forming the cross-linked structure. The tension and three-point bending of DGEBA-epoxy resin cured by DDM were also studied. As a result, the dependencies of the resin's physical-mechanical properties (modulus of elasticity, strength, elongation at break, glass temperature) on the concentrations of the amine and epoxy groups ratio and the curing temperature were obtained [58]. In the same area, more practical cases were also investigated, e.g., epoxy resin glass fibre-reinforced composite. It was found that the epoxy resin: curing agent ratio did indeed influence both interfacial and thermal properties [59]. Interfacial Shear Stress (IFSS), using the test for the Thermal Mechanical Analyser (Araldite 506 epoxy resin and triethylenetetramine (TETA) hardener), was also analysed for fibre-reinforced polymers [60]. There are also known attempts at investigating the behaviour of a composite based on industrial hemp fibre and epoxy resin in terms of the ratio of the epoxy resin to the hardener. It was found that alkali treatment increased the interfacial shear strength, the composite tensile strength, the Young's modulus and the elongation at break. The highest tensile strength was obtained with an epoxy-resin-to-curing-agent ratio of 1:1, while the best Young's modulus was achieved with a resin-to-agent ratio of 1:1.2 [61].

The results obtained in the study presented in this paper confirm the influence of the inaccuracy of the amine-to-epoxy (a/e) stoichiometric ratio in the structure of adhesives on the strength of joints made with them. An excess of as little as 10% hardener was found to cause a statistically significant deterioration in strength (regardless of the type of adhesive tested), but, interestingly, a small deficiency of this hardener led to an increase in strength. A statistically significant increase in strength was shown to be particularly evident for joints tested at elevated temperatures, regardless of whether the adhesive was of a rigid or flexible type. By comparing the two adhesives tested, it can be seen that when the adhesive bonds are working at elevated temperatures, they are far more sensitive to changes in the epoxy-to-resin ratio than they are when working at ambient temperature. It is worth noting that the more rigid Epidian 57/PAC, when subjected to heat curing, showed less sensitivity to hardener excess, as it only registered a loss in strength characteristics at greater than 10% excess amine hardener. It should be remembered, however, that strength is one, but not the only, important property in the context of the operation of adhesive joints. In the characteristics of the final epoxy adhesives, the curing process is of key importance. The conditions during which the curing of the epoxy resin will take place will affect the material characteristics of the adhesive obtained, such as density [62], glass transition temperature [63,64], thermal behaviour [65] or the strength of the material itself [66], but also the quality of the adhesive effect to the material to be bonded [67,68]. Dramatic effects on the dynamic mechanical behaviour of flexible epoxy are shown, for example, in [69], where the dynamic mechanical behaviour of a new kind of flexible epoxy FE-1, which was crosslinked under four different thermal crosslink conditions, was examined. The results obtained in the research described in this paper confirm this phenomenon, with heat-cured joints, in most cases, being significantly stronger than unheated ones (by 30%, on average, when tested at ambient temperature and 15% when tested at elevated temperatures), as mentioned before. In one case, however, a deterioration in strength was observed after heat curing for joints made with the more rigid Hysol 9492 adhesive with an excessive proportion of a curing agent—only when tested at elevated temperatures (70 °C).

Despite the many benefits of epoxy resins and adhesives formed on their basis, one of their disadvantages is a certain sensitivity to operating conditions, precisely elevated temperatures [70–75], elevated pressures [76] and other interdependent factors, such as chemical agents typical of aerospace operations: water, jet fuel, hydraulic fluid and fuel additive [77,78]. The problem of the thermal degradation of epoxy adhesives has particularly been given much attention. Isotropic conductive adhesives (ICAs) (eco-friendly alternatives to lead solder in surface mount technology (SMT)) have been studied during two different environmental tests: a thermal cycle from −40 to 125 °C and a humid exposure of 85 °C at 85% RH. The development of some additional defects at the joint interface, such as microcracks and layers of Sn oxide, resulted in the interfacial degradation of the assembled chip components. [79]. The electrically conductive adhesives (ECAs), e.g., silver-epoxy ECAs, can also be affected due to environmental ageing (heat/humidity ageing (85 °C/85% relative humidity) and accelerated thermal cycling (−40 to 125 °C)), as it affects their electrical conductivity [80]. Due to the need for the joint to operate at a specific temperature, research was also carried out to analyse potential adhesives made of two compositions—one highly resistant to low temperatures and the other resistant to high temperatures—and to build models to estimate the strength of such a bi-component adhesive. Numerical analysis was conducted using finite element models to investigate the distribution of stresses in the mixed adhesive joint in order to find the optimal design for titanium/um/titanium and titanium/composite double lap joints. For the joint with non-similar adhesives, the mixed double adhesive joint was shown to produce an increased load capacity for the considered temperature range compared to the application of a high-temperature adhesive itself. [81]. As shown, the epoxy adhesives tested in this work also showed significant sensitivity to the elevated temperature conditions, which always weakened the joints analysed.

Further research in the topic is planned to analyse the influence of other parameters on the strength of adhesive joints, including: the method of surface preparation, the use of admixtures of materials that modify the properties of the adhesive, the degradation of joints due to aggressive environmental conditions and variable cyclic fluctuations in the temperature of the working environment of the joint [82–85]. To this end, statistical tools will be used to support mathematical modelling, and it is also planned to apply an analytical approach and use finite element numerical analysis software [86–90]. All this with the aim of obtaining even more durable and reliable adhesive joints that will be able to complement but also compete with other types of joints to an even greater extent.

5. Conclusions

The inaccuracy of the amine-to-epoxy stoichiometric ratio and the fact that heat is used to assist in curing the adhesive (resin cross-linking) are important factors in the design of adhesive joints. By using experimental tests on various types of adhesives and valid statistical methods (post-hoc tests), the influence of the variation in the above parameters was assessed, and a statistically significant variation in strength depending on the parameters analysed was demonstrated. As little as a 10% excess hardener, in the case of a more rigid adhesive, will, in most cases, lead to a significant deterioration in the strength of the joint. The flexible Hysol 9492, on the other hand, is resistant to up to a 30% excess hardener. On the other hand, a shortage of the hardener either did not adversely affect the strength of the joints or even led to an improvement. However, one should be cautious, as this may be apparent, and the modification of the adhesive may be followed by other negative changes of a non-strength nature.

It was shown that the heat curing operation at the stage of forming the butt joint leads to an increase in its static strength at ambient temperature, as well as in the case of joints made with adhesives with a deviation from the initial resin/hardener ratio considered optimal. It has been proven, however, that the heat curing operation cannot be regarded as universally improving the strength of the joint in all operational cases of the joint (e.g., at elevated temperatures of joint operation).

A reasonably good fit was achieved between simple mathematical models and the results obtained, which makes it possible to use them to predict other intermediate cases of variation in the parameters analysed, without the need for further destructive tests on adhesive joints. As shown in the paper, the second-order polynomial model was sufficient to describe the variation in strength from the inaccuracy of the epoxy/amine hardener ratio quite well. In addition, higher even-order models were tested, but it was not considered that a slight increase in fit would justify the use of these types of curves.

In the research described in this paper, the adhesive joint was analysed in one of several possible joint loading arrangements, i.e., butt joint, axially tensile. Classical test methods for adhesives and adhesive joints often also pay a lot of attention to lap, shear-operated joints, as this is one of the more favourable loading options for adhesive joints. It thus seems reasonable to continue the tests that have been undertaken for other joint conditions to see whether the observed patterns will be confirmed in other cases. In addition, it is worth including chemical and thermal analyses in the study, which could show in a practical way how an inaccurately made adhesive cross-links and which cross-linking state corresponds to which strength characteristics of the adhesive and the joints made with it.

Author Contributions: Conceptualization, J.S. and R.K.; data curation, J.S.; formal analysis, J.S., R.K. and M.F.; funding acquisition, J.J.; investigation, J.S. and R.K.; methodology, J.S. and R.K.; project administration, J.J. and M.F.; resources, J.S.; software, J.S.; supervision, J.J. and M.F.; validation, J.S. and R.K.; visualization, J.S.; writing—original draft preparation, J.S. and R.K.; writing—review and editing, J.S., R.K. and M.F. All authors have read and agreed to the published version of the manuscript.

Funding: This research project was financed in the framework of the Lublin University of Technology–Regional Excellence Initiative project, funded by the Polish Ministry of Science and Higher Education (contract no. 030/RID/2018/19).

Institutional Review Board Statement: Not applicable.

Informed Consent Statement: Not applicable.

Data Availability Statement: The data presented in this study are available from the corresponding authors upon request.

Conflicts of Interest: The authors declare no conflict of interest. The funders had no role in the design of the study; in the collection, analyses or interpretation of data; in the writing of the manuscript; or in the decision to publish the results.

References

1. Fay, P.A. A History of Adhesive Bonding. In *Adhesive Bonding*; Elsevier: Amsterdam, The Netherlands, 2021; pp. 3–40. ISBN 978-0-12-819954-1.
2. Chandrasekaran, N.K.; Arunachalam, V. State-of-the-art Review on Honeycomb Sandwich Composite Structures with an Emphasis on Filler Materials. *Polym. Compos.* **2021**, *42*, 5011–5020. [CrossRef]
3. Comyn, J. *Adhesion Science*, 2nd ed.; Royal Society of Chemistry: London, UK, 2021; ISBN 978-1-78801-888-3.
4. Rudawska, A.; Worzakowska, M.; Bociąga, E.; Olewnik-Kruszkowska, E. Investigation of Selected Properties of Adhesive Compositions Based on Epoxy Resins. *Int. J. Adhes. Adhes.* **2019**, *92*, 23–36. [CrossRef]
5. Rudawska, A.; Szabelski, J.; Miturska, I.; Doluk, E. Strength of Solder and Adhesive Joints of Copper Sheets. In Proceedings of the 9th International Conference on Fracture, Fatigue and Wear, Ghent, Belgium, 2–3 August 2021; Abdel Wahab, M., Ed.; Lecture Notes in Mechanical Engineering. Springer: Singapore, 2022; pp. 85–95, ISBN 9789811688096.
6. Doluk, E.; Rudawska, A.; Stančeková, D.; Mrázik, J. Influence of Surface Treatment on the Strength of Adhesive Joints. *Manuf. Technol.* **2021**, *21*, 585–591. [CrossRef]
7. Ozun, E.; Ceylan, R.; Özgür Bora, M.; Çoban, O.; Kutluk, T. Combined Effect of Surface Pretreatment and Nanomaterial Reinforcement on the Adhesion Strength of Aluminium Joints. *Int. J. Adhes. Adhes.* **2022**, *114*, 103274. [CrossRef]
8. Miturska, I.; Rudawska, A.; Müller, M.; Valášek, P. The Influence of Modification with Natural Fillers on the Mechanical Properties of Epoxy Adhesive Compositions after Storage Time. *Materials* **2020**, *13*, 291. [CrossRef]
9. Park, S.-J. Thermal Stability and Fracture Toughness of Epoxy Resins Modified with Epoxidized Castor Oil and Al_2O_3 Nanoparticles. *Bull. Korean Chem. Soc.* **2012**, *33*, 2513–2516. [CrossRef]
10. Kumar, R.; Mohanty, S.; Nayak, S.K. Thermal Conductive Epoxy Adhesive Composites Filled with Carbon-Based Particulate Fillers: A Comparative Study. *J. Adhes. Sci. Technol.* **2020**, *34*, 807–827. [CrossRef]
11. Kim, H.-J.; Lim, D.-H.; Hwang, H.-D.; Lee, B.-H. Composition of Adhesives. In *Handbook of Adhesion Technology*; da Silva, L.F.M., Öchsner, A., Adams, R.D., Eds.; Springer International Publishing: Cham, Switzerland, 2018; pp. 319–343. ISBN 978-3-319-55410-5.
12. Frigione, M.; Lettieri, M. Recent Advances and Trends of Nanofilled/Nanostructured Epoxies. *Materials* **2020**, *13*, 3415. [CrossRef]
13. Rudawska, A.; Miturska, I. Impact Study of Single Stage and Multi Stage Abrasive Machining on Static Strength of Lap Adhesive Joints of Mild Steel. *MATEC Web Conf.* **2018**, *244*, 02006. [CrossRef]
14. Miturska, I.; Rudawska, A.; Müller, M.; Hromasová, M. The Influence of Mixing Methods of Epoxy Composition Ingredients on Selected Mechanical Properties of Modified Epoxy Construction Materials. *Materials* **2021**, *14*, 411. [CrossRef]
15. Godzimirski, J.; Komorek, A. The Influence of Selected Test Conditions on the Impact Strength of Adhesively Bonded Connections. *Materials* **2020**, *13*, 1320. [CrossRef] [PubMed]
16. Doluk, E.; Rudawska, A.; Brunella, V. The Influence of Technological Factors on the Strength of Adhesive Joints of Steel Sheets. *Adv. Sci. Technol. Res. J.* **2020**, *14*, 107–115. [CrossRef]
17. Savvilotidou, M.; Vassilopoulos, A.P.; Frigione, M.; Keller, T. Development of Physical and Mechanical Properties of a Cold-Curing Structural Adhesive in a Wet Bridge Environment. *Constr. Build. Mater.* **2017**, *144*, 115–124. [CrossRef]
18. Petrie, E.M. *Handbook of Adhesives and Sealants*, 3rd ed.; McGraw-Hill: New York, NY, USA, 2021; ISBN 978-1-260-44044-7.
19. Packham, D.E. (Ed.) *Handbook of Adhesion*, 2nd ed.; John Wiley: Hoboken, NJ, USA, 2005; ISBN 978-0-471-80874-9.
20. *Standard PN-EN ISO 10365:2022*; Adhesives—Designation of Main Failure Patterns. ISO: Warszawa, Poland, 2022.
21. Gallas, S.; Devriendt, H.; Croes, J.; Naets, F.; Desmet, W. Experimental Model Update for Single Lap Joints. In *Nonlinear Structures & Systems, Volume 1*; Brake, M.R.W., Renson, L., Kuether, R.J., Tiso, P., Eds.; Conference Proceedings of the Society for Experimental Mechanics Series; Springer International Publishing: Cham, Switzerland, 2023; pp. 177–180. ISBN 978-3-031-04085-6.
22. Rozylo, P. Stability and Failure of Compressed Thin-walled Composite Columns Using Experimental Tests and Advanced Numerical Damage Models. *Int. J. Numer. Methods Eng.* **2021**, *122*, 5076–5099. [CrossRef]
23. Jonak, J.; Karpiński, R.; Wójcik, A.; Siegmund, M.; Kalita, M. Determining the Effect of Rock Strength Parameters on the Breakout Area Utilizing the New Design of the Undercut/Breakout Anchor. *Materials* **2022**, *15*, 851. [CrossRef]
24. Falkowicz, K.; Debski, H.; Wysmulski, P. Effect of Extension-Twisting and Extension-Bending Coupling on a Compressed Plate with a Cut-Out. *Compos. Struct.* **2020**, *238*, 111941. [CrossRef]
25. Rozylo, P. Experimental-Numerical Study into the Stability and Failure of Compressed Thin-Walled Composite Profiles Using Progressive Failure Analysis and Cohesive Zone Model. *Compos. Struct.* **2021**, *257*, 113303. [CrossRef]
26. Liu, P.F.; Liu, J.W.; Wang, C.Z. Finite Element Analysis of Damage Mechanisms of Composite Wind Turbine Blade by Considering Fluid/Solid Interaction. Part II: T-Shape Adhesive Structure. *Compos. Struct.* **2022**, *301*, 116211. [CrossRef]
27. Wysmulski, P.; Debski, H.; Falkowicz, K.; Rozylo, P. The Influence of Load Eccentricity on the Behavior of Thin-Walled Compressed Composite Structures. *Compos. Struct.* **2019**, *213*, 98–107. [CrossRef]
28. Falkowicz, K.; Debski, H. The Post-Critical Behaviour of Compressed Plate with Non-Standard Play Orientation. *Compos. Struct.* **2020**, *252*, 112701. [CrossRef]

29. Romero, A.; Galvín, P.; Domínguez, J. 3D Non-Linear Time Domain FEM–BEM Approach to Soil–Structure Interaction Problems. *Eng. Anal. Bound. Elem.* **2013**, *37*, 501–512. [CrossRef]

30. Romero, A.; Galvín, P.; Tadeu, A. An Accurate Treatment of Non-Homogeneous Boundary Conditions for Development of the BEM. *Eng. Anal. Bound. Elem.* **2020**, *116*, 93–101. [CrossRef]

31. Machrowska, A.; Karpiński, R.; Jonak, J.; Szabelski, J.; Krakowski, P. Numerical Prediction of the Component-Ratio-Dependent Compressive Strength of Bone Cement. *Appl. Comput. Sci.* **2020**, *16*, 87–101. [CrossRef]

32. Machrowska, A.; Szabelski, J.; Karpiński, R.; Krakowski, P.; Jonak, J.; Jonak, K. Use of Deep Learning Networks and Statistical Modeling to Predict Changes in Mechanical Parameters of Contaminated Bone Cements. *Materials* **2020**, *13*, 5419. [CrossRef] [PubMed]

33. Rogala, M.; Gajewski, J.; Górecki, M. Study on the Effect of Geometrical Parameters of a Hexagonal Trigger on Energy Absorber Performance Using ANN. *Materials* **2021**, *14*, 5981. [CrossRef]

34. Szabelski, J.; Karpiński, R.; Machrowska, A. Application of an Artificial Neural Network in the Modelling of Heat Curing Effects on the Strength of Adhesive Joints at Elevated Temperature with Imprecise Adhesive Mix Ratios. *Materials* **2022**, *15*, 721. [CrossRef]

35. Bauer, L.; Stütz, L.; Kley, M. Black Box Efficiency Modelling of an Electric Drive Unit Utilizing Methods of Machine Learning. *ACS* **2021**, *17*, 5–19. [CrossRef]

36. Dsouza, K.J.; Ansari, Z.A. Histopathology Image Classification Using Hybrid Parallel Structured DEEP-CNN Models. *Appl. Comput. Sci.* **2022**, *18*, 20–36.

37. Jonak, J.; Karpiński, R.; Siegmund, M.; Machrowska, A.; Prostański, D. Experimental Verification of Standard Recommendations for Estimating the Load-Carrying Capacity of Undercut Anchors in Rock Material. *Adv. Sci. Technol. Res. J.* **2021**, *15*, 1–15. [CrossRef]

38. Świć, A.; Gola, A. Theoretical and Experimental Identification of Frequency Characteristics and Control Signals of a Dynamic System in the Process of Turning. *Materials* **2021**, *14*, 2260. [CrossRef]

39. Berzins, R.; Merijs Meri, R.; Zicans, J. Different Epoxide Compound Influence on Two Component Silyl-Terminated Polymer/Epoxide Systems Mechanical, Rheological and Adhesion Properties. *IOP Conf. Ser. Mater. Sci. Eng.* **2019**, *500*, 012032. [CrossRef]

40. Szabelski, J.; Karpiński, R.; Krakowski, P.; Jojczuk, M.; Jonak, J.; Nogalski, A. Analysis of the Effect of Component Ratio Imbalances on Selected Mechanical Properties of Seasoned, Medium Viscosity Bone Cements. *Materials* **2022**, *15*, 5577. [CrossRef] [PubMed]

41. Karpinski, R.; Szabelski, J.; Maksymiuk, J. Analysis of the Properties of Bone Cement with Respect to Its Manufacturing and Typical Service Lifetime Conditions. *MATEC Web Conf.* **2018**, *244*, 01004. [CrossRef]

42. Karpiński, R.; Szabelski, J.; Maksymiuk, J. Seasoning Polymethyl Methacrylate (PMMA) Bone Cements with Incorrect Mix Ratio. *Materials* **2019**, *12*, 3073. [CrossRef] [PubMed]

43. Szabelski, J. Effect of Incorrect Mix Ratio on Strength of Two Component Adhesive Butt-Joints Tested at Elevated Temperature. *MATEC Web Conf.* **2018**, *244*, 01019. [CrossRef]

44. Rabiej, M. *Analizy Statystyczne z Programami Statistica i Excel*; Wydawnictwo Helion: Gliwice, Poland, 2018; ISBN 978-83-283-3922-4.

45. Ross, S.M. *Introduction to Probability and Statistics for Engineers and Scientist*, 6th ed.; Elsevier/Academic Press: London, UK, 2021; ISBN 978-0-12-824346-6.

46. Navidi, W.C. *Statistics for Engineers and Scientists*, 3rd ed.; McGraw-Hill: New York, NY, USA, 2011; ISBN 978-0-07-337633-2.

47. Devore, J.L.; Farnum, N.R.; Doi, J. *Applied Statistics for Engineers and Scientists*, 3rd ed.; Cengage Learning: Stamford, CT, USA, 2014; ISBN 978-1-133-11136-8.

48. Heilio, M.L.; Lähivaara, T.; Laitinen, E.; Mantere, T.; Merikoski, J.; Pohjolainen, S.; Raivio, K.; Silvennoinen, R.; Suutala, A.; Tarvainen, T.; et al. *Mathematical Modelling*; Springer: Berlin/Heidelberg, Germany, 2018; ISBN 978-3-319-80226-8.

49. Rudawska, A.; Miturska-Barańska, I.; Doluk, E.; Olewnik-Kruszkowska, E. Assessment of Surface Treatment Degree of Steel Sheets in the Bonding Process. *Materials* **2022**, *15*, 5158. [CrossRef]

50. Miturska-Barańska, I.; Rudawska, A. Effect of Surface Abrasive Treatment on the Strength of Galvanised Steel Sheet Adhesive Joints. *Adv. Sci. Technol. Res. J.* **2022**, *16*, 75–84. [CrossRef]

51. Miturska-Barańska, I.; Rudawska, A. Influence of Adhesive Compound Viscosity on the Strength Properties of 1.0503 Steel Sheets Adhesive Joints. *Adv. Sci. Technol. Res. J.* **2022**, *16*, 196–205. [CrossRef]

52. Rudawska, A.; Zaleski, K.; Miturska, I.; Skoczylas, A. Effect of the Application of Different Surface Treatment Methods on the Strength of Titanium Alloy Sheet Adhesive Lap Joints. *Materials* **2019**, *12*, 4173. [CrossRef]

53. Kim, D.; Nutt, S.R. Processability of DDS Isomers-Cured Epoxy Resin: Effects of Amine/Epoxy Ratio, Humidity, and out-Time: Processability of DDs Isomers-Cured Epoxy Resin: Effects of Amine/Epoxy Ratio, Humidity, and Out-Time. *Polym. Eng. Sci.* **2018**, *58*, 1530–1538. [CrossRef]

54. Ladaniuc, M.A.; Hubca, G.; Gabor, R.; Andi-Nicolae, C.; Sandu, T. Study on the Reactants Molar Ratio Influence on the Properties of Standard Epoxy Resin/Glycols-Modified Epoxy Resin Compounds. *Mater. Plast.* **2015**, *52*, 433–436.

55. Odagiri, N.; Shirasu, K.; Kawagoe, Y.; Kikugawa, G.; Oya, Y.; Kishimoto, N.; Ohuchi, F.S.; Okabe, T. Amine/Epoxy Stoichiometric Ratio Dependence of Crosslinked Structure and Ductility in AMINE-CURED Epoxy Thermosetting Resins. *J. Appl. Polym. Sci.* **2021**, *138*, 50542. [CrossRef]

56. Wang, H.; Li, S.; Yuan, Y.; Liu, X.; Sun, T.; Wu, Z. Study of the Epoxy/Amine Equivalent Ratio on Thermal Properties, Cryogenic Mechanical Properties, and Liquid Oxygen Compatibility of the Bisphenol A Epoxy Resin Containing Phosphorus. *High Perform. Polym.* **2020**, *32*, 429–443. [CrossRef]

57. Fernandez-Nograro, F.; Valea, A.; Llano-Ponte, R.; Mondragon, I. Dynamic and Mechanical Properties of DGEBA/Poly(Propylene Oxide) Amine Based Epoxy Resins as a Function of Stoichiometry. *Eur. Polym. J.* **1996**, *32*, 257–266. [CrossRef]

58. Garifullin, N.O.; Komarov, B.A.; Kapasharov, A.T.; Malkov, G.V. Dependence of the Physical-Mechanical Properties of Cured Epoxy-Amine Resin on the Ratio of Its Components. *KEM* **2019**, *816*, 146–150. [CrossRef]

59. Minty, R.F.; Thomason, J.L.; Yang, L. The Role of the Epoxy Resin: Curing Agent Ratio on Composite Interfacial Strength and Thermal Performance. In Proceedings of the European Conference on Composite Materials-ECCM 17, Munich, Germany, 26–30 June 2016.

60. Minty, R.; Thomason, J.; Petersen, H. The Role of the Epoxy Resin: Curing Agent Ratio in Composite Interfacial Strength by Single Fibre Microbond Test. In Proceedings of the 20th International Conference on Composite Materials, Copenhagen, Denmark, 19–24 July 2015.

61. Islam, M.S.; Pickering, K.L. Influence of Alkali Treatment on the Interfacial Bond Strength of Industrial Hemp Fibre Reinforced Epoxy Composites: Effect of Variation from the Ideal Stoicheometric Ratio of Epoxy Resin to Curing Agent. In *Advanced Materials Research*; Trans Tech Publications Ltd.: Stafa, Switzerland, 2007; Volume 29–30, pp. 319–322. ISBN 978-0-87849-466-8.

62. Pang, K.P.; Gillham, J.K. Anomalous Behavior of Cured Epoxy Resins: Density at Room Temperature versus Time and Temperature of Cure. *J. Appl. Polym. Sci.* **1989**, *37*, 1969–1991. [CrossRef]

63. Michel, M.; Ferrier, E. Effect of Curing Temperature Conditions on Glass Transition Temperature Values of Epoxy Polymer Used for Wet Lay-up Applications. *Constr. Build. Mater.* **2020**, *231*, 117206. [CrossRef]

64. Chandrathilaka, E.R.K.; Gamage, J.C.P.H.; Fawzia, S. Effects of Elevated Temperature Curing on Glass Transition Temperature of Steel/CFRP Joint and Pure Epoxy Adhesive. *EJSE* **2018**, *18*, 1–6. [CrossRef]

65. Musa, A.; Alamry, K.A.; Hussein, M.A. The Effect of Curing Temperatures on the Thermal Behaviour of New Polybenzoxazine-Modified Epoxy Resin. *Polym. Bull.* **2020**, *77*, 5439–5449. [CrossRef]

66. Jin, N.J.; Yeon, J.; Seung, I.; Yeon, K.-S. Effects of Curing Temperature and Hardener Type on the Mechanical Properties of Bisphenol F-Type Epoxy Resin Concrete. *Constr. Build. Mater.* **2017**, *156*, 933–943. [CrossRef]

67. Rudawska, A. The Influence of Curing Conditions on the Strength of Adhesive Joints. *J. Adhes.* **2020**, *96*, 402–422. [CrossRef]

68. Rudawska, A.; Czarnota, M. Selected Aspects of Epoxy Adhesive Compositions Curing Process. *J. Adhes. Sci. Technol.* **2013**, *27*, 1933–1950. [CrossRef]

69. Dai, P.; Wang, Y.; Huang, Z. Effect of Thermal Crosslink Conditions on Dynamic Mechanical Behaviors of Flexible Epoxy. *J. Wuhan Univ. Technol.-Mat. Sci. Edit.* **2008**, *23*, 825–829. [CrossRef]

70. Goda, Y.; Sawa, T.; Himuro, K.; Yamamoto, K. Impact Strength Degradation of Adhesive Joints Under Heat and Moisture Environmental Conditions. In *Volume 9: Mechanics of Solids, Structures and Fluids*; ASMEDC: Vancouver, BC, Canada, 2010; pp. 29–36.

71. Ocaña, R.; Arenas, J.M.; Alía, C.; Narbón, J.J. Evaluation of Degradation of Structural Adhesive Joints in Functional Automotive Applications. *Procedia Eng.* **2015**, *132*, 716–723. [CrossRef]

72. Cui, M.; Zhang, L.; Lou, P.; Zhang, X.; Han, X.; Zhang, Z.; Zhu, S. Study on Thermal Degradation Mechanism of Heat-Resistant Epoxy Resin Modified with Carboranes. *Polym. Degrad. Stab.* **2020**, *176*, 109143. [CrossRef]

73. Rudawska, A.; Abdel Wahab, M.; Szabelski, J.; Miturska, I.; Doluk, E. The Strength of Rigid and Flexible Adhesive Joints at Room Temperature and After Thermal Shocks. In Proceedings of the 1st International Conference on Structural Damage Modelling and Assessment, Online, 4–5 August 2021; Abdel Wahab, M., Ed.; Lecture Notes in Civil Engineering. Springer: Singapore, 2021; Volume 110, pp. 229–241, ISBN 9789811591204.

74. Rudawska, A.; Miturska, I.; Szabelski, J.; Abdel Wahab, M.; Stančeková, D.; Čuboňová, N.; Madleňák, R. The Impact of the Selected Exploitation Factors on the Adhesive Joints Strength. In Proceedings of the 13th International Conference on Damage Assessment of Structures, Porto, Portugal, 9–10 July 2019; Wahab, M.A., Ed.; Lecture Notes in Mechanical Engineering. Springer: Singapore, 2020; pp. 899–913, ISBN 9789811383304.

75. Wang, Q.; Min, Z.; Li, M.; Huang, W. Mechanical Behavior and Thermal Oxidative Aging of Anhydride-Cured Epoxy Asphalt with Different Asphalt Contents. *J. Mater. Civ. Eng.* **2022**, *34*, 04022245. [CrossRef]

76. Simenou, G.; Le Gal La Salle, E.; Bailleul, J.-L.; Bellettre, J. Heat Transfer Analysis during the Hydrothermal Degradation of an Epoxy Resin Using Differential Scanning Calorimetry (DSC). *J. Therm. Anal. Calorim.* **2016**, *125*, 861–869. [CrossRef]

77. Sakaguchi, K. *Estimation of the Mechanism to Suppress Water Degradation of 1K Heat-Curing Epoxy Adhesive with High Durability*; SAE: Warrendale, PA, USA, 2020; p. 2020-01-0227.

78. Campbell, R.A.; Pickett, B.M.; Saponara, V.L.; Dierdorf, D. Thermal Characterization and Flammability of Structural Epoxy Adhesive and Carbon/Epoxy Composite with Environmental and Chemical Degradation. *J. Adhes. Sci. Technol.* **2012**, *26*, 889–910. [CrossRef]

79. Kim, S.S.; Kim, K.S.; Suganuma, K.; Tanaka, H. Degradation Mechanism of Ag-Epoxy Conductive Adhesive Joints by Heat and Humidity Exposure. In Proceedings of the 2008 2nd Electronics Systemintegration Technology Conference, Greenwich, UK, 1–4 September 2008; pp. 903–908.

80. Ivanov, V.; Wolter, K.-J. Environmental Ageing Effects on the Electrical Resistance of Silver-Epoxy Electrically Conductive Adhesive Joints to a Molybdenum Electrode. In Proceedings of the 2013 IEEE 15th Electronics Packaging Technology Conference (EPTC 2013), Singapore, 11–13 December 2013; pp. 44–47.
81. Da Silva, L.F.M.; Adams, R.D. Joint Strength Predictions for Adhesive Joints to Be Used over a Wide Temperature Range. *Int. J. Adhes. Adhes.* **2007**, *27*, 362–379. [CrossRef]
82. Rudawska, A. Mechanical Properties of Epoxy Compounds Based on Bisphenol a Aged in Aqueous Environments. *Polymers* **2021**, *13*, 952. [CrossRef]
83. Miturska, I.; Rudawska, A.; Niedziałkowski, P.; Szabelski, J.; Grzesiczak, D. Selected Strength Aspects of Adhesive Lap Joints and Butt Welded Joints of Various Structural Materials. *Adv. Sci. Technol. Res. J.* **2018**, *12*, 135–141. [CrossRef]
84. Rudawska, A.; Miturska-Barańska, I.; Doluk, E. Influence of Surface Treatment on Steel Adhesive Joints Strength—Varnish Coats. *Materials* **2021**, *14*, 6938. [CrossRef] [PubMed]
85. Rudawska, A. Experimental Study of Mechanical Properties of Epoxy Compounds Modified with Calcium Carbonate and Carbon after Hygrothermal Exposure. *Materials* **2020**, *13*, 5439. [CrossRef]
86. Wysmulski, P. The Analysis of Buckling and Post Buckling in the Compressed Composite Columns. *Arch. Mater. Sci. Eng.* **2017**, *85*, 35–41. [CrossRef]
87. Falkowicz, K. Numerical Investigations of Perforated CFRP Z-Cross-Section Profiles, under Axial Compression. *Materials* **2022**, *15*, 6874. [CrossRef] [PubMed]
88. Wysmulski, P. The Effect of Load Eccentricity on the Compressed CFRP Z-Shaped Columns in the Weak Post-Critical State. *Compos. Struct.* **2022**, *301*, 116184. [CrossRef]
89. Jonak, J.; Karpiński, R.; Wójcik, A. Numerical Analysis of Undercut Anchor Effect on Rock. *J. Phys.: Conf. Ser.* **2021**, *2130*, 012011. [CrossRef]
90. Jonak, J.; Karpiński, R.; Wójcik, A. Numerical Analysis of the Effect of Embedment Depth on the Geometry of the Cone Failure. *J. Phys. Conf. Ser.* **2021**, *2130*, 012012. [CrossRef]

 *materials* 

Article

Mechanical Performance of Adhesive Connections in Structural Applications

Marco Lamberti [1,2,*], Aurélien Maurel-Pantel [1,*] and Frédéric Lebon [1]

[1]   Aix-Marseille University, CNRS, Centrale Marseille, LMA, 13453 Marseille, France
[2]   ENEA, Brasimone Research Center, 40032 Camugnano, Italy
*   Correspondence: lamberti@lma.cnrs-mrs.fr or marco.lamberti@enea.it (M.L.); maurel@lma.cnrs-mrs.fr (A.M.-P.)

Abstract: Adhesive bonding is an excellent candidate for realising connections for secondary structures in structural applications such as offshore wind turbines and installations, avoiding the risk and associated welding problems. The strength of the adhesive layer is an important parameter to consider in the design process it being lower than the strength capacity of the bonding material. The presence of defects in the adhesive materials undoubtedly influences the mechanical behaviour of bonded composite structures. More specifically, the reduction in strength is more pronounced as the presence of defects (voids) increases. For this reason, a correct evaluation of the presence of defects, which can be translated into damage parameters, has become essential in predicting the actual behaviour of the bonded joints under different external loading conditions. In this paper, an extensive experimental programme has been carried out on adhesively bonded connections subjected to Mode I and Mode II loading conditions in order to characterise the mechanical properties of a commercial epoxy resin and to define the damage parameters. The initial damage parameters of the adhesive layer have been identified according to the Kachanov–Sevostianov material definition, which is able to take into account the presence of diffuse initial cracking.

Keywords: adhesive bonding; damage; porosity; Kachanov–Sevostianov's material

Citation: Lamberti, M.; Maurel-Pantel, A.; Lebon, F. Mechanical Performance of Adhesive Connections in Structural Applications. *Materials* **2023**, *16*, 7066. https://doi.org/10.3390/ma16227066

Academic Editors: Ricardo J. C. Carbas and Chih-Chun Hsieh

Received: 25 September 2023
Revised: 27 October 2023
Accepted: 4 November 2023
Published: 7 November 2023

Copyright: © 2023 by the authors. Licensee MDPI, Basel, Switzerland. This article is an open access article distributed under the terms and conditions of the Creative Commons Attribution (CC BY) license (https://creativecommons.org/licenses/by/4.0/).

1. Introduction

In the last two decades, the use of adhesive materials in the field of mechanical and civil engineering has grown exponentially due to their capacity to easily and quickly connect several types of materials to each other such as metals, composite, concrete and masonry.

The use of bonding techniques in various industries has increased significantly due to the growing demand for the design of lightweight structures in the mechanical field, such as aircraft and vehicle frames. For this reason, the use of adhesive bonding to join advanced lightweight materials that are dissimilar, coated, and difficult to weld have been widely studied in recent years [1].

Although bonding has been used as a traditional joining method for many centuries, it is only in the last seventy years that the scientific results and the technology of the bonding technique have advanced significantly [2–7]. In addition to civil engineering, the adhesive bonding technique has been increasingly used in structural strengthening and reinforcement of concrete elements by adding FRP sheets, both in fully composite structures such as pedestrian bridges and in buildings where pultruded profiles have been matched to form complex and structured cross-sections [8]. Furthermore, these types of joints are particularly suitable for the realization of secondary structures such as parapets, stairs and railings in various types of structures such as buildings, cooling towers and offshore installations.

Among the factors which have limited the spread and development of adhesive connections for marine and offshore structures, there is the long-term durability of joints in critical environments.

Nowadays, in offshore installations, most connections are made using the welding technique. However, the welding technique does not represent the optimal solution for safety and building technique reasons. Avoiding the presence of high welding temperatures leads to safer construction in marine environments. In addition, it will positively contribute to the preservation of and improvement in the quality of the environment by reducing the amount of welding slag created.

Adhesive bonding in the marine environment for offshore applications is still very much in its infancy despite some successes. However, it is still needed to establish this joining process as a standard process considering the design, fabrication, and modification of offshore structures.

It is important to emphasize that the choice of thickness geometry must derive from on-site feasibility assessments, considering that thin and uniform adhesive thicknesses are easily made in a specialized laboratory using skilled workers, otherwise it becomes difficult to make them on site.

For these reasons, the scientific and industrial communities have become interested in providing tools to describe and simulate the behaviour of adhesively bonded joints.

The mechanical behaviour of an adhesive joint is influenced not only by the geometry of the joint, but also by various boundary conditions.

Several approaches and theories have been formulated in the literature to describe material characteristics to investigate different types of applications using analytical, mechanical, or finite element analyses. Among them, damage modelling is increasingly used to simulate debonding processes and fractures in adhesive connections.

One of the most important characteristics is undoubtedly the stiffness of the adhesive layer, which, if properly defined, allows a realistic evaluation of the displacements exhibited after the application of loads that could act during the life of the structure.

Damage modelling techniques are distinguished into local or continuous approaches. In the first, the continuous approach, damage is implemented over a finite region, while in the second, the local approach, damage is located to zero-volume lines leading it to be referred as the cohesive zone model [9].

The cohesive zone model [10] has received considerable attention over the past two decades and has been used to predict interlaminar failure of composite materials. Fractures in bonded materials particularly affect the machined zone in front of the macrocrack tip, where microcracks or cavities form, grow and coalesce. This process region can be modelled by assuming that the material along the crack path follows the established tensile separation laws of an appropriate cohesive region model. There are a large number of cohesion laws in the literature, ranging from exponential to trapezoidal laws.

One of the earliest theories of the elastic contact model for flat metal surfaces was formulated by Greenwood and Williamson [11].

The model proposed was based on the existence of elastic contact hardness, a composite quantity that is a function of the elastic properties and topography, considering a statistical distribution of asperities that do not interact with each other.

Subsequently, Yoshioka and Scholz [12] developed a theory for predicting the behaviour of contacting surfaces focused on micromechanics under elastic and non-slip conditions, opening up a new way of understanding the behaviour of contacting surfaces.

A few years later, Sherif and Kossa [13], using the theories of Greenwood and Williamson [11], carried out a theoretical analysis to calculate the normal and tangential contact stiffnesses between two elastic flat surfaces, giving an interpretation of the experimental results obtained founded on the evaluation of the natural frequencies at the contact region. Following the same strategy, Krolikowski and Szczepek [14], based on the Green–Wood–Williams model and the Hertz–Mindlin theory [15], provided an analytical description of the normal and tangential contact stiffness between rough surfaces with spherical properties. Contact stiffness has also been measured using an ultrasonic method focused on the measure of the reflection coefficient of ultrasonic waves at the interface.

The definition of contact stiffness has been carried out by several experimental studies that can be found in the literature.

In addition to the cases mentioned above [13,14], Gonzalez-Valadez et al. [16] proposed the use of a simple spring model influenced by the amount, shape and distribution of the contact asperities, relating the interfacial stiffness to the reflection of ultrasound obtained in a rough contact.

Finally, a new approach has been proposed by Kachanov et al. [17].

The Kachanov theory consists of considering the presence of initial cracks in the interior of an adhesive material. The main assumptions of the microcracked adhesive are based on the absence of interaction among the several cracks, constant stress vector along the crack and finally the absence of effect due to the presence of the crack edge in the stress field. Furthermore, the peculiarity of this model is that it considers some of the most important variabilities of the adhesive, such as thickness variation, porosity and initial damage [18,19].

The Kachanov-type model has previously been successfully applied to aluminium foam alloy [18], composite materials [19] and also other types of structures.

The accuracy of this approach, which is a function of the density of the cracks, is satisfactory up to fairly small distances between the cracks. The distances between the cracks is much smaller than their width. For linear cracks, Kachanov's model includes a global parameter indicated as crack density, which is attributable to the number and length of all cracks.

In this work, an extensive experimental programme was carried out to determine the properties of the undamaged material. The experimental programme consists of static tensile tests performed under Mode I and Mode II loading conditions on bonded specimens using an Arcan-modified apparatus and double lap shear-bonded joints.

The bonded joints were realised with different sizes of thickness and surface area of the adhesive layer in order to provide a better comprehension of the damage parameters. Finally, an imperfect interface model, obtained thanks to the homogenisation technique and the asymptotic approach, was used to reproduce the global response of the adhesive joints in Mode I and II loading conditions, using the initial damage parameters evaluated experimentally.

Since it is essential that the adhesive connections must be able to guarantee long-term properties and sufficient mechanical strength in order to propose reliable solutions, the effects of the aging conditions will be investigated as a perspective of the present investigation.

2. Experimental Program

To characterise the mechanical behaviour of adhesive connections under normal and tangential forces, two types of specimens were produced and tested: cylindrical and double lap shear joints.

In addition, several adhesive thicknesses and diameters for the cylindrical specimens were experimentally tested under static loading conditions in order to evaluate the variation in damage parameters as a function of adhesive volume.

The adhesive used in the current investigation is available on the market and is named Sicomin Isobond SR 5030/SD 503x [20]. Specifically, Sicomin Isobond is a two-component epoxy paste designed for structural bonding and fillet joints, with high mechanical strength and high thixotropy for good behaviour on vertical surfaces. The maximum strength of the adhesive is reached after a curing time of 24 h at 23 °C or 10 h at 70 °C.

The mechanical properties declared by the manufacturer are summarized in Table 1.

Table 1. Mechanical properties of Sicomin Isobond SR 5030/SD 503x.

Property	Unit Measure	Value
Modulus of Elasticity	N/mm^2	4500
Tensile strength	N/mm^2	62
Elongation at break	%	2.9
Shear strength	N/mm^2	13.4

2.1. Cylindrical Adhesive Joints

Cylindrical aluminium specimens were used to make the adhesive joints. The mechanical properties of the aluminium material are provided in Table 2. Each specimen has a straight surface which allows a homogeneous adhesive layer to be produced at the interface.

Table 2. Mechanical properties of aluminium cylinder.

Property	Unit Measure	Value
Modulus of Elasticity	N/mm^2	70,000
Yield strength	N/mm^2	210
Poisson ratio	-	0.3

The cylindrical adhesive samples are realized using a steel device made by an upper and lower horizontal element connected to each other by means of two vertical columns; in this way, a constant total height of the adhesive layer is performed. The total height, Ht, of 64 mm is due to the height of two half-specimens plus the adhesive thickness of the adhesive. In more detail, three different thicknesses, ta, are tested: 1, 2.5, and 5 mm and three-cylinder diameters considered, dc: 18, 14 and 10 mm. Further geometrical details are reported in Figure 1. The samples were cured at room temperature for 24 h.

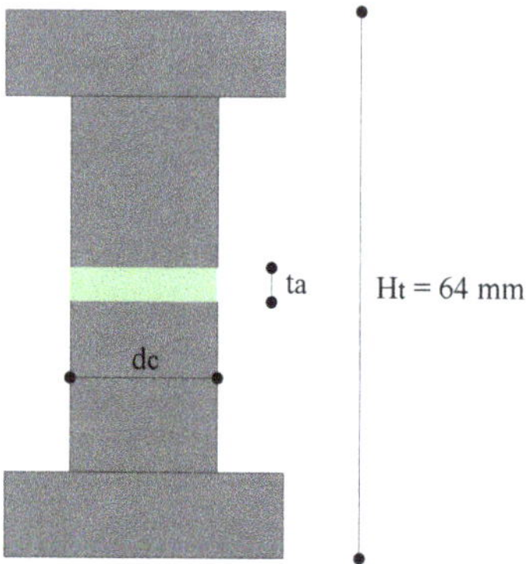

Figure 1. Dimension of cylinder adhesive connections.

To ensure the effectiveness of the bonding, the surfaces were well-cleaned using acetone. Surfaces may be contaminated with dust or micro-particles and may have poorly adhering surface layers, which affects the effectiveness of bonding and may lead to premature failure.

2.2. Double Lap Shear Joints

A total of 6 double lap shear joints were manufactured to investigate the shear strength of the epoxy adhesive. The specimens were realized in accordance with the standard code ASTM D3528-96 [21] using rectangular S275 steel elements. More specifically, the specimens were made up of two rectangular plates of 112 mm, 26 mm wide and 4 mm thick, and two other rectangular plates of 50 mm, 26 mm wide and 2 mm thick (some details in

Figure 2). The steel elements were bonded together with four 20×26 mm rectangular adhesive layers.

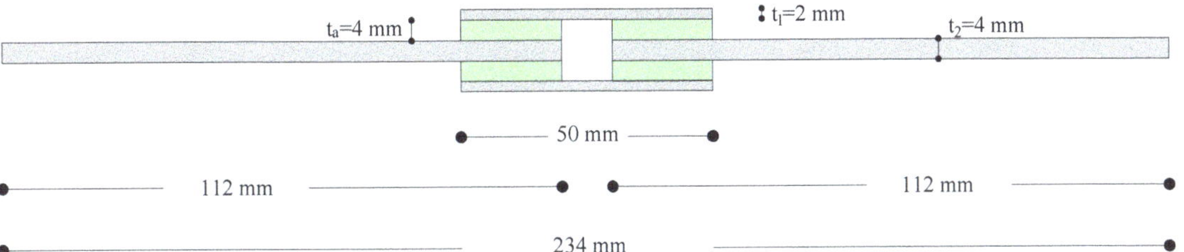

Figure 2. Double lap shear adhesive joint dimension.

The mechanical properties of the steel plate are reported in Table 3.

Table 3. Mechanical properties of steel plate.

Property	Unit Measure	Value
Modulus of Elasticity	N/mm^2	210,000
Yield's strength	N/mm^2	275
Poisson ratio	-	0.3

Figure 3 shows the double lap shear adhesive specimens realized.

Figure 3. Double lap shear adhesive specimens.

3. Experimental Tests

All the tests were performed at the Laboratory of Mechanics and Acoustics in Marseille using the universal testing machine characterized by a load capacity of 100 kN.

As is well known in the literature, in fracture mechanics which are concerned with the study of crack propagation in materials, the force is divided into its components. This process leads to the definition of the following two modes: Mode I, also known as the "opening mode" where a tensile stress is applied perpendicular to the plane of the crack, and Mode II, also known as the "sliding mode" where a shear stress is applied parallel to the plane of the crack.

In the current investigation, the specimens were subjected to both Mode I and II loading conditions.

The experimental set-up and results are described and discussed in the following sections.

3.1. Cylindrical Adhesive Joints in Mode I

A total of 37 cylindrical adhesive joints with different surfaces and volumes of adhesive layer have been programmed and realized.

The cylindrical adhesive joints were subjected to a vertical displacement by means of an Arcan-modified device at a rate of 1 mm/min. The experimental test set-up is depicted in Figure 4.

Figure 4. Experimental set-up of adhesive connections subject to Mode I loading by means of an Arcan-modified device.

3.2. Double Lap Shear Joints in Mode II

The double lap shear joints were placed directly in the clamps of the universal testing machine, as shown in Figure 5. All the specimens were tested in displacement control at a rate of 1 mm/min.

Figure 5. Experimental set-up of double lap shear tests.

4. Experimental Test Results

In this section, the experimental results are analyzed and discussed for both aluminium cylindrical and double lap shear adhesive joints tested in Mode I and II loading conditions.

4.1. Cylindrical Adhesive Joints Test Results

The experimental results of cylindrical adhesive joints are evaluated in this section. As mentioned above, in the current investigation, several diameters were considered equal to 18, 14 and 10 mm, respectively, and for each of them the adhesive thicknesses equal to 1, 2.5 and 5 mm have been investigated.

The experimental data are summarized in the following tables in terms of ultimate force, F_u, ultimate stress, σ_u, corresponding displacement u_{max}, and global stiffness K.

In more detail, the experimental results of cylindrical adhesive joints characterized by a diameter of 18 mm at different adhesive thicknesses are reported in Table 4.

Table 4. Results of adhesive joints with a diameter equal to 18 mm.

Test ID	F_u (N)	$F_{u,av}$ (N)	σ_u (MPa)	$\sigma_{u,av}$ (MPa)	u_{max} (mm)	$u_{max,av}$ (mm)	K (N/mm^3)	K_{av} (N/mm^3)
D18T1#1	2719		10.69		0.123		87	
D18T1#2	4538		17.83		0.172		104	
D18T1#3	4358		17.13		0.188		91	
D18T1#4	3230		12.69		0.146		87	
D18T1#5	2882	3399 ± 781	11.32	13.36 ± 3.07	0.128	0.139 ± 0.03	89	96 ± 9
D18T1#6	3586		14.09		0.126		112	
D18T1#7	2369		9.31		0.094		99	
D18T1#8	4086		16.06		0.171		94	
D18T1#9	2824		11.10		0.106		104	
D18T2.5#1	2877		11.31		0.142		80	
D18T2.5#2	2051	2794 ± 704	8.06	10.98 ± 2.77	0.116	0.146 ± 0.03	70	75 ± 5
D18T2.5#3	3453		13.57		0.179		76	
D18T5#1	2087		8.20		0.136		60	
D18T5#2	3676	2028 ± 953	14.17	10.04 ± 3.59	0.207	0.126 ± 0.05	68	63 ± 4
D18T5#3	1968		7.74		0.117		66	

In Tables 5 and 6, the experimental data of adhesive joints of diameter equal to 14 and 10 mm at different adhesive thicknesses are reported, respectively.

Table 5. Results of adhesive joints with a diameter equal to 14 mm.

Test ID	F_u (N)	$F_{u,av}$ (N)	σ_u (MPa)	$\sigma_{u,av}$ (MPa)	u_{max} (mm)	$u_{max,av}$ (mm)	K (N/mm^3)	K_{av} (N/mm^3)
D14T1#1	2515		16.34		0.139		117	
D14T1#2	1412	1770 ± 645	9.17	11.50 ± 4.19	0.091	0.103 ± 0.03	101	111 ± 9
D14T1#3	1384		8.99		0.078		116	
D14T2.5#1	1915		12.44		0.106		118	
D14T2.5#2	1927		12.52		0.112		112	
D14T2.5#3	1718	1703 ± 326	11.16	11.06 ± 2.12	0.076	0.100 ± 0.02	147	111 ± 21
D14T2.5#4	1283		8.34		0.095		88	
D14T2.5#5	1326		8.61		0.090		96	
D14T2.5#6	2046		13.29		0.124		107	
D14T5#1	1158		7.52		0.062		122	
D14T5#2	1613	1261 ± 228	10.48	8.95 ± 1.48	0.118	0.092 ± 0.03	89	97 ± 25
D14T5#3	1365		8.87		0.122		73	

Table 6. Results of adhesive joints with a diameter equal to 10 mm.

Test ID	F_u (N)	$F_{u,av}$ (N)	σ_u (MPa)	$\sigma_{u,av}$ (MPa)	u_{max} (mm)	$u_{max,av}$ (mm)	K (N/mm^3)	K_{av} (N/mm^3)
D10T1#1	875	875	11.15	11.15	0.079	0.079	141	141
D10T2.5#1	648		8.25		0.059		140	
D10T2.5#2	568	764 ± 293	7.23	10.72 ± 3.73	0.053	0.076 ± 0.04	136	133 ± 34
D10T2.5#3	1207		15.37		0.139		111	
D10T2.5#4	944		12.02		0.063		192	
D10T5#1	517		6.59		0.050		132	
D10T5#2	768		9.77		0.067		146	
D10T5#3	811	763 ± 241	10.33	9.72 ± 3.07	0.082	0.078 ± 0.02	126	126 ± 17
D10T5#4	584		7.44		0.074		100	
D10T5#5	1135		14.45		0.115		126	

It is important to note that the choice of high thicknesses is due to their feasibility on site by workers for the realization of adhesive connection for secondary structures in civil and mechanical engineering construction. However, the value of standard deviation reported in Tables 4–6 can be explained by the presence of initial defects inside the adhesive layer.

On the other hand, Figures 6–8 show the bar charts in terms of ultimate normal stress for the adhesive joints under monotonic loading conditions for each cylinder diameter and adhesive thickness.

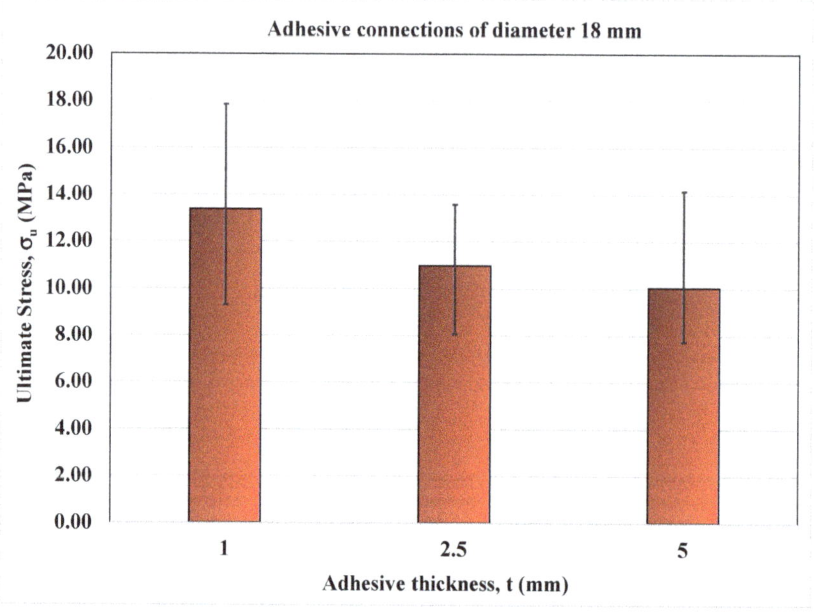

Figure 6. Bar-chart of adhesive connections of diameter equal to 18 mm.

As can be seen, the strength is higher at lower adhesive thicknesses and similar between the thicknesses of 2.5 and 5 mm.

For each specimen, the failure has occurred after the initiation of a crack in the adhesive layer and its instantaneous propagation, resulting the separation of the bonded metallic elements (see Figure 9). After the failure, some of the adhesive layer remains on the two cylindrical surfaces (cohesive failure).

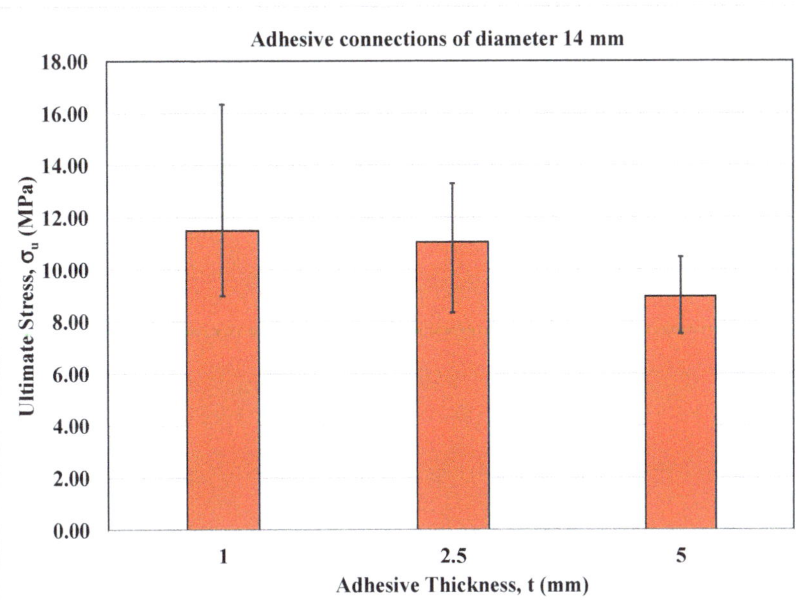

Figure 7. Bar-chart of adhesive connections of diameter equal to 14 mm.

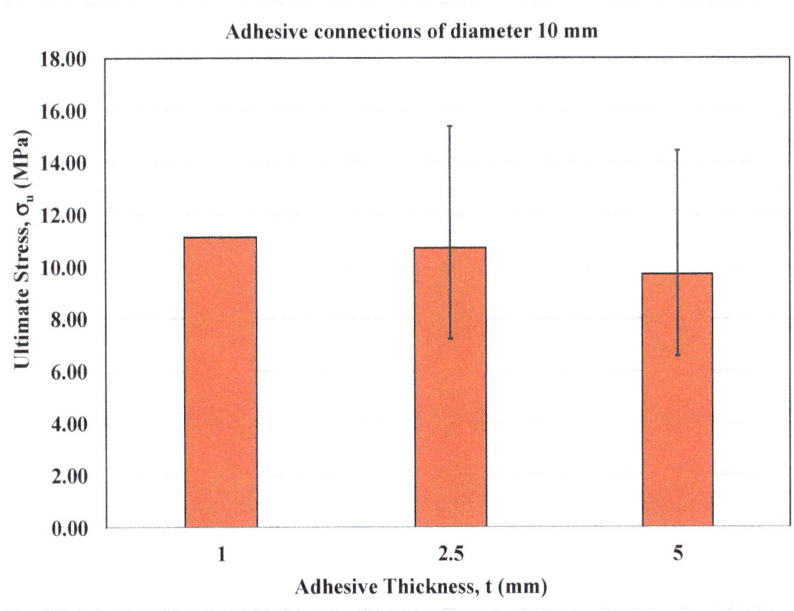

Figure 8. Bar-chart of adhesive connections of diameter equal to 10 mm.

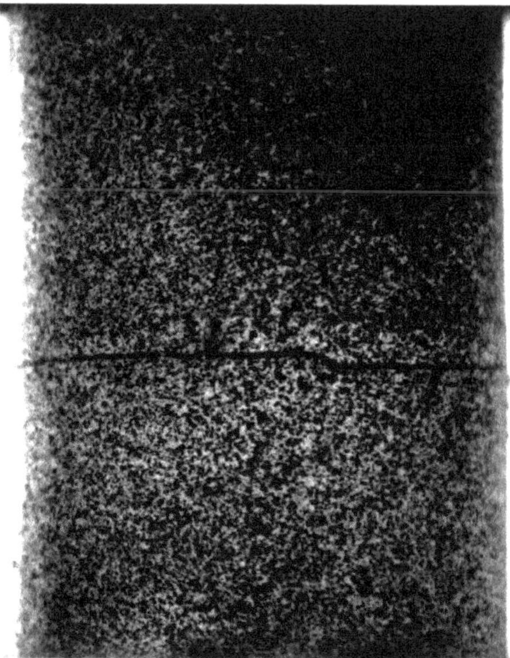

Figure 9. Cohesive failure recorded by the camera for the specimen D18T2.5#2.

4.2. Double Lap Shear Test Results

The experimental results of double lap shear adhesive tests in terms of ultimate force, F_u; average shear stress, τ_m; displacement at failure, u_{max}; and global stiffness, K, are summarized in Table 7.

Table 7. Experimental results of double lap shear joints.

Test ID	F_u (N)	$F_{u,av}$ (N)	τ_m (MPa)	$\tau_{m,av}$ (MPa)	u_{max} (mm)	$u_{max,av}$ (mm)	K (N/mm³)	K_{av} (N/mm³)
DLSJ4#1	6010		5.78		0.113		51	
DLSJ4#2	5707		5.49		0.104		53	
DLSJ4#3	7735	5780 ± 1014	7.44	5.49 ± 0.97	0.139	0.103 ± 0.02	53	54 ± 2
DLSJ4#4	5257		5.06		0.091		56	
DLSJ4#5	4793		4.61		0.081		57	
DLSJ4#6	4744		4.56		0.087		52	

Figure 10 shows the global mechanical response in terms of force versus displacement of the double lap shear adhesive joints.

The failure occurs after the initiation of a crack in the adhesive layer and the consequent instantaneous propagation, which leads to the separation of the bonded metallic adherents.

Figure 11 shows the picture recorded by the camera at the failure instant for the specimen DLSJ4#3.

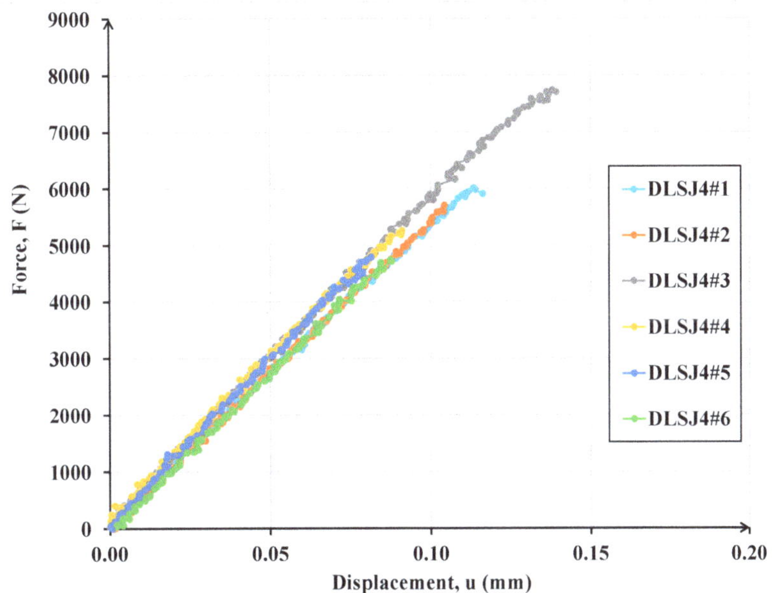

Figure 10. Global response of double lap shear tests.

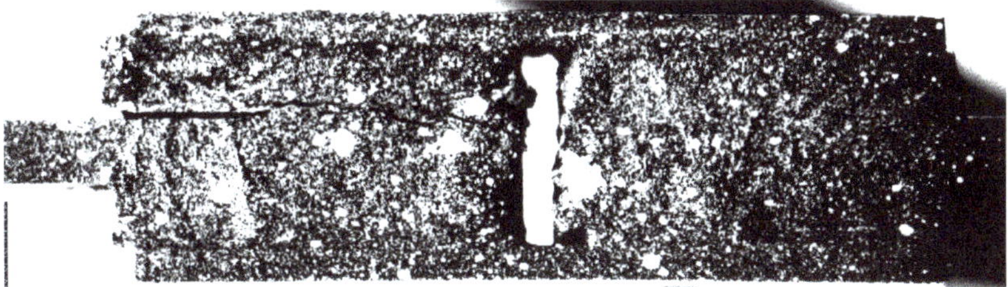

Figure 11. Detailed view of the failed adhesive area for the specimen DLSJ4#3.

5. Damage Parameters

The results of the experimental investigations made it possible to evaluate the initial damage properties of a material defined according to Kachanov–Sevostianov's material theory.

In Equations (1) and (2), the stiffness in Mode I and II, respectively, is defined by the function of the adhesive Young's Moduli, E, in the normal and tangential direction; the Poisson ratio, ν; the geometric dimension of the adhesive, S; and initial damage length, l_0. Note that the stiffness is directly dependent on the length l_0.

In this approach, the crack density $\rho(l_0)$ is defined in Equation (3) [17–19]. In more detail, the crack density can be evaluated as the ratio between the cubic length of the crack l_0 and the elementary volume in this way is able to describe the material at the microscale. Note that the crack density is therefore inversely proportional to the thickness of the adhesive interface:

$$K_N(l_0) = \frac{3E_N S}{16 l_0^3 (1 - \nu^2)} \tag{1}$$

$$K_T(l) = \frac{3E_T S (2 - \nu)}{32 l_0^3 (1 - \nu^2)} \tag{2}$$

$$\rho(l_0) = \frac{l_0^3}{V} \tag{3}$$

Finally, using the experimental results in terms of stiffness in Equations (1) and (2), the initial damage lengths l_0 are calculated according to Kachanov–Sevostianov's theory for Mode I and II and are summarized in Table 8a,b, respectively.

Table 8. Comparison between the initial damage length evaluated by means of mathematical formulation for Mode I and II.

(a) Mode I				
Thickness t (mm)	Surface S (mm^2)	Volume V (mm^3)	Porosity Rate $\rho(l_0)$	Initial Damage Length l_0 (mm)
	254	254	9.91	13.61
1	154	154	8.58	10.97
	79	79	6.77	8.10
	254	636	5.08	14.79
2.5	154	385	3.43	10.97
	79	196	2.86	8.25
	254	3185	1.21	15.66
5	154	1924	0.78	11.47
	79	982	0.51	8.41
(b) Mode II				
Thickness t (mm)	Surface S (mm^2)	Volume V (mm^3)	Porosity Rate $\rho(l_0)$	Initial Damage Length l_0 (mm)
4	460	1840	3.70	18.94

As highlighted in Table 8, the initial crack length assumes higher values as the adhesive volume increases due to a higher presence of defects or voids. As expected, the initial crack length as a damage parameter is a function of the volume of the adhesive layer.

6. Imperfect Interface Model

In this section, the steps are illustrated that led to the formulation of the imperfect interface model. The theoretical model is obtained by homogenization techniques and by asymptotic methods in the context of small perturbation coupling of unilateral contact and damage [18,19,22–26].

The approach of the considered damage behaviour is introduced in [24,27]: a thin adhesive interphase is located between two elements (adherents) and is assumed to be a microcracked material undergoing a degradation process. Further details can be found in [26,27].

Each step of the proposed procedure is described below:

(1) The microstructure of the glue layer incorporates multiple families of randomly arranged and distributed microcracks. The family of parallel microcracks is chosen as the only representative of the macroscopic behaviour of the adhesive and is indicated as the equivalent length l of the family of microcracks. Furthermore, the direction of the crack is considered to be parallel to the adherent surface.

(2) The actual mechanical properties of theoretical microcracked elements are obtained through the Kachanov-type homogenization mathematical technique [18,19], based on the Eshelby problem. The consequent elastic properties depend on the microcrack density ρ, whose three-dimensional formula is $\rho = \frac{l^3}{V}$, where V is the volume of the representative element. It is emphasized that the equivalent length l of a family of microcracks can be characterized experimentally, as illustrated in Section 5. The Young's modulus E_N is defined in the normal direction to the adhesive joint surface and is equal to $\frac{E_0}{1+C\rho}$ where E_0 is the initial Young's modulus and C is calculated as $\frac{16(1-v_0^2)}{3}$ where v_0 is the Poisson's ratio. Note the subscript 0 indicates the undamaged material. The peculiarity of the present model is centred in the definition of the

crack density. This crack density changes over the time and therefore represents a damage parameter. Similarly to the crack density ρ, the equivalent length l is defined by an evolution law. In fact, the time variation of l has to be related to a dissipative pseudo-potential ϕ, which is given by the sum of quadratic term (rate dependent) and a positively homogeneous functional (rate-independent) [27]. The dissipative pseudo-potential equation is defined in Equation (4), where η indicates a positive viscosity parameter function of the adhesive layer thickness and I_B denotes the indicator function of a set B, in particular $I_B = 0$ if $x \in B$ and $I_B = \infty$ otherwise.

$$\phi\left(\dot{l}\right) = \frac{1}{2}\eta\dot{l}^2 + I_{]0,\infty[}\left(\dot{l}\right) \tag{4}$$

Additionally, the indicator function term $I_{]0,\infty[}$ forces the length of cracks to acquire a positive value. This causes the crack length to increase over time, making the adhesive degradation process irreversible. Furthermore, damage will only begin when the elastic work is greater than a certain value which depends on the adhesive geometry (thickness) [27]. It has been proven that Kachanov-type materials are soft materials, this means that, for example, the stiffness of the glue is of the same order of its thickness. However, in order to force one-sided contact, which implies non-penetrating conditions during asymptotic expansion, the adhesive is considered a soft material only under tension (Equation (4)).

(3) The homogenised material is employed to implement a thin adhesive interphase. As aforementioned, the interface is located between two elements or adherents. The adhesion between the interface and element's surface is considered perfect, which means that the continuity of interface separation and of stress vectors is always verified.

(4) Using an appropriate asymptotic expansion [22], due to the small thickness of the glue layer, it is possible to obtain at the limit that the interphase volume of the adhesive is substituted by an interface named S of normal unit n. The equations between the two adherents that link the stress vector n and the interface separation [u] across the surface S are obtained:

$$\sigma n = K(l)[u]_+ + \tau n \text{ on } S \tag{5}$$

$$\tau[u].n = 0, \ \tau \leq 0, \ [u].n \geq 0 \text{ on } S \tag{6}$$

$$\overline{\eta}\dot{l} = \left(\overline{\omega} - \frac{1}{2}K_{,l}(l)[u]_+.[u]_+\right)_+ \text{ on } S \tag{7}$$

For more clarity, the symbol $()_{,l}$ denotes the partial derivate in l, $()_+$ that represent the positive part of a function, i.e., $[u]_+ = [u]$ if $[u].n \geq 0$, $[u]_+ = [u] - [u].n$ if $[u].n \leq 0$. The quantity $\overline{\eta}$ is the limit of $\eta\varepsilon$ for $\varepsilon \to 0$ as well as the limit of $\omega\varepsilon$ (further details on the application of asymptotic expansion can be found in [22]). Moreover, the term K represents the interface stiffness. As it is possible to note, the interface constitutive law provided in Equations (5)–(7) is a spring-like nonlinear interface model characterized by a nonlinear damage evolution. It is important to emphasize that the interface stiffness K remembers the mechanical properties of the initial interphase (mechanical properties, geometry, and damage).

7. Finite Element Simulation

The imperfect model presented in Section 6, was implemented in the commercial finite element software, COMSOL Multiphysics 5.6 [28], to verify the reliability of the estimated damage parameters. Both experimental tests on specimens are simulated. Finally, the comparison between numerical and experimental investigation is performed.

Validation of the Model on Adhesive Tests

In the current model, aluminium and steel substrates assume an isotropic linear elastic behaviour. The adherents' mechanical properties are reported in Tables 2 and 3, for aluminium and steel, respectively.

The interface adhesive model offers the link between the two substrates and implements the mechanical behaviour of the adhesive layer which mechanical properties are collected in Table 1 and by integrating the effect of damage parameters evaluated in Table 8a,b. Thanks to the presence of a symmetry plane only a quarter of the cylindrical specimen and only half part of the double lap shear joint have been modelled.

The mesh of the specimen is depicted in Figures 12 and 13.

Figure 12. Mesh details of cylindrical joints.

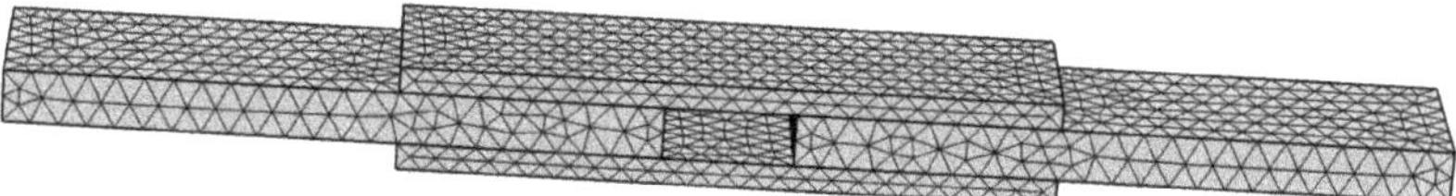

Figure 13. Mesh details of double lap shear joints.

After a mesh sensitivity analysis involved on the elastic response of the adhesive connection, a fine mesh size (minimal 0.1 mm) of elements is implemented. In more detail, for meshing the entire cylindrical geometry, triangular elements were used while for what concerns the double lap shear joint, tetrahedral elements were employed.

Boundary conditions correspond to the experimental set-up for both tests: the specimen is embedded at one surface and on the opposite extremity a displacement along the vertical axis is applied.

The value of normal and tangential stiffnesses have been evaluated by means of the experimental data extrapolated by the global mechanical response and reported in Tables 4 and 7.

Figure 14 shows the comparison between the experimental results of the cylindrical joints of 18 mm diameters connected by an adhesive layer of thickness equal to 1 mm.

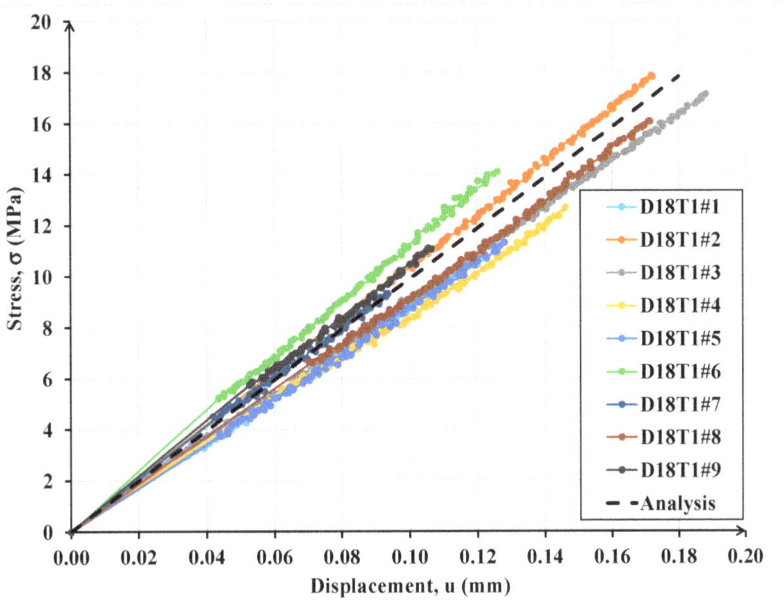

Figure 14. Comparison between numerical and experimental results for the cylindrical specimens with diameter equal to 18 mm and adhesive thickness equal to 1 mm.

In Figure 15, the comparisons between the numerical and experimental results for the double lap shear joints tests are depicted.

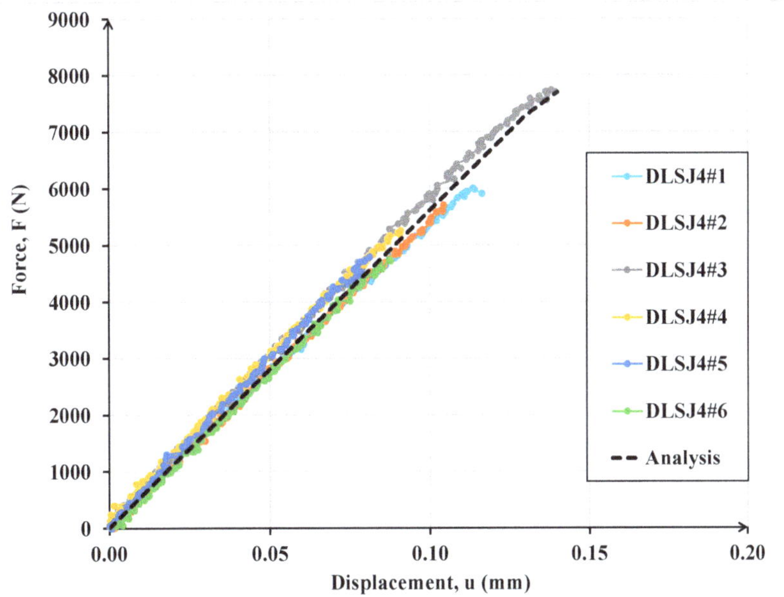

Figure 15. Comparison of numerical and experimental results for the double lap shear adhesive specimens.

As it is possible to note in both cases (Figures 14 and 15), the numerical curves are collocated in the dispersion of experimental evidence. The dispersion of experimental results in terms of ultimate resistance is due to the presence of defects that influence the mechanical behaviour of adhesive connections especially in the case of high thicknesses (double lap shear joints). The numerical data are in good agreement with the experimental ones underlying the power of the interface imperfect model.

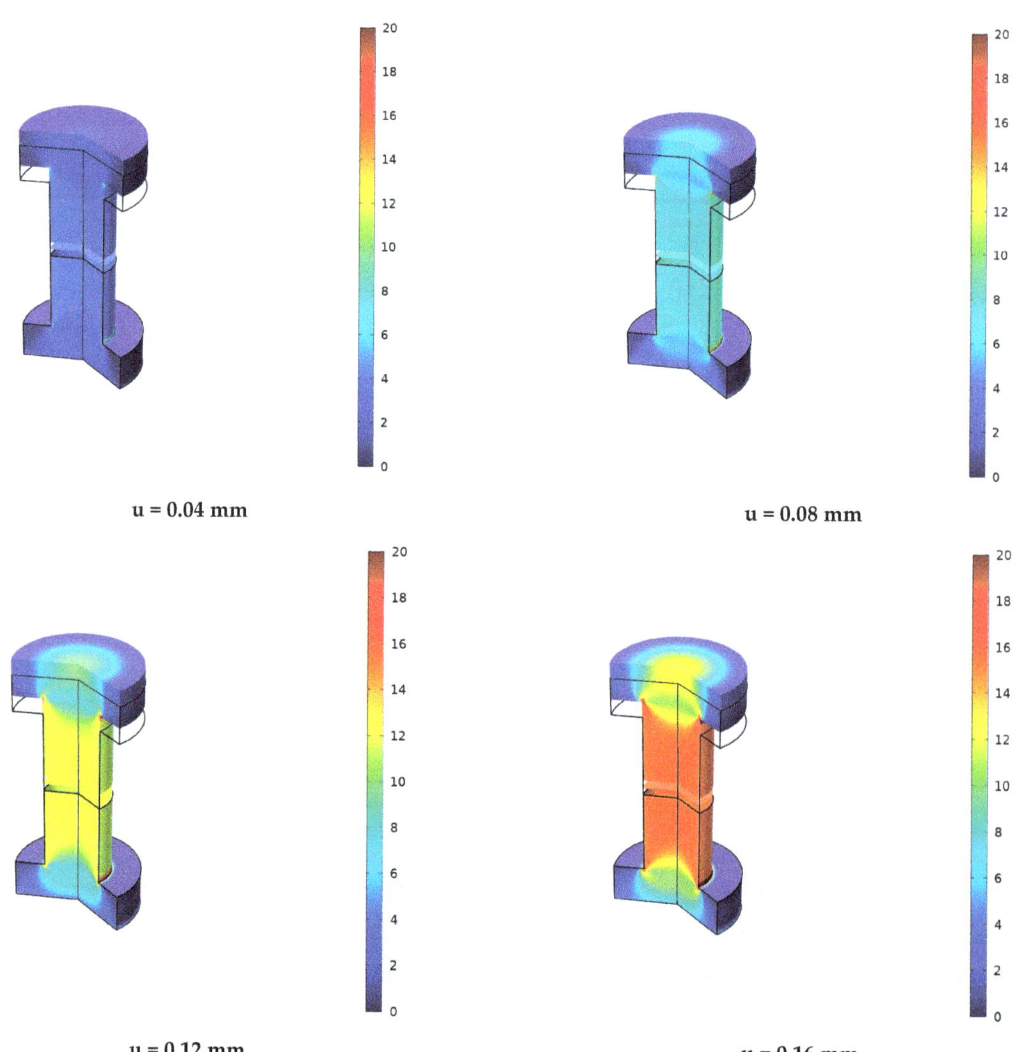

Figure 16. Normal stress distribution in MPa at different displacement levels for cylindrical adhesive joint.

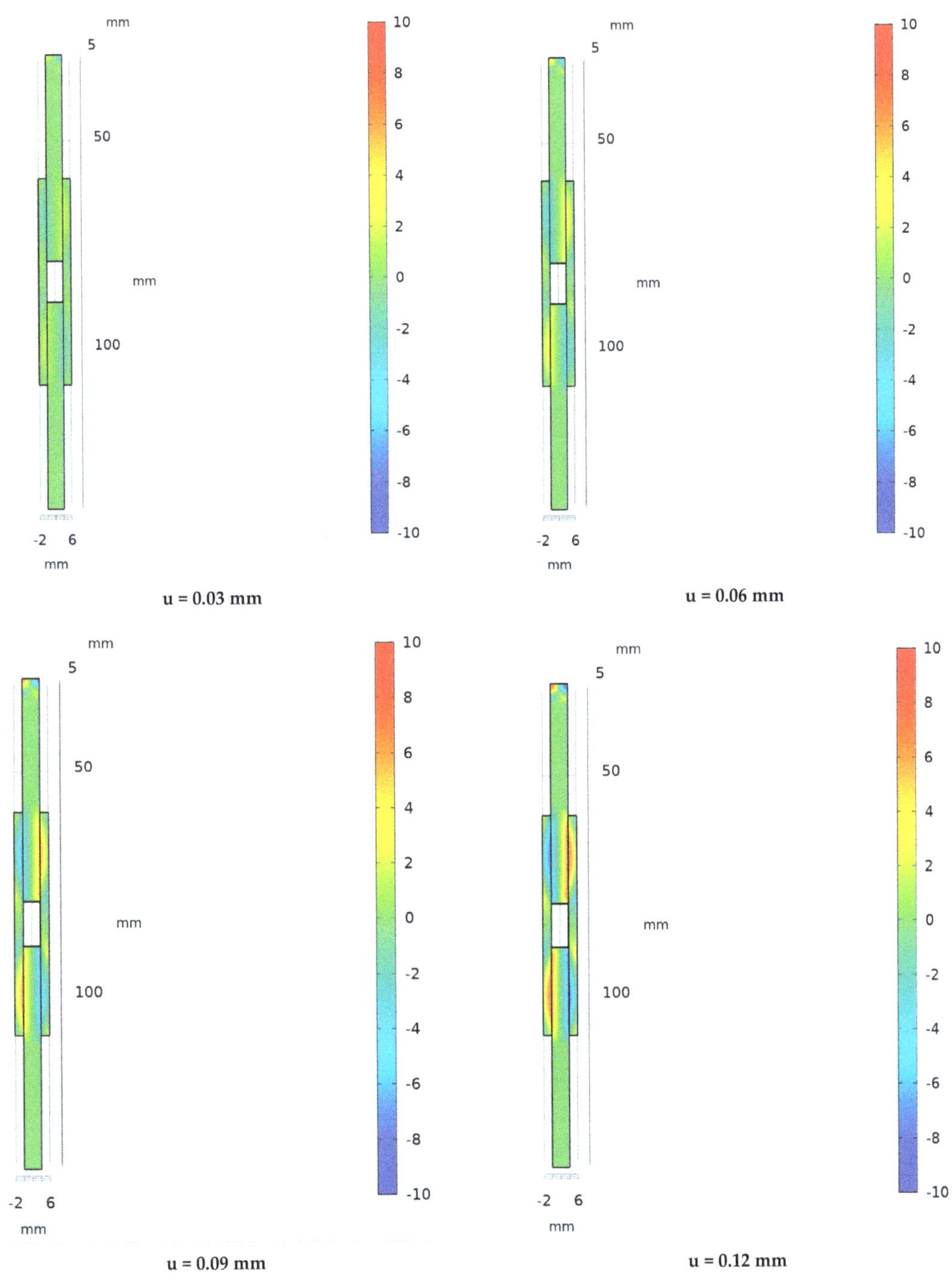

Figure 17. Shear stress distribution in MPa at different displacement levels for double lap shear joint.

Figure 16 shows the contour plots obtained by the revolution of the 2D numerical model in terms of normal stress at varying of displacement applied, 0.04, 0.08, 0.12 and 0.16 mm, respectively, for the numerical analysis of cylindrical adhesive joint.

Finally, in Figure 17, the contour plots in terms of shear stress at varying of displacement applied, 0.03, 0.06, 0.09 and 0.12 mm, respectively, are presented for the numerical analysis of double lap shear joint.

8. Conclusions

An extensive experimental programme has been conducted at the Laboratory of Mechanics and Acoustics in Marseille to evaluate the performance of mechanical adhesive connections for secondary members in structural applications such as offshore installation. In particular, the damage parameters have been evaluated experimentally in order to improve the design process of such connections taking into account the presence of defects or voids.

The results discussed support the following conclusions:

(1) The volume of the adhesive layer influences the mechanical strength of the joints as experimentally observed.
(2) The damage parameters are functions of the volume of the adhesive layers; a greater quantity of adhesive increases the probability of having defects or voids.
(3) The damage parameters evaluated under Mode I and Mode II loading conditions implemented in the imperfect interface model allow the reproduction of the experimental tests with high accuracy.
(4) The imperfect interface model has been shown to be powerful enough to be used and considered as a design process for adhesive connections.

Author Contributions: Conceptualization, M.L., A.M.-P. and F.L.; methodology, M.L., A.M.-P. and F.L., software, M.L.; validation, M.L.; formal analysis, M.L.; investigation, M.L.; data curation, M.L.; writing—original draft preparation, M.L.; writing—review and editing, M.L., A.M.-P. and F.L.; visualization, M.L.; supervision, M.L., A.M.-P. and F.L.; project administration, M.L.; funding acquisition, M.L. All authors have read and agreed to the published version of the manuscript.

Funding: This project has received funding from the European Union Horizon 2020 research and innovation programme under the Marie Skłodowska-Curie grant agreement No. 843218-ASSO (Adhesive Connection for Secondary Structures in Offshore wind installations).

Institutional Review Board Statement: Not applicable.

Informed Consent Statement: Not applicable.

Data Availability Statement: The data required to reproduce these findings are all reported in the paper.

Acknowledgments: This project has received funding from the European Union Horizon 2020 research and innovation programme under the Marie Skłodowska-Curie grant agreement No. 843218-ASSO (Adhesive Connection for Secondary Structures in Offshore wind installations). The authors thank Messrs Luca Giuliani and Mohamed Amine Ouelhazi, master's students at the Department of Mechanical Engineering at the University of Ferrara, for their assistance during the preparation and testing of the specimens. The authors would also thank the industry BW Ideol for its financial support.

Conflicts of Interest: The authors declare no conflict of interest.

References

1. He, X. A review of finite element analysis of adhesively bonded joints. *Int. J. Adhes. Adhes.* **2011**, *31*, 248–264. [CrossRef]
2. Adams, R.D.; Comyn, J.; Wake, W.C. *Structural Adhesive Joints in Engineering*; Chapman and Hall: London, UK, 1998.
3. Lamberti, M.; Ascione, F.; Napoli, A.; Razaqpur, G.; Realfonzo, R. Nonlinear Analytical Procedure for Predicting Debonding of Laminate from Substrate Subjected to Monotonic or Cyclic Load. *Materials* **2022**, *15*, 8690. [CrossRef]
4. Chataigner, S.; Caron, J.F.; Diaz, A.; Aubagnac, C.; Benzarti, K. Non-linear failure criteria for a double lap bonded joint. *Int. J. Adhes. Adhes.* **2010**, *30*, 10–20. [CrossRef]
5. Cognard, J.Y.; Devaux, H.; Sohier, L. Numerical analysis and optimization of cylindrical adhesive joints under tensile loads. *Int. J. Adhes. Adhes.* **2010**, *30*, 706–719. [CrossRef]

6. Freddi, F.; Sacco, E. Mortar joints influence in debonding of masonry element strengthened with FRP. *Key Eng. Mater.* **2015**, *624*, 197–204. [CrossRef]

7. Carrara, P.; Freddi, F. Statistical assessment of a design formula for the debonding resistance of FRP reinforcements externally glued on masonry units. *Compos. Part B Eng.* **2014**, *66*, 65–82. [CrossRef]

8. Zhao, L.; Burguono, R.; Rovere, H.L.; Seible, F.; Karbhari, V. *Preliminary Evaluation of the Hybrid Tube Bridge System*; Technical Report No. TR-2000/4–59AO032; California Department of Transportation: Sacramento, CA, USA, 2000.

9. Banea, M.D.; da Silva, L.F.M. Adhesively bonded joints in composites materials: An overview. *Proc. Inst. Mech. Eng. Part L-J. Mater. Des. Appl.* **2009**, *223*, 1–18. [CrossRef]

10. Haran-Nogueira, A.; Kasaei, M.M.; Akhavan-Safar, A.; Carbas, R.J.C.; Marques, E.A.S.; Kim, S.K.; da Silva, L.F.M. Development of hybrid bonded-hole hemmed joints: Process design and joint characterization. *J. Manuf. Process.* **2023**, *95*, 479–491. [CrossRef]

11. Greenwood, J.A.; Williamson, J.B. Contact of nominally flat surfaces. *Proc. R. Soc. Lond. Ser. A* **1966**, *295*, 300–319.

12. Yoshioka, N.; Scholz, C.H. Elastic properties of contacting surfaces under normal and shear loads: 1 Theory. *J. Geophys. Res. Solid Earth* **1989**, *94*, 17681–17690. [CrossRef]

13. Sherif, H.A.; Kossa, S.S. Relationship between normal and tangential contact stiffness of nominally flat surfaces. *Wear* **1991**, *151*, 49–62. [CrossRef]

14. Krolikowski, J.; Szczepek, J. Assessment of tangential and normal stiffness of contact between rough surfaces using ultrasonic method. *Wear* **1993**, *160*, 253–258. [CrossRef]

15. Mindlin, R.D. Compliance of elastic bodies in contact. *J. Appl. Mech.* **1949**, *71*, 259–268. [CrossRef]

16. Gonzalez-Valadez, M.; Baltazar, A.; Dwyer-Joyce, R.S. Study of interfacial stiff-ness ratio of a rough surface in contact using a spring model. *Wear* **2010**, *268*, 373–379. [CrossRef]

17. Kachanov, M. Elastic solids with many cracks and related problems. In *Advances in Applied Mechanics*; Hutchinson, J., Wu, T., Eds.; Academic Press: New York, NY, USA, 1993; Volume 30, pp. 259–445.

18. Tsukrov, I.; Kachanov, M. Effective moduli of an anisotropic material with elliptical holes of arbitrary orientational distribution. *Int. J. Solids Struct.* **2000**, *69*, 5919–5941. [CrossRef]

19. Sevostianov, I.; Kachanov, M. On some controversial issues in effective field approaches to the problem of the overall elastic properties. *Mech. Mater.* **2014**, *69*, 93–105. [CrossRef]

20. Sicomin Isobond SR 5030/SD 503x Technical Data Sheet. 2018. Available online: https://sicomin.com/datasheets/product-pdf1 254.pdf (accessed on 4 December 2018).

21. *ASTM D3528-96*; Standard Test Method for Strength Properties of Double Lap Shear Adhesive Joints by Tension Loading. ASTM: West Conshohocken, PA, USA, 2002; Volume 10, p. 1996.

22. Lebon, F.; Rizzoni, R. Asymptotic behavior of a hard thin linear elastic interphase: An energy approach. *Int. J. Solids Struct.* **2011**, *49*, 441–449. [CrossRef]

23. Rekik, A.; Lebon, F. Homogenization methods for interface modeling in damaged masonry. *Adv. Eng. Softw.* **2012**, *46*, 35–42. [CrossRef]

24. Maurel-Pantel, A.; Lamberti, M.; Raffa, M.L.; Suarez, C.; Ascione, F.; Lebon, F. Modelling of a GFRP adhesive connection by an imperfect soft interface model with initial damage. *Compos. Struct.* **2020**, *239*, 112034. [CrossRef]

25. Lamberti, M.; Maurel-Pantel, A.; Lebon, F.; Ascione, F. Cyclic behaviour modelling of GFRP adhesive connections by an imperfect soft interface model with damage evolution. *Compos. Struct.* **2022**, *279*, 114741. [CrossRef]

26. Lamberti, M.; Maurel-Pantel, A.; Lebon, F. Experimental characterization and modelling of adhesive bonded joints under static and non-monotonic fracture loading in the mode II regime. *Int. J. Adhes. Adhes.* **2023**, *124*, 103394. [CrossRef]

27. Bonetti, E.; Bonfanti, G.; Lebon, F.; Rizzoni, R. A model of imperfect interface with damage. *Meccanica* **2017**, *52*, 1911–1922. [CrossRef]

28. COMSOL. *COMSOL Multiphysics®v. 5.6*; COMSOL: Stockholm, Sweden, 2020.

Disclaimer/Publisher's Note: The statements, opinions and data contained in all publications are solely those of the individual author(s) and contributor(s) and not of MDPI and/or the editor(s). MDPI and/or the editor(s) disclaim responsibility for any injury to people or property resulting from any ideas, methods, instructions or products referred to in the content.

MDPI AG

Grosspeteranlage 5

4052 Basel

Switzerland

Tel.: +41 61 683 77 34

Materials Editorial Office

E-mail: materials@mdpi.com

www.mdpi.com/journal/materials

Disclaimer/Publisher's Note: The title and front matter of this reprint are at the discretion of the . The publisher is not responsible for their content or any associated concerns. The statements, opinions and data contained in all individual articles are solely those of the individual Editors and contributors and not of MDPI. MDPI disclaims responsibility for any injury to people or property resulting from any ideas, methods, instructions or products referred to in the content.

www.ingramcontent.com/pod-product-compliance
Lightning Source LLC
LaVergne TN
LVHW071250170726
843515LV00006BA/1373